W9-DIM-992

WITHDRAWN

F V

St. Louis Community College

Library

5801 Wilson Avenue
St. Louis, Missouri 63110

THE OMNI
SPACE ALMANAC

THE

OMNI

SPACE ALMANAC

A Complete Guide to the Space Age

NEIL McALEER

AN OMNI BOOK

WORLD ALMANAC
AN IMPRINT OF PHAROS BOOKS • A SCRIPPS HOWARD COMPANY
NEW YORK

Cover and text design by Elyse Strongin
Typeset on Atex by Charles Castillo

First published in 1987.
Distributed in the United States by Ballantine
Books, a division of Random House, Inc., and in
Canada by Random House of Canada, Ltd.
Library of Congress Catalog Card Number 86-050445
Pharos Books ISBN 0-88687-280-4
Ballantine Books ISBN 0-345-34395-6

Printed in the United States of America

World Almanac
An Imprint of Pharos Books
A Scripps Howard Company
200 Park Avenue
New York, NY 10166

10 9 8 7 6 5 4 3 2 1

To my father, Jim and my wife, Constance

. . . Our civilization is no more than the sum of all the dreams that earlier ages have brought to fulfillment. And so it must always be, for if men cease to dream, if they turn their backs upon the universe, the story of our race will end.

—ARTHUR C. CLARKE

There can be no thought of finishing, for "aiming at the stars," both literally and figuratively, is a problem to occupy generations, so that no matter how much progress one makes, there is always the thrill of just beginning.

—ROBERT HUTCHINGS GODDARD

So many worlds, so much to do.
So little done, such things to be.

—ALFRED, LORD TENNYSON

CONTENTS

THE OMNI SPACE ALMANAC

INTRODUCTION

The year 1987 marks the fortieth anniversary of my historic flight in the X-1 rocket plane, when I became the first man to fly at supersonic speed and break the sound barrier. While I'm certainly proud of this claim to fame, I'm even happier about the fact that I lived to tell about it; there were a lot of people who didn't think I would. Although I flew rocket-powered planes dozens of times during my test-pilot days in the late 1940s and early 1950s, that breakthrough flight on October 14, 1947 is something I remember like yesterday.

Once I had my "Glamorous Glennis" leveled off at 42,000 feet (almost 8 miles high), and had lit up the third rocket chamber, I knew I was going all the way. And suddenly I was traveling faster than sound in smooth silence: the X-1 reached Mach 1.07 or 700 miles per hour, and I held her there for about 20 seconds. The daytime sky before me turned to dark purple and the Moon and stars came out. I was at the edge of space ten years before the Russians put up *Sputnik 1* in 1957—the date now considered the official beginning of the Space Age.

As the years fly by, it's easy to lose perspective, but this book helped me regain it. In reading through *The Omni Space Almanac*, I'm amazed what we've accomplished in just a few short decades, despite some painful setbacks, and where we're headed as we approach the 21st century. In 1947, I had no idea that my flight would lay the groundwork for everything to come. We have probed the edge of space with our flying skills and our rocket power since the end of World War II. We've explored the space frontier above planet earth for four decades now, and we're still at it. Space is a boundless frontier. We'll never run out of new territory to explore, new adventures to beckon and challenge us. Each time we pit our spacefaring skills against the gravitational bonds of Earth, we make a new contribution to space exploration. Make no mistake; great and exciting discoveries await us as we continue to explore the other

planets and moons and eventually colonize the solar system and beyond.

This *Almanac*, with thousands of fascinating facts and more than 100 photographs, tells the story of the progress that has been made in space in these forty years. It covers everything—from those crude Mercury spaceflights in the 1960s (we called them "Spam-in-a-can" missions at Edwards Air Force Base) to the enormous industrial and economic opportunities that the high frontier promises to our free-enterprise system.

The Omni Space Almanac covers the hardware, the missions, the people, and the future of the Space Age. An entire chapter is devoted to the development of rockets and spacecraft. Another chapter covers the unmanned robot ships that have sent back their spectacular images of the planets and moons of our solar system. And a full chapter is devoted to the space station we will be building in orbit in another few years. But what stands out for me as I was reading this book are the people—the astronauts, cosmonauts, engineers, and technicians, many of whom I've known personally—who have turned humanity's recurring dream of space travel into reality.

The men and women who leave earth know from day one that there are no guarantees in this line of work. Pilots never think of *any* flight as routine—a word that NASA has recently dropped from its vocabulary. The pioneers of the Space Age are heroes precisely because they're willing to risk their lives on every mission they fly. These men and women are doing nothing less than creating our future.

The loss of seven courageous men and women in the explosion of the space shuttle *Challenger* put a black cloud over NASA and the United States space program. The world's worst single space disaster has completely changed the future scenario of space exploration for decades to come. This, in the long run, will be positive. *The Omni Space Almanac* reminds us that triumphs follow tragedies and tells us what our future triumphs in space are likely to be. It was only two-and-a-half years after the terrible launchpad fire in 1967 killed three of America's best that two men walked on the Moon.

There will, of course, always be risks, but accidents should never interfere with what I see as the natural destiny of Americans to go and explore space. Now that we know what went wrong, the mechanical and decision-making flaws will be corrected. Before the end of this decade, the *Challenger* explosion will be seen as a temporary setback on the course we started in the 1940s.

While I've broken many speed and altitude records in my day, and was one of the first pilots to see the earth's curvature, I still haven't orbited this planet. I must confess that I envy the first astronauts who saw our blue planet through the portholes of their space capsules. But who knows? I might get into orbit yet. Perhaps I can fly a shuttle to the space station and celebrate the fiftieth anniversary of my X-1 flight in the fall of 1997. I'm ready to go.

Anyone who reads *The Omni Space Almanac* will experience the thrills that the pilots and astronauts first felt when they headed into space. I hope all of us will help to keep the dream alive—the same dream that motivated the pioneers of my generation. This book certainly does. It's filled with dreams that have come true and dreams that are still being chased. Every step the test pilots and astronauts took in the early days was a ''first,'' which only meant that there was always so much more to come. Just what is to come is described right here in the pages that follow. The wonders will not cease.

CHUCK YEAGER
CEDAR RIDGE, CALIFORNIA

ACKNOWLEDGEMENTS

Two people deserve special credit for making *The Omni Space Almanac* possible: Hana Umlauf Lane, editor-in-chief of Pharos Books; and Bob Weil, editor of Omni Books.

For information and photographs, I wish to thank the following people who serve NASA at various centers in the United States: Don Bane, Vera Buescher, Lynn Cline, Carter Dove, Terry Eddleman, Jim Elliott, Susan Fruchter, Azeez Jaffer, Tom Jaqua, Karen Kleinsorge, Bob Marshall, Ed Medal, Joyce Milliner, Bill O'Donnell, Maurice Paker, Janet Ross, Patricia Ross, Lee Saegesser, Barbara Selby, Larry Thomas, Lisa Vazquez, Terry White, and Don Zylstra. Without exception, every one of my NASA contacts has been helpful, knowledgeable, and pleasant.

I also wish to thank the following individuals and their companies or institutions for providing print resources and photo materials: Dick Barton, Robert Howard, and Joyce Lincoln, Rockwell International; Roger Beall, Lockheed Missiles and Space Company, Inc.; Margot Bellman, the British Embassy; Odile Burton and Louis Laidet, Ambassade de France; Kenneth Carter, Department of Defense; Dave Christensen, Wyle Laboratories; Don Dixon, Spacescapes; Paul Horowitz, Harvard University; Chu Ishida, National Space Development Agency of Japan; Janis Kreiser and James L. Work, LTV Aerospace and Defense Company; Ed Collins, Perkin-Elmer; Walt Cooper, Martin Marietta Aerospace; Steve Eames, IBM; Bill Ennis, TRW; Jeff Fister and Susan Flowers, McDonnell Douglas; Jack Isabel, General Dynamics; Donna Mikov and Bill Rice, Boeing Aerospace Company; Don Norton, RCA; Pat Rawlings, Eagle Engineering; Miriam Reid and Lois Lovisolo, Grumman Aerospace Corporation; Bill Standing, Geostar Corporation; Lieutenant Andree Swanson, U.S. Air Force Space Division; Ruth Thomas, General Electric Company; and Phyllis Wiepking, EROS Data Center.

Finally, I express gratitude to my wife, Connie, who understands what writers do and go through because she too is a writer.

PHOTOS ON THE PRECEDING PAGES:

(Pp. ii-iii) In the 1990s, the space shuttle, pictured here, will carry unmanned free-flying spacecraft called orbit maneuvering vehicles to orbit to help build the space station. Courtesy TRW. (Pp. iv-v) The world's first space salvage operation occurred during a Discovery *mission in November 1984, when two satellites were retrieved, secured in the cargo bay, and brought back to earth. Here the* Westar VI, *with astronaut Joe Allen's help, is secured to the RMS robot arm and brought into the cargo bay. Courtesy NASA. (Pp. vi-vii) Before the shuttle* Challenger *disaster in January 1986, the fleet of four spaceships had twenty-four successful launches. Only after the tragic explosion did most people realize the shuttle's awesome power. Courtesy NASA. (P. viii) Close-up images of the moons of Uranus taken by* Voyager 2, *including those of Miranda (the closest view) and Ariel, revealed dramatic fractures, cliffs, and craters for the first time. Courtesy Jet Propulsion Laboratory.*

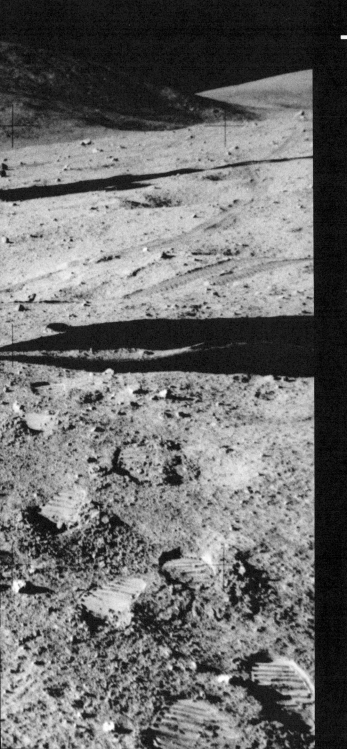

1

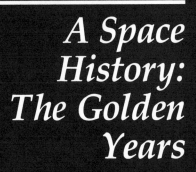

A Space History: The Golden Years

THE DREAMS BEFORE SPUTNIK

Luminous Mysteries

Human dreams of soaring to the moon, the stars, and the wandering bright planets no doubt began countless millennia ago, as eyes turned upward during ancient nights, dark and clear, long before written history began. As night followed night, the moon waxed and waned before these curious eyes. Most-bright Venus, before it had a name, cast shadows at times on the night earth when the moon was down. Now and then, perhaps, a star burst forth, brightening to change the sky, or a long-tailed wisp of comet light arched across the bowl of night. More eyes looked up, some more curious than others, and the celestial lights stirred their wonder and ancient dreams. And some stretched out their arms and fingers and tried to touch these luminous mysteries. These were the beginning dreams of human spaceflight.

Dreams to Quill and Ink

During the last few thousand years, these fantasies of spaceflight manifested themselves in the written record, and some have survived. The Greek Lucian of Samosata wrote a satire in A.D. 160 entitled *Vera Historia* (True History), which poked fun at absurd fictions put forth as truth. In this story a ship is lifted to the moon by a powerful whirlwind, and the author describes the mysterious celestial place as "a great countrie in the aire, like a shining island."

In a second story by Lucian, *Icaro-Menippus*, the hero, Menippus, launches himself from Mount Olympus to the moon with only the power of two wings—one a vulture's and one an eagle's.

Centuries later, the great astronomer Johannes Kepler wrote *Somnium* (Dream), which takes the reader to the moon in a dream voyage. This work was posthumously published in 1634. And in England in 1638, Francis Godwin, the Bishop of Hereford, published *The Man in the Moone: Or a Discourse of a Voyage Thither* under the pseudonym Domingo Gonsales. The hero flies to the moon on a chair pulled by thirty or forty trained wild swans. Even France's Cyrano de Bergerac wrote two fictional tales of voyages to the moon: *Voyage dans la Lune* (Voyage to the Moon; 1649) and *Histoire Comique des États et Empires du Soleil* (Comic History of the States and Empires of the Sun; 1662).

In all this early moon travel literature, it is important to note that the authors never present their stories in a serious vein. They are either written as satires or couched in supernatural or dream-state terms to avoid censure because of prevalent religious dogma. Such speculations were often presented as laughing matters by the author for self-protection.

If there ever was a turning point at which dreams of space travel began their long and arduous journey to reality, it was with two classics of Jules Verne, published in 1865 and 1870: *De la Terre à la Lune* (From the Earth to the Moon) and *Autour de la Lune* (Around the Moon). These books eventually found their way into the hands of a new generation, men who would begin solving the real problems of how to leave the gravity of earth and travel in space. All this from an ancient gleam in ancestral eyes.

FACING PAGE: *Three men and their pioneering work brought the Space Age to the twentieth century:* TOP, *the Russian, Konstantin Tsiolkovsky.* BOTTOM LEFT, *the American, Robert Goddard.* BOTTOM RIGHT, *and the German, Hermann Oberth.* Courtesy Smithsonian Institution. PRECEDING PAGES: *Charles Duke of* Apollo 16 *salutes the flag with Stone Mountain in the background. While on the surface, he had an inside-the-helmet mishap, when his orange drink dispenser squirted by mistake. "I wouldn't give you two cents for that orange juice as hair tonic," Duke said later.* Courtesy NASA.

From Prophets to Pioneers

Three men and their pioneering work are generally recognized for bringing the Space Age to the twentieth century: Konstantin Eduardovich Tsiolkovsky (1857-1935); Robert Hutchings Goddard (1882-1945); and Hermann Julius Oberth (1894-   ). All three men acknowledged the influence of Jules Verne's writings on their formative years. If there must be only *one* "father" of space travel—and all three have been called this—then the title should go to Tsiolkovsky, on the basis of his birthdate alone. Still, all three men contributed immensely to transforming an old dream into a twentieth-century reality.

Konstantin E. Tsiolkovsky. A Russian school teacher, Tsiolkovsky spent much of his spare time pursuing his dream of spaceflight and taught himself higher mathematics, physics, and astronomy. An early article submitted for publication in 1898, "Exploration of Cosmic Space by Means of Reaction Devices," presented the basic theory of rocket propulsion. This article was finally published in 1903. Tsiolkovsky continued to write technical papers as well as science fiction for the next thirty years. In 1926 he suggested the use of artificial earth satellites, space stations, and multistage rockets—a concept he described as rocket trains.

Although Tsiolkovsky was a great pioneer of space travel, he never received the recognition he deserved in his lifetime. He was a deaf and impoverished school teacher, and it would be decades before his great space literature was distributed and appreciated in the West. Even his own country was late in understanding his importance. But because his work eventually inspired Russian rocketeers such as Fridrikh Tsander, he undoubtedly deserves substantial credit for the beeping Russian surprise of 1957. He was finally honored in 1959 by having a crater on the far side of the moon named after him, a crater photographed for the first time by *Luna 3* in the fall of that year.

Robert H. Goddard. America's loner rocket man, Goddard did not live to see *Apollo 11's Eagle* land on the dry Sea of Tranquillity. He died in 1945, the year that the first atomic bomb was detonated. His widow, Esther Goddard, however, spoke for her dead husband: "That was his dream, sending a rocket to the moon . . . . He would just have glowed."

Where the Russian Tsiolkovsky theorized and wrote, the American Goddard wrote and tinkered. He too was a visionary, but he actually built and flew rockets. After earning his doctorate from Clark University, he became a professor of physics there to earn his living. But rocketry was his first love, and his great accomplishment can be traced back to a near-mystical experience he had as a boy still in high school.

On October 19, 1899, young Goddard climbed a cherry tree in his backyard in Worcester, Massachusetts, to trim the branches. As he looked down on the meadow, he imagined the possibility of creating a centrifugal-force device that would lift upward, spinning faster and faster, ascending to the planet Mars. The vision was imagined on a small scale from his tree perch, and his life was changed forever. He would work to fulfill that vision for the rest of his life.

Goddard's famous paper "A Method of Reaching Extreme Altitudes" was published by the Smithsonian in 1920. In the main, it was a summary of his experiments and findings, but he did mention the idea of an unmanned rocket that would carry to the moon some explosive, which would flash a signal on impact. The press picked this up out of context and ran with it, labeling him "the moon rocket man." He was a shy person, and this ridicule of his work and dream pained him terribly. He did continue his studies, but he remained somewhat reclusive to avoid the humiliation of such insensitive distortions of his work.

He achieved his greatest joy and earned his place in history, however, on March 16, 1926, when he successfully flew the world's first liquid-propellant rocket from his Aunt Effie's cabbage patch in Auburn, Massachusetts. The rocket's fuel was liquid oxygen and gasoline, and it thrust his dream up 41 feet (12.5 meters) in the air for 2.5 seconds before it crashed to the ground 184 feet (56 meters) away. It was the beginning.

"There can be no thought of finishing," he wrote later in his life, "for 'aiming at the stars', both literally and figuratively, is a problem to occupy generations, so that no matter how much progress one makes, there is always the thrill of just beginning."

The sixtieth anniversary of Goddard's famous rocket flight was celebrated in March 1986, when a replica rocket was flown from the Goddard Space Flight Center. Twelve men had walked and driven on the moon, and earth-made robots had left the solar system in those sixty years.

Hermann J. Oberth. He was born at Hermannstadt, Transylvania (now Rumania), in 1894. At the age of twelve, in 1906, Oberth's mother gave him a book to read; it was *De la Terre à la*

All three fathers of the Space Age acknowledged their debt to the writings of Jules Verne, especially De la Terre a la Lune *(From the Earth to the Moon), which was first published in 1865.* Courtesy Smithsonian Institution.

Lune by Jules Verne. The boy read the book six times, and it was a powerful influence on the rest of his life.

Oberth was a theorist, not an inventor. In 1930 in Berlin, he was given the chance to build a rocket for a film producer, but it failed. After the fact, he did learn the practical trades of the mechanic and locksmith. Had he known them earlier, he believed, his rocket would have flown.

In 1917, Oberth produced a design for a long-range liquid fuel rocket. Although his Ph.D. thesis of 1922 was rejected as too speculative, he turned this work into his classic book *Die Rakete zu den Planetenr¨aumen* (The Rocket into Interplanetary Space), which contained the first detailed discussion of orbiting space stations. The book, published in 1923, also examined such space travel problems as space food, space suits, space walks, and probable missions for both space stations and interplanetary journeys. Oberth had read a newspaper account of Goddard's work in 1919.

German publication and distribution of Oberth's work goaded the Russians into paying more attention to their own space pioneer Tsiolkovsky, and as a result, the Russian was read and studied. The same year that Lindbergh flew the Atlantic, 1927, Oberth joined the German Rocket Society, and the following year he went to Berlin as a technical adviser for the film classic *Frau im Mond* (Woman in the Moon). Oberth's major work, *Wege zur Raumschiffahrt* (The Way to Space Travel), appeared in 1929, just three years after Goddard's famous rocket launch.

Perhaps most important to the launching of the Space Age was Oberth's encounter and life-long relationship with another space visionary. When Oberth's first book was published, he received a letter from a young man who wrote that he could not understand the equations—this young man was Wernher von Braun. More than thirty years later (1955), Oberth would join von Braun's rocket team at Redstone Arsenal in the United States for three years . . . and the moon loomed large.

TWELVE YEARS TO TRANQUILLITY BASE

First Above Earth

The nations of the world were astounded when the Soviet Union successfully launched the first artificial satellite—*Sputnik 1*—on October 4, 1957. This event marked the dawning of the Space Age. An aluminum sphere 23 inches (58.4 centimeters) in diameter, *Sputnik 1* weighed 184.3 pounds (83.6 kilograms) and was outfitted with four metal-rod antennae 96 to 116 inches (244 to 295 centimeters) in length. These antennae transmitted radio signals to earth, indicating the satellite's on-board temperature, for twenty-one days.

A Soviet radio announcer spoke of *Sputnik 1*'s "beep-beep" radio signal in October 1957: "Until two days ago, that sound had never been heard on this earth. Suddenly it has become as much a part of twentieth-century life as the whir of your vacuum cleaner . . . . The satellite is still maintaining a speed of 18,000 miles an hour, a dozen times faster than any man has ever flown."

(In a BBC interview after the launch of *Sputnik 1*, a London woman said, "I guess the American people are alarmed that a foreign country, especially an enemy country, can do this . . . . We fear they have something that the majority of people don't know about.")

The Russian feat, however, was not a complete surprise to American space experts. As early as 1954, a proposal called Project Orbiter was put before the National Science Foundation. It proposed that a Redstone rocket place an artificial satellite into earth orbit. A formal statement of intent to launch an artificial earth satellite as part of the contribution to the 1957-1958 International Geophysical Year was announced in late 1954. Soviet representatives at the interna-

tional meeting listened quietly, and early in 1955 Radio Moscow predicted that a Soviet satellite in orbit was imminent. Russia, at least, knew a race was on.

Army, Navy, and the Red Star

The Naval Research Laboratory was responsible for Project Vanguard, begun in 1955, which intended to launch the first U.S. satellite into earth orbit. But the Russians knew it was coming, and they beat it off the launchpad with *Sputnik 2*, which carried a Russian dog as its passenger.

After that, the Vanguard program promised only to restore some U.S. technological self-esteem, but instead it delivered another blow to American pride: The first launch exploded on the launchpad in December 1957, and the second mission broke up in flight in February 1958. The British press referred to Vanguard as "Kaputnik and Flopnik."

While Vanguard gave the United States only embarrassing failures, the von Braun rocket team, urged to openly compete with Vanguard, rushed to prepare the Army's Jupiter C missile (to be known as *Juno 1*) and another satellite for launch. It was this team that, after the first Vanguard failure, succeeded in propelling the first U.S. satellite—*Explorer 1*—into orbit on January 31, 1958. The slender projectile-shaped satellite, 80 inches (203 centimeters) in length and weighing almost 31 pounds (14 kilograms), orbited at its highest point some 984 miles (1,583 kilometers) above the earth, higher than *Sputnik 1* and 2. As a result, it discovered the Van Allen radiation belts, belts of high energy particles that surround the earth like a giant doughnut.

Explorer 1 weighed about 1/36th the weight of the Soviet *Sputnik 2* that was launched three months earlier. There was every reason for T. Keith Glennan, NASA's first administrator, to say in October 1958, when the agency was first formed (one year after *Sputnik 1* achieved orbit): "Let's get on with it." Army may have beaten Navy to a successful launch, but Russia's red star still led in the space race.

The Booster Learning Curve

Launch vehicles were still unpredictable in the infant years of the Space Age. In 1958, the first full year of launches by the United States and the Soviet Union, there were nine failures out of eighteen launches on both sides—a 50 percent failure rate. Vanguards blew up, Jupiter Cs lost thrust and dived into the ocean, and Thor-Ables fizzled.

There was a slight improvement in the success rate in 1959: Only seven out of seventeen launches during that year failed. And the rocket folk continued to improve during the 1960s and 1970s.

Some Early Orbits

1958. *Vanguard 1*, launched in March 1958, not quite six months after *Sputnik 1*, was actually the third attempt to place a Vanguard satellite into earth orbit. The first two attempts, both designated as Vanguard TV 3s (the "TV" was an abbreviation for Test Vehicle), were very visible and embarrassing failures—they exploded! The number "1" was no doubt saved by the Navy for the first successful Vanguard launch, which came after *Explorer 1*. An ad hoc committee had obviously backed the wrong launch vehicle. Some experts contend that, had the Von Braun Army team been given the go ahead earlier, they could have beat the Russians into orbit by almost a year.

Launch vehicles aside for the moment, the Vanguard satellite itself—even though it weighed only 3 pounds (1.4 kilograms), less than one-sixtieth of *Sputnik 1*'s 184 pounds (83.5 kilograms)—contained the same reliable electronics and quality engineering that would eventually put U.S. astronauts on the moon before the Russians. The satellite's tiny radio continued to transmit data for six years. It was a precise tracking of *Vanguard 1* by means of its radio signal that allowed geophysicists to determine for the first time the precise shape of the earth—a slightly pear-shaped planet, with the neck of the pear near the North Pole. As a scientific package, *Vanguard 1* was sophisticated and worth waiting for. As a means of boosting shaken U.S. national pride, however, it failed.

1959. The *Explorer 6* satellite, launched on August 7, 1959, in a Thor-Able rocket, took the first crude television image of the cloud-covered earth. Its elongated elliptical orbit swung it out to 16,373 miles (26,344 kilometers) above the earth and gave earthlings the first Space-Age view of planet Earth.

1960. Less than eight months later, *TIROS 1*, the first of the prototype weather satellites, was launched into orbit. From April 1 to June 17, 1960, *TIROS 1* transmitted 22,952 photos back to earth.

In the summer of 1960, a Delta rocket blasted off from Cape Canaveral and placed into orbit a payload which then inflated into a huge, 100-foot (30.5 meter) aluminum-coated balloon. This was *Echo 1*, the first passive communications satellite in earth orbit to relay voice and TV signals from one ground station to another. A week after launch, the Bell Telephone Laboratories announced that they had made the first transatlantic wireless code transmission via *Echo 1* from Holmdel, New Jersey, to Issyles-Moulineaux, France.

But even more important to the future of the Space Age was the fact that millions of people throughout the world began to observe what appeared to be a bright star moving across the night sky. This was dramatic visual proof that the Space Age had arrived, and this bright beacon also heralded the coming communications revolution that has changed our world.

1964. *Echo 2*, a somewhat larger inflated sphere (135 feet, 41 meters, in diameter) with a more rigid skin than *Echo 1*, was launched on January 25, 1964. This satellite was used to exchange data with the Soviet Union and thus became the first cooperative effort in space between the United States and Russia.)

The First Transatlantic TV

After the Echo balloon satellites (passive reflection mirrors for radio signals, which imitated stars passing overhead), there were rapid developments in communications satellites. *Telstar 1*, built by the American Telephone and Telegraph Company, was launched on July 10, 1962, by NASA and became the first Fortune-500-company satellite. Telstar was also the first active-repeater satellite, meaning that it received, amplified, and retransmitted radio signals. This solar-cell-studded satellite, which somewhat resembled the "Death Star" of *Star Wars*, represented the beginning of a revolution in global communications. *Telstar 1* carried the first transatlantic television broadcast between the United States and Europe. The signal was beamed from a ground station in Andover, Maine, to *Telstar 1* high over the Atlantic, which then retransmitted the signal to Europe. Today the impact of important world events is felt immediately because of the speed-of-light images transmitted through the present generation of communications satellites.

A Stationary High

Echo, Telstar, and other medium-altitude communications satellites of the early 1960s moved above the earth's surface and were able to communicate only when they were in the line of sight of two ground stations. Their orbits were never higher than 5,000 miles (8,045 kilometers), and this meant that there would have to be thirty to fifty of these medium-altitude satellites to achieve continuous global communications.

NASA tackled this problem in 1963 and 1964 by experimenting with geostationary (also called geosynchronous) orbits with a series of satellites called Syncom, which stood for synchronous communications. It was Arthur C. Clarke who first described such stationary orbits high above the earth in a now famous 1945 technical paper. In such an orbit, a satellite's orbital period around the earth is about twenty-four hours—the same as the earth's rotation period. Such a geostationary satellite circles above the earth's equator at a distance of some 22,300 miles (35,880 kilometers) and remains over the same point on earth. Three such geostationary satellites would provide planet-wide communications service.

In 1962, however, scientists were not sure that such orbits could be achieved and maintained or that the long distance over which the radio waves were transmitted would not degrade the signal's quality or cause a time lag or echo.

After one launch failure in early 1963, *Syncom 2* was launched into a synchronous orbit in late July of the same year. The satellite traced an elongated figure 8 with the cross-over point on the equator, and it proved to scientists that a synchronous orbit was practical. Then, on August 19, 1964, the world's first geostationary satellite, *Syncom 3*, was successfully put into its high orbital frontier. *Syncom 3* laid the basis for today's sophisticated and ever-growing global communications satellite system.

All Creatures on High

What did Laika, Belka, Strelka, Ptsyolka, Mushka, Chernushka, and Zvezdochka all have in common? They were canine heroes, Russian space dogs who flew various Sputnik missions during the early space days and prepared the way for human spaceflight. Three of these seven dogs, including Laika, the first living creature to orbit the earth in *Sputnik 2*, died in orbit or during reentry of their spacecraft, making the ultimate sacrifice for their Soviet masters.

Sputnik 2's Laika (in Russian, Laika means Barker) was a mongrel female dog who lived inside a cylindrical compartment for ten days. She died in orbit before the 1,121 pound (508.5 kilogram) satellite burned up as it plunged through the atmosphere. Her ten-day survival proved for the first time that living organisms could survive in outer space.

Laika's canine comrades Pchelka (Little Bee) and Mushka (Little Fly), of *Sputnik 6*, died during descent, when the spacecraft reentered the atmosphere at an incorrect angle and burned up. The dogs of *Sputnik 5*, Belka (Squirrel) and Strelka (Little Arrow), were recovered in the capsule after eighteen orbits, and Chernushka (Blackie) of *Sputnik 9* and Zvezdochka (Little Star) of *Sputnik 10* were both recovered after one orbit. These dog days of spaceflight took place from November 3, 1957 (Laika, *Sputnik 2*), to March 25, 1961 (Zvezdochka, *Sputnik 10*).

While the Russians were orbiting dogs, the United States had its own monkeys-in-space program. In fact, although the Russians had been sending animals on short up-and-down suborbital flights for several years before *Sputnik 1* beeped into history, it was the United States that accomplished the first successful spaceflight for a live creature. On September 20, 1951, a monkey and eleven mice were put atop an Aerobee sounding rocket, which blasted straight up before falling back to earth. The creatures were recovered alive. Before this flight, in the late 1940s, two monkeys (Albert 1 and 2) died in the nose cones of V-2s during test flights. Another monkey and several mice perished in early 1951, before the successful follow-up launch, when their parachute failed to open.

It was not until November 29, 1961—months after the suborbital Mercury-Redstone missions of Alan Shepard and Virgil Grissom—that the United States rocketed a chimp into orbit above the earth and recovered him alive after two circuits. The chimp's name was Enos, and he went up to be sure that an orbiting Mercury capsule was a safe place for Lieutenant Colonel John H. Glenn of the United States Marine Corps, soon to become the first American to orbit the earth in his Mercury *Friendship 7* (February 20, 1962). Both the sacrificial chimp and astronaut Glenn returned safely to earth.

Just over two weeks before *Apollo 11*'s historic voyage to the moon, *Biosatellite 3* was launched, carrying a small monkey to determine how long-duration spaceflight would affect the biology of primates. The flight was planned for thirty days, but it was cut short after nine days, and the capsule was brought back to earth on July 7, 1969, because the monkey became ill in orbit. Shortly after the capsule was recovered and opened, the monkey died. It was later determined that the monkey died because of body fluid loss. This finding confirmed that the replacement of body fluids lost during spaceflight is vital to the well-being of astronauts.

Did the *Apollo 11* crew—Armstrong, Aldrin, and Collins—know about the death of their fellow primate? It is doubtful. Considering the tremendous pressures the crew was under on the eve of this historic moon-landing mission, no one would have wanted to upset them with a dead space monkey story just days before they were to blast off to the moon. It is likely, however, that they did get the word from their physician to drink plenty of liquids on their voyage to the dry Sea of Tranquillity.

Yuri in Orbit

The first man in history to orbit the earth and witness two sunrises in just under two hours was the Soviet cosmonaut Yuri Alekseyevich Gagarin. Born in 1934, Gagarin was selected as a candidate for the U.S.S.R. cosmonaut program in early 1960. Less than a year and a half later, on April 12, 1961, he was strapped into the *Vostok 1* spacecraft perched atop the powerful A-1 rocket to be thrust into orbit on a tower of fire.

Sir Bernard Lovell, a British astronomer interviewed by the BBC at the time, said: "I think this is one of the greatest achievements in the history of mankind. It's remarkable when one realizes that this success has been achieved by a nation which a generation ago was largely illiterate."

The one orbit around planet Earth took 108 minutes (1.8 hours), and the cosmonaut was weightless for about 89 of those minutes. While in orbit, air force major Gagarin practiced drinking and eating from a small food container. He also practiced writing down his observations in his weightless state. His handwriting did not change, he said, but he had to hold on to his writing tablet so that it would not float away.

Gagarin manually piloted his spacecraft at

times while he was in orbit, but the descent, which began over Africa, and the landing were fully automatic. At an altitude of about 23,000 feet (7,000 meters), the cosmonaut was ejected from the capsule and landed separately from the *Vostok 1*. Descending under a huge parachute in his orange spacesuit and helmet, Gagarin landed in a field near the farming village of Smelovaka. A woman and small girl watched his descent to earth, and farm workers rushed to the scene, but a helicopter appeared and quickly took him away for a debriefing. The *Vostok 1* spacecraft, still hot and charred, landed some distance away from the cosmonaut's landing site.

The first human to orbit the planet Earth died a little more than a year before Americans walked on the moon. On March 27, 1968, Gagarin's jet trainer crashed while he was training for another space mission.

Life Above Earth: The First Year

The year that men from earth first flew in the void of space was 1961. After Gagarin made his historic voyage around the world in less than two hours, two of the original seven U.S. Mercury astronauts—Alan B. Shepard, Jr., and Virgil I. (Gus) Grissom—flew suborbital flights to test the structural integrity of their Mercury spacecraft and its Redstone rocket booster.

First, on May 5, 1961, Shepard piloted *Mercury-Redstone 3 (Freedom 7)* on a fifteen-and-a-half minute suborbital flight that reached an altitude of 116.5 miles (187.5 kilometers). The astronaut and capsule were recovered some 304 miles (489 kilometers) downrange. After the flight, Shepard recalled that he had only about thirty seconds to look out of the window.

While Shepard's successful flight demonstrated that the United States was continuing to build momentum in its space race with the Russians, it nevertheless appeared to be a modest accomplishment when measured against Gagarin's historic flight, which flew almost 25,000 miles (around the earth), compared with Shepard's 304 miles (489 kilometers) of up-and-

down distance. The cosmonaut's *Vostok 1* spacecraft also weighed almost five times more than the Mercury capsule.

Yuri Gagarin belittled the U.S. suborbital missions by saying that the Soviet Union had put some dogs up and down just like Alan Shepard. There was truth behind his sarcasm—a United States astronaut would not orbit the earth for another nine months.

After Alan Shepard became America's first man in space, Virgil "Gus" Ivan Grissom flew a similar suborbital mission—in *Mercury-Redstone 4*'s *Liberty Bell 7* on July 21, 1961—which reached a few miles higher and lasted a few seconds longer. But his mission had an unexpected dramatic finale that made it unique.

Grissom's *Liberty Bell 7* splashed down in the Atlantic some 303 miles (487.5 kilometers) southeast of Cape Canaveral. Then, unexpectedly, the hatch cover blew off prematurely. Water began pouring into the interior cabin of the spacecraft, and Grissom abandoned the Mercury capsule, then floated nearby it, buoyed up by his space suit. As Grissom watched the recovery helicopter hovering above, he realized that his flight suit was shipping water through an open vent. He was sinking! The increasing weight dragged him down in the Atlantic. He was going under, swallowing water. The astronaut struggled to grab a sling that was lowered from the helicopter, but the downwash from the copter's blades kept pushing him away.

Finally, after a struggle, Grissom grasped the sling and was hoisted to safety. The *Liberty Bell 7*, however, had taken on too much water for the helicopter to lift it out of the water and carry it to the recovery ship. The decision was made to abandon it, and the spacecraft sank to the bottom of the sea, some 16,500 feet (5,029 meters) below.

Grissom's July 1961 flight was to be the last American manned spaceflight for that year. But the Russians would end the historic year as they began it—with another space feat that the Americans could not match in the near future. In August of 1961, Soviet air force colonel Gherman Stepanovich Titov, the backup cosmonaut for Gagarin's first orbital flight, orbited the earth seventeen times in *Vostok 2*. The mission duration was 25.3 hours—more than a day in space. This was a fitting record to set on the last manned mission of 1961—the first year of human life in space.

God Speed John Glenn

The day had come for the United States to let the world know that it was definitely an up-and-comer in the space race. It was February 20, 1962, and Lieutenant Colonel John Herschel Glenn, Jr. was ready to be launched into space by an Atlas D rocket and orbit the earth three times in his Mercury spacecraft, *Friendship 7*. America's hopes depended on his success.

Glenn squeezed through the hatch of *Friendship 7* a few minutes after 6:00 A.M. that morning, almost four hours before lift-off. His custom-made space suit had some unique features, including fingertip lights on the gloves to provide light during periods of darkness in orbit. A mirror mounted on the chest reflected the instrument panel for in-flight photographs and a mirror on each wrist allowed Glenn to view any area of the capsule interior with minimum movement.

After the initial bumpy ride, when the entire rocket shook during powered flight, Glenn began to feel his weight increasing. During the first minutes of flight, the rocket burned more fuel than a jetliner needs to fly coast to coast across North America. At 1 minute and 56 seconds into the flight, Glenn weighed close to half a ton because of the g forces against his body. Three minutes later, he felt himself lifting slightly out of his chair and told the ground: "Zero g, and I feel fine." The ground soon gave Glenn the good news: "Seven, you have a go, at least seven orbits." The truth was that, according to Goddard Space Flight Center computers in Maryland, *Friendship 7* was good for almost one hundred orbits, although the mission plan called for only three.

Nineteen minutes after lift-off, Glenn reported seeing a beautiful view of the African coast, looking back over 900 miles (1,448 kilometers) of the Spanish Sahara. At 40 minutes into the flight, Glenn was over the Indian Ocean, watching the first of four sunsets. During the first night of Glenn's voyage, when he passed over western Australia, the people of Perth turned on all their lights as an earthbound beacon for *Friendship 7* before it went out over the Pacific Ocean during darkness. The spacecraft's cabin clock read 1 hour, 4 minutes, and the flight plan called for some food. Glenn opened his visor and squeezed a tube that squirted applesauce in his mouth.

At 1 hour, 37 minutes into the flight, the spacecraft was back over the Cape and the first orbit was complete: two more times to go around the world. During his second pass over Florida, at 3 hours, 12 minutes, he told the Bermuda tracking station: "I can see the whole state of Florida just laid out like a map. It's beautiful. I can see clear back along the Gulf Coast."

One hour and 10 minutes later, John Glenn was preparing for retrofire and his return to earth. When he was over Hawaii, the capsule communicator told Glenn: "We have been reading an indication on the ground of . . . landing bag deploy. We suspect this is erroneous." But to suspect something is not enough when a spacecraft's exterior is about to become a fireball as it plunges through the dense atmosphere during reentry with an astronaut inside. And not 3 feet (1 meter) away from the astronaut's back is the shock wave from the friction-heated air, with a temperature five times that of the sun's surface. According to the signal at the telemetry control console, the spacecraft heat shield and the compressed landing bag were not locked in position. If the signal was true, then the lifesaving heat shield was being held onto the capsule only by the straps of the retro-rocket package, which was to be jettisoned prior to reentry. What to do? This was a life-and-death decision.

Glenn was not aware of his potential danger, but he became suspicious when one tracking site after the other asked him to be sure that the deploy switch was off. There was a debate going on in the control center: One group believed the retropack should be jettisoned; the other group urged that it be retained. Christopher Kraft and Walter Williams made the final decision.

"Keep your retropack on until you pass Texas," Walter Schirra advised Glenn as *Friendship 7* passed over California.

"That's affirmative," Glenn responded. The astronaut was thinking hard now. This had never come up in any of the simulated flights. "This is *Friendship 7*. What is the reason for this? Do you have any reason?"

"Not at this time," was the reply. "This is the judgment of Cape Flight."

A few minutes later, over the Cape, ground control told him the problem and the reasons for their decision. Then there was a communications blackout as the spacecraft started its searing descent.

Glenn, cut off from the world, saw outside the window an orange glow, which grew brighter. Next he heard an explosion and saw a retropack strap hanging in front of his window. The retropack must have broken off, he thought. Then large, flaming pieces of metal came rushing back past the window. Because he thought the retropack was gone, he feared for a moment that his spacecraft was burning and breaking up. He expected at any moment to feel the first heat as the intense fire burned through *Friendship*'s walls. But it did not happen.

Back in the atmosphere, past the 100,000-foot (30,480-meter) mark, communications were reestablished.

"Seven, this is Cape . . . . Are you feeling pretty well?"

"My condition is good, but that was a real fireball, boy! I had great chunks of that retropack breaking off all the way through."

John Glenn was falling safely back to the surface of Planet earth. *Friendship 7* would splash down in the Atlantic Ocean, 4 hours, 55 minutes, and 23 seconds after lift-off, and America's manned space program was finally on its way to the moon.

Glenn's adventure, the flight of *Friendship 7*, was reported to the world by more than 1,500 news correspondents from all over the world—the largest concentration of journalists and pho-tographers in history. And some 100 million Americans viewed portions of the historic event on television.

From the Rose Garden, President John F. Kennedy expressed the sentiments of a nation—relief, pride, and thanks to all who participated. "We have a long way to go in this space race," the President said. "But this is the new ocean, and I believe the United States must sail on it and be in a position second to none."

Fireflies in Orbit: The "Glenn Effect"

When *Friendship 7* passed from its short nighttime over the Pacific, just after crossing the international date line, Glenn saw his first sunrise and exclaimed, "Oh, the sun is coming up behind me in the periscope—a brilliant, brilliant red." The sun was blinding through the scope, and he placed a dark filter over it.

Then, 1 hour and 15 minutes into flight, he turned from the periscope to the window and noticed something unusual: Was the window filled with stars? He turned away to check the instruments. All was normal. Looking back, Glenn saw that the brilliant specks were not stars at all, but luminous particles—yellowish-green in color—that drifted slowly by the window.

"It is as if I were walking though a field of fireflies," he told the Canton Island ground station. Close to the window in the shade, the particles appeared white, like snowflakes.

No one was sure what Glenn was seeing, but as the sunrise continued and the earth became brighter, the thousands of luminous particles became only a few and more difficult to see.

These "fireflies" were referred to by scientists as the "Glenn effect." They remained a space mystery until Scott Carpenter, during the follow-up three-orbit flight in *Aurora 7* on May 24, 1962, tapped *Aurora*'s interior wall with his hand. This action released hundreds of the so-called fireflies and allowed experts to determine their source: frost from the spacecraft's reaction control jets sparkling during the orbital dawns.

After-Splashdown Facts

- WHY HAD John Glenn's *Friendship 7* splashed into the Atlantic about 40 miles (64 kilometers) short of the predicted area? Apparently, those who worked out the retrofire calculations had not factored in the spacecraft's weight loss due to consumables, especially the thruster and retrofire fuels. That this rudimentary consideration was neglected forcibly demonstrates to everyone the high risk involved in the 1962 flight of John Glenn in *Friendship 7*.

- ASTRONAUT GLENN lost 5 pounds, 5 ounces from his preflight weight, which averages out to just over 1 pound for every hour of flight (from lift-off to splashdown—4 hours, 55 minutes and 23 seconds).

- A METAL fragment from the Atlas booster rocket that thrust Glenn into orbit landed on a farm in South Africa about three hours after splashdown. This meant that the spent Atlas rocket stayed in orbit for about eight hours before its fiery plunge to earth.

- FEBRUARY 26, 1962, was proclaimed John Glenn Day in Washington, D.C. It included a White House reception, a parade down the rainy Washington streets before an estimated crowd of 250,000, and finally Glenn's informal address to a joint session of Congress.

- BUT IT was New York City's John Glenn Day on March 1 that brought out the crowd. An estimated 4 million people lined the streets of lower Manhattan to honor America's astronaut heroes—Glenn, Grissom, and Shepard.

- COLONEL JOHN Glenn retired from NASA and the Marine Corps two years after his historic flight, in 1964. He joined the Royal Crown Cola Company and entered politics. A decade later, in 1974, Glenn was elected U.S. senator from the state of Ohio. In 1984, he was defeated in his bid for the Democratic nomination for President of the United States.

- JOHN GLENN traveled 83,450 miles (134,271 kilometers) on three orbits around the earth. His average speed was, therefore, 17,031 miles (27,403 kilometers) per hour. His flight pay for three circuits around the planet was $245, which came out to less than one third of a U.S. penny per mile traveled.

- IF JOHN Glenn's orbit were shown to scale on a 12-inch (30.5-centimeter) globe, it would circle only .25 (.64 cm) inch above the globe's surface.

Other Mercury Flights

After John Glenn's flight in early 1962, three more Mercury-Atlas missions were flown before Project Mercury was completed. In total, two suborbital and four orbital missions were successfully flown in the Mercury program in the early sixties. Together, they represented America's initiation rites into manned spaceflight, the first series of test flights to build the knowledge, men, and hardware necessary for the Apollo voyages to the moon.

Here are some highlights of the other Mercury flights:

Mercury-Atlas 7: **May 24, 1962.** Scott Carpenter's flight aboard the *Aurora 7* in the spring of 1962 basically duplicated John Glenn's flight, thereby confirming its success. But there was one change for the better: The astronaut had more than applesauce to eat. The Pillsbury Company supplied the flight with three high-protein cereal snacks, The Nestlé Company sent along some "bone bones" made of high-protein cereals with raisins and almonds—space granola squares. This food, Carpenter soon learned, crumbled badly. The floating crumbs were not just annoying; they were also hazardous to his breathing.

Like Glenn, Carpenter heard hissing sounds

as *Aurora 7* blazed along its return path to earth, but there was no dramatic fireball because his retropack had been jettisoned. He saw instead a light green glow.

All the way through the strong g forces that built up during reentry, Carpenter kept talking, but it became difficult to squeeze words out. During the peak g period, it took Carpenter a forceful breath to utter anything.

Aurora 7 overshot its predicted impact point in the Atlantic near Puerto Rico by some 250 miles (402 kilometers). This was because of the spacecraft's position at the time of retrofire and because there was a three-second delay of retrofire. These factors resulted in *Aurora 7* and Carpenter bobbing in the Atlantic ocean for three hours—the same amount of time it took him to go two times around the world. He was out of radio contact for forty-one minutes, and his whereabouts were unknown. This made a lot of NASA people very nervous. Finally, after he signaled a nearby aircraft with a mirror, the world found Carpenter again.

Mercury-Atlas 8: **October 3, 1962.** *Sigma 7,* flown by Walter "Wally" Marty Schirra for six orbits around the earth, had a textbook flight from launch to recovery. The splashdown occurred only 5 miles (8 kilometers) away from the prime recovery ship, the U.S.S. *Kearsarge.*

During Schirra's fourth circumnavigation, about six hours into the flight and above California, the first-ever telecast was beamed back from orbit. John Glenn was Wally's on-the-earth interviewer, and the images from space were relayed via the Telstar satellite to television audiences in Western Europe.

One of Schirra's unique and memorable phrases was: "I'm in chimp configuration," which told the ground that all the spacecraft's systems were on automatic and working well.

Mercury-Atlas 9: **May 15, 1963.** Astronaut L. Gordon Cooper, Jr. piloted his *Faith 7* spacecraft on an orbital marathon May 15 and 16, 1963. The twenty-two-orbit mission took 34 hours, 19 minutes, and 49 seconds, and it was the last

MERCURY MANNED SPACEFLIGHT SUMMARY

Spacecraft Name	Crew	Date	Flight Time (Hrs., Min., Sec.)	Revolutions	Remarks
Freedom 7	Alan B. Shepard, Jr.	5/5/61	00:15:22	Suborbital	America's first manned spaceflight.
Liberty Bell 7	Virgil I. Grissom	7/21/61	00:15:37	Suborbital	Evaluated spacecraft fuctions.
Friendship 7	John H. Glenn, Jr.	2/20/62	04:55:23	3	America's first manned orbital spaceflight.
Aurora 7	M. Scott Carpenter	5/24/62	04:56:05	3	Initiated research experiments to further future space efforts.
Sigma 7	Walter M. Schirra, Jr.	10/3/62	09:13:11	6	Developed techniques and procedures applicable to extended time in space.
Faith 7	L. Gordon Cooper, Jr.	5/15-16/63	34:19:49	22	Met the final objective of the Mercury program—spending one day in space.

Courtesy NASA.

Mercury flight of the U.S. space program.

Cooper successfully deployed a strobe light experiment during the third orbit and later, on the night side of the fourth orbit, was able to see the flashing beacon. Cooper therefore became the first man ever to launch a satellite while in orbital flight.

Because there was some concern that water in the system was responsible for a short circuit that caused a panel light to malfunction, flight directors decided that Cooper should reenter the atmosphere in the manual mode rather than trust in an automatic system that could malfunction at a critical phase of the mission. He thus became the first American to reenter the atmosphere in the manual mode while at the mercy of powerful g forces.

Cooper slept and napped more than any other astronaut before him, and this is natural because of the duration of his flight. Still, what is amazing is that he took his first nap early on, when he was still on the launchpad waiting to blast off into orbit!

Red Star Still Rising

As the U.S. Mercury project came to an end with the flight of Cooper's *Faith 7* in mid-1963 after five years and a cost of $384 million, the Russian space program continued to make dramatic gains. Both nations knew that rendezvous in earth orbit between two spacecraft was an essential technique to master for a lunar flight. This was one of the main objectives of the upcoming Project Gemini, but as usual, even five years into the Space Age, the Russians were there first: In August 1962, they put two spacecraft into orbit at the same time to test their rendezvous techniques.

Vostok 3 (call sign *Falcon*) was launched on August 11, 1962, and piloted by Maj. Gen. Andrian Gregoryevich Nikolayev of the Soviet air force. The next day, *Vostok 4* (call sign *Golden Eagle*) rocketed into orbit with cosmonaut Pavel Romanovich Popovich at the controls.

The two spacecraft came within 4 miles (6.4 kilometers) of one another, but they did not link up because they were not equipped to do so. Western experts believed that these earth-orbited rendezvous techniques were needed to assemble moon rocket components—one of the basic strategies of a manned flight to the moon.

Vostok 3 and *Vostok 4* both landed by parachute on August 15, just a short distance apart, and Nikolayev in *Vostok 3* had set a new space endurance record with sixty-four orbits in 94.4 hours.

Russian Rendezvous Two

The second two-spacecraft Russian mission, in June 1963, marked the end of the Vostok spaceflight program as well as the beginning for women in space, with the first flight of a female cosmonaut, Valentina Tereshkova, a twenty-five year-old former textile mill worker whose hobby was parachute jumping. Her spaceship, *Vostok 6*, was launched June 16 and came to within 3 miles (4.8 kilometers) of her mission companion spacecraft, *Vostok 5*, piloted by Lt. Col. Valery Fyodorovich Bykovsky. The spacecraft did not dock, however. The *Vostok 5* was launched two days earlier than *6*, on June 14, 1963. Both spaceships landed safely on June 19 in the region of Karaganda, a city about 1,300 miles (2,092 kilometers) southeast of Moscow, after eighty-one orbits for *Vostok 5* and forty-eight orbits for *Vostok 6*.

A Vostok Marriage

A few months after her seventy-hour flight in *Vostok 6*, on November 3, 1963, the first woman cosmonaut, Valentina Tereshkova, married cosmonaut Andrian Nikolayev, the pilot of *Vostok 3*. The next year the couple gave birth, on June 7, 1964, to a daughter named Yelana, who weighed in at 6 pounds, 13 ounces (3 kilograms). There was speculation that the marriage and pregnancy were encouraged and sanctioned by the Soviet government, even though the couple had a genuine relationship.

Was it really an experiment of the Soviet

space program to learn how orbital flight affected human reproduction? No one in the West knew for sure. Valentina Tereshkova had a normal, healthy pregnancy. Two-plus days of space travel appeared to have no effect on ovulation, conception, or fetal gestation.

Gemini: Two by Two

Twenty-two months passed between the end of Project Mercury and the first two-man mission of Project Gemini. It was the mid-1960s. President Kennedy was dead, and the American people still mourned. But his vision and goal to put men on the moon before the end of the 1960s was well defined, and the program was building momentum.

The Gemini manned space program was the junior year of America's commitment in space, the time when skills were acquired and hard decisions were made. The accomplishments of this program would determine if Americans would walk on the moon before 1970.

The twelve flights of Gemini, two early unmanned flights and ten manned missions, had to answer three basic questions. Could men and equipment remain in space and function well up to two weeks? Could two spacecraft launched at different times rendezvous and dock with each other in earth orbit and then change orbit together? Could astronauts perfect methods of reentry through the earth's atmosphere and attain precision-controlled landings? These very basic space faring skills had to be mastered before a moon voyage could be attempted.

Planning began in 1961 for an earth-orbital rendezvous program to follow Project Mercury, and it was originally described as Mercury Mark II. The name Gemini was chosen in early 1962 because it means "twins" in Latin and is a constellation made up of twin stars Castor and Pollux. It was appropriate because the Gemini spacecraft would fly a two-man crew and would rendezvous or dock with another spacecraft. Also, as a space project it was number two, after Mercury, in the American space program.

At first glance, a Gemini spacecraft looks like a larger Mercury capsule, but this is a completely false impression. Gemini, while only 20 percent larger than Mercury, had twice the cabin space. It weighed 8,400 pounds (3,810 kilograms)—two and a half times the weight of a Mercury capsule. Storage batteries supplied the electrical power to Mercury, but an innovative fuel-cell system in Gemini supplied power to the spacecraft by mixing hydrogen and oxygen, with water as a by-product, and chemically producing electrical current.

The real difference between these first- and second-generation spacecraft, however, was in the controls—Gemini carried what was called an orbital attitude maneuvering system (OAMS), which consisted of sixteen thrusters of varying power. The Mercury capsule, however, emphasized automatic controls, so that even Enos, a chimpanzee, could orbit the earth and land safely. But Gemini was built for astronaut pilots, who could exercise much more control over the spacecraft, even during the critical reentry, than astronauts could over their Mercury capsules. Gemini could be maneuvered in space in any direction. The orbital path could be made higher or lower by acceleration or deceleration with the thrusters. The spacecraft could be maneuvered sideways or up or down or made to pitch, roll, and yaw to any degree. The Gemini astronauts were true spacefaring pilots, and their missions proved it.

Besides building the skills for moon voyages, Gemini contributed to man's new view of the home planet Earth. Some 1,400 color photographs taken of the earth from various altitudes taught us much about our planet. Scientists realized that photographs of the earth could serve as a valuable tool to help identify and husband the planet's dwindling resources. Perhaps future historians will see this aspect of the Gemini program as its most lasting contribution.

Gemini 3 (Molly Brown). The first manned flight of the Gemini Project (March 23, 1965) was the last one to carry the crew's choice of an unofficial nickname—*Molly Brown.* Project Mercury and the original seven astronauts had started this practice, and it appeared at first that

Ed White became the first U.S. space walker and the world's first self-propelled astronaut on Gemini 4 *in 1965. When he had to return to his spacecraft, he said, "It was the saddest moment of my life."* Courtesy NASA.

this spacecraft-naming tradition would continue. Gordon Cooper, however, had had trouble selling NASA his choice of *Faith 7* for the last spacecraft in the Mercury program. The crew of *Gemini 3*, Virgil I. Grissom and John W. Young, had had the same difficulty with *Molly Brown*, a choice that implied the spacecraft was "unsinkable" (a reference to the heroine of the Broadway stage hit), unlike Grissom's Mercury capsule, which sank to the bottom of the Atlantic. NASA management thought the name lacked dignity—a typical bureaucratic response—but they grudgingly consented when they heard Grissom's second choice: *Titanic.* After the Grisson-Young *Gemini 3* flight, NASA announced that all forthcoming Gemini flights would use a single official designation for the mission. A little pizzazz left the program during the other Gemini flights.

Did this Gemini spacecraft do what it was designed and engineered to do? Almost. The mission was for only three earth orbits—4 hours, 53 minutes—and the all-important spacecraft orbital changes were performed. By firing the 100-pound (45.3 kilogram) thrusters for 74 seconds, Grissom and Young were able to make the orbit more circular; the highest point dropped about 34 miles (55 kilometers). This was a historic first. Another orbital change was executed on the second orbit. On the third and final revolution, the crew attempted to fly the *Molly Brown* to a splashdown area in the Atlantic after the reentry, but the descent control was off, and they splashed down 58 miles (93 kilometers) short of the projected impact area. But this was a shakedown flight for Gemini, and *Gemini 3*'s major objectives were carried out. The smuggled-in corned beef sandwich that Young had gotten from Schirra and had given to Grissom to eat during the flight caused an uproar in Congress, however. More stringent rules were the result, and astronauts had strict limitations as to what they could take on board their Gemini spacecraft.

Gemini 4. This mission blasted off on June 3, 1965, for a four-day flight in orbit, almost twenty times longer than the first manned Gemini flight. Millions of people around the world viewed the launch because the scene was broadcast to twelve European nations via the Early Bird satellite. Also, for the first time in the U.S. space program, the new Mission Control Center in Houston was the nerve center for the flight. Add to these factors the planned extravehicular activity of the mission, and the flight of *Gemini 4* generated a high public interest that was never again matched in the Gemini program.

Astronaut James A. McDivitt was the command pilot, and astronaut Edward H. White II was pilot. It was White who took the United States' first "space walk," an extravehicular (outside the spacecraft) activity, or EVA, that lasted twenty-two minutes.

Their first task in orbit was to rendezvous with the second stage of the Titan 2 rocket that had thrust them into orbit. Their target had a flasher beacon to make it easier to spot. This first attempt at orbital navigation proved more difficult than expected, however, and *Gemini 4* never got as close to its booster as the flight plan had called for. But again, astronauts were learning how to do something for the first time—the start of the learning curve. The plan for flying in formation ("station-keeping") with the booster was abandoned when half the thruster fuel was spent.

Next it was Edward White's 6,000-mile (9,654-kilometer) "walk" in space. The Gemini cabin was depressurized over Australia; after that, all White had to do was open the hatch above him and float or push himself into outer space. As *Gemini 4* orbited high above Hawaii, he left the spacecraft. With a hand-held portable gas thruster, nicknamed a Zot gun, White propelled himself to the end of his gold-plated lifeline tether and continued to practice his weightless acrobatics traveling almost 18,000 miles (28,926 kilometers) an hour high above the earth. He became the world's first self-propelled astronaut.

White was more of a space "swimmer" than a space "walker," and he remained on his EVA as he swam over California, Mexico, and Texas, toward Bermuda. Once his shoulder brushed the spacecraft's window, and McDivitt exclaimed,

"You smeared up my windshield, you dirty dog!"

The EVA time was almost over, but White was euphoric, and he didn't want to stop his above-earth swim. It was, he said, "the saddest moment of my life." But he was finally persuaded by McDivitt and the ground communicators to return to *Gemini 4* before it entered darkness.

After completing sixty-two orbits, the crew splashed down in the Atlantic about 50 miles (80 kilometers) short of the planned target, on June 7, 1965.

Gemini 5. Astronauts Gordon Cooper and Charles "Pete" Conrad were blasted into orbit on August 21, 1965, and their spacecraft was the first to fly with fuel cells to produce electrical power on their 8-day, 120-orbit flight. On the sixth day, Cooper and Conrad surpassed the Russian *Vostok 5* flight record of June 1963. The mission established a new world record.

Much of the schedule was devoted to evaluating the navigation and guidance system for future Gemini rendezvous with target vehicles launched by the Atlas-Agena D rocket. This was, after all, the raison d'être for the program. But problems with the fuel-cell power system occupied the crew, and a deployed radar pod ran out of power before it could be used in rendezvous exercises. A simulated Agena rendezvous—which involved catching an imaginary point in orbit by executing four complex maneuvers using the orbit attitude and maneuver system—was conducted. This was a complete success, and many mission planners began to sleep better. While many problems plagued this mission, most of them were solved. An example of one of the easiest: When the spacecraft was powered down to conserve fuel, it tumbled through space, and the astronauts were bothered by the stars spinning around outside the window. What did they do? They blocked out the dizzying view by covering the windows.

Gemini 5, after eight days in space, splashed down safely on August 29, 1965.

Gemini 7 and *Gemini 6:* The Spirit of "Gemini 76." Late 1965 was a busy time for the Gemini program. Astronauts Walter M. Schirra, Jr. and Thomas P. Stafford were on the launchpad ready to go on the morning of October 25, but their mission was scrubbed when the rendezvous and docking target vehicle—an Agena atop an Atlas ICBM—failed to achieve orbit. The main goal of their mission was, all of a sudden, impossible, even before they got off the ground. Their flight was rescheduled for December, which put *Gemini 7* into orbit before *Gemini 6*. The two spacecraft would eventually fly in orbital formation together. NASA control room personnel referred to the overlapping missions as "Gemini 76."

About six weeks after the scrub, on December 4, 1965, *Gemini 7* was launched with Frank Borman and James A. Lovell, Jr. at the controls. They would fly just short of two full weeks (13 days, 18 hours, and 35 minutes), set a new world's record for manned spaceflight endurance, and just barely stand up on rubbery legs on their return to earth.

Edwin Aldrin performed the first work in space on his Gemini 12 *EVA in late 1966 before becoming the second man to walk on the moon in 1969.* Courtesy NASA.

On December 15, *Gemini 6* was finally launched, and Schirra and Stafford began to chase *Gemini 7*, which was acting as a rendezvous target. During the third orbit, Schirra in *Gemini 6* told the ground that his radar had locked on to *Gemini 7*. The spacecraft were 246 miles (396 kilometers) apart. About an hour later, Schirra exclaimed, "My gosh, there is a real bright star out there. That must be Sirius." It turned out to be *Gemini 7*, reflecting sunlight from 62 miles (100 kilometers away). Later still that day, after chasing its twin craft a few times around the earth and executing several crucial maneuvers, *Gemini 6* eased to within 1 foot (.3 meter) of *Gemini 7* and later flew around it. A sign in the portal of *Gemini 6* read *BEAT ARMY*. All crew members were either Air Force or Navy.

The two spacecraft orbited together for another three and a half turns around the world before separating. On December 15, 1965, the two spacecraft and their four astronauts had accomplished for the first time a complex rendezvous high above the earth, just as it would have to be done by the Apollo missions around the moon. Also, the first photograph of another orbiting spacecraft was taken: one of *Gemini 7* from *Gemini 6*.

Gemini 6 reentered the atmosphere and returned to earth on December 16, leaving the orbital marathoners, Borman and Lovell, to continue for nearly three more days to reach their record time of almost two weeks. These last days would be hard on the *Gemini 7* crew. They did some reading to pass the time. Borman read some of Mark Twain's *Roughing It*, and Lovell some of Walter D. Edmond's *Drums Along the Mohawk*, two books that had *nothing* to do with space. They toughed it out for the last two days and proved that astronauts could hack a moon voyage lasting up to two weeks.

The year 1965 had been a fabulous one for manned space flight. It was a Merry Christmas for the U.S. space program.

Gemini 8. Scheduled to last three days, this mission was aborted after the spacecraft was in orbit, but the crew duo, Neil A. Armstrong and

David R. Scott, returned safely to earth on March 16, 1966, after a mission time of 10 hours and 41 minutes.

Even though this was the first emergency landing of a manned U.S. spacecraft, the mission nonetheless fulfilled one of its two primary objectives—to rendezvous and dock with the Gemini Agena target vehicle that was launched into orbit about two hours earlier on the same day.

After insertion into orbit, *Gemini 8* completed nine orbital-change maneuvers before it rendezvoused with the Agena target, after about six hours, during its third orbit. In another half hour, a hard docking between the two spacecraft was completed—a world's first in manned spaceflight.

"Flight, we are docked!" Armstrong radioed the ground. "It's ... really a smoothie—no noticeable oscillations at all." But Armstrong had spoken too soon. In less than half an hour, the *Gemini 8* spacecraft would be completely out of control and the crew would be fighting for their lives. (For an account of this dangerous period of the *Gemini 8* mission, read Chapter 5, "Space Disasters and Close Calls.")

Following the emergency landing of Armstrong and Scott in *Gemini 8*, the Agena target rocket remained in orbit and ground control used it to conduct tests. The main engine was fired nine times, and some five thousand commands were received and executed by Agena's control system—five times the number required by the contract between NASA and the manufacturer. Finally, during its tenth day in orbit, the electrical power went out and the Agena was placed into a circular decay orbit. The Agena test added to the rapidly growing store of knowledge needed for the Apollo program, as did the aborted *Gemini 8* mission. It was, after all was said and done, a dangerous intersection of the learning curve that was heroically negotiated by the man who would be the first to walk on the moon.

Gemini 9. After the tragic deaths of the primary crew, Elliot See and Charles A. Bassett II, in a T-38 aircraft crash at Lambert Field in St. Louis on February 28, 1966, NASA assigned Thomas P. Stafford and Eugene A. Cernan to pilot the mission.

Rendezvous was the main goal of the *Gemini 9* flight. Second, the astronaut maneuvering unit (AMU) was to be tested during an EVA outside the spacecraft. But the Agena target vehicle failed to orbit in May, and the mission was scrubbed. A makeshift target vehicle was then substituted and successfully launched on June 1, 1966. Two days later, *Gemini 9* followed it into orbit, and on its third revolution, it rendezvoused with the 12-foot (3.7-meter) target. But the target was not receptive to docking; it was, instead, an "angry alligator," its protective shroud still attached, the two halves opened like jaws. Docking was impossible, even though the rendezvous skills were practiced.

Astronaut Cernan began his space walk on June 5. In struggling to don his backpack maneuvering unit, Cernan breathed so hard that his faceplate became fogged. His space suit's air conditioner could not handle the moisture load, and his vision was limited. After about an hour, it was realized that no useful activities could be performed under these conditions, and Cernan reentered *Gemini 9*.

Gemini 9 gave NASA more problems to solve. For one, they learned that EVA work would not be easy. That was important to know. But at least *Gemini 9* splashed down to one of the most precise landings of the Gemini program—less than 2 miles (3.2 kilometers) from their recovery ship, the *Wasp*. And after three days and forty-five orbits, the two astronauts returned safely to earth—the true measure of a successful mission.

Gemini 10. The crew of this mission, John W. Young and Michael Collins, would become famous later in the space program, Young as the commander of the first *Columbia* space shuttle mission and Collins as the command module pilot for the historic *Apollo 11* mission. Their three-day mission included two rendezvous with two different Agena target vehicles—their own and the *Gemini 8* Agena. A space walk to retrieve an experiment package from the *Gemini 8* Agena target was also planned. It was the most ambitious manned U.S. mission to date.

Rendezvous with their Agena target took place just 5 hours and 21 minutes after the launch, on July 18, 1966. After the spacecraft was mated and docked with the Agena, they fired the powerful Agena propulsion system, which sent them into a new 458-mile (737-kilometer) high orbit—a new altitude record for astronauts. At the time of the rocket thrust, Young and Collins were thrown forward in the cockpit. "We got a tremendous thrill on our way to apogee," Young said. The crew and their two joined spacecraft would remain in the record-high orbit overnight.

At sunset, during the fifteenth orbit, both hatches were opened. Collins then "stood" into space, by "standing" in his seat, and carried out an astronomy experiment that called for photographing the stars in ultraviolet light—something impossible to do on earth because of the atmosphere. He aimed the camera at the southern Milky Way and exposed twenty-two frames. At daylight, the astronaut began some color photography, but this was interrupted when an irritant in the oxygen system caused both astronauts' eyes to fill with tears and they were temporarily blinded, as if tear gas had found its way into their space helmets. They immediately closed the hatches and repressurized the cabin. With new oxygen in the cabin, the crew recovered. No physical evidence of the irritant was found after the flight, but the flight surgeon, Dr. Charles Berry, believed that it may have been the antifogging mixture used to keep the faceplates clear of moisture.

After thirty-nine hours, *Gemini 10* undocked from its target Agena, and on the third day of the mission, Young and Collins maneuvered their spacecraft to a rendezvous with the *Gemini 8* Agena in another orbit. Here Collins opened the hatch and propelled himself with the handheld Zot gun over the Agena, where he collected two scientific packages that scientists wanted to evaluate—a canister of microorganisms (T-1 bacteriophage), some of which had survived the hostile environment for four months, and a micrometeoroid collector package. Collins, the first astronaut to retrieve a scientific package from

orbit, returned to the spacecraft after an EVA lasting thirty-nine minutes. With the long umbilical lifeline coiling all over the cabin, Young had to help Collins untangle it. Young said it made "the snakehouse at the zoo look like a Sunday school picnic."

After a sleep period, the crew prepared for reentry, which ended with a pinpoint landing only 3.4 miles (5.5 kilometers) off their target. The sailors on the *Guadalcanal* recovery ship watched the spacecraft hit the water and cheered.

Gemini 11. Launched during a two-second launch window on September 12, 1966, this second-to-last Gemini mission achieved the altitude record for the program when their Agena target spacecraft boosted them 853 miles (1,372 kilometers) above the earth while they were over Australia. From that altitude, Charles "Pete" Conrad and Richard F. Gordon could view more than 17 million square miles (10 billion square acres) of the earth's surface—all of Australia, Borneo, and Southeast Asia spread out beneath them.

Even more important to the Apollo moon missions to follow, however, was that Conrad and Gordon caught up to and docked with their Agena target between Hawaii and Texas during their first orbit, just one and a half hours after launch. This simulated a lunar orbit rendezvous, which was essential for a moon landing.

On the second day of the flight, during the fifteenth orbit, Gordon began an EVA, and he floated over Houston and watched lightning flashes below. Then he hitched a 100-foot (30.5-meter) nylon rope from *Gemini 11* to the Agena to determine how the tethered vehicles would react after they undocked. But the physical exertion of the task proved too much—he was breathing 40 times a minute, and his heart rate soared to 102 beats a minute—a high pulse for him in zero gravity. To compound his problem, his faceplate was fogged up, and, like Cernan of *Gemini 9*, he could not see. The EVA was terminated after forty-four minutes, less than half the planned time, and the cabin was repressurized.

Later, the crew performed some tests to determine how the two tethered vehicles would perform. Once the tether was taut and the resulting spacecraft oscillations had damped down, the joined vehicles slowly rotated around each other once every nine minutes. A camera that had been floating in the cabin moved back against the bulkhead. This proved that gravity had been created—not much, about 1.5-thousandths of our normal earth gravity, but artificial gravity nonetheless—the first time it had every been done. Larger space stations of the future, such as the famous torus wheel design of Arthur C. Clarke's *2001*, may create artificial gravity by spinning.

After a sleep and another practice rendezvous and docking with the Agena, they cast off from their orbital companion for the last time. "We were sorry to see that Agena go," Gordon said. "It was very kind to us."

The flight plan of *Gemini 11* called for an automatic reentry, a hands-off splashdown, unlike the other Gemini missions. On September 15, 1966, during the forty-fourth orbit around the earth, the retro-rockets fired, and Conrad and Gordon watched the computer make its decisions and adjustments all the way down to the ocean waves. The computer had done itself proud. *Gemini 11* came down only 2.8 miles (4.5 kilometers) from the recovery ship, the *U.S.S. Guam*. As it made its final descent, John Young in Houston told the crew: "You're on TV now." The Space Age had come to America's living rooms.

Gemini 12. On Veterans Day, November 11, 1966, the last Project Gemini spacecraft was launched skyward. James A. Lovell and Edwin E. "Buzz" Aldrin were the last two Gemini astronauts before the Apollo years began. As they shuffled up the ramp on pad 17, they each wore a sign on their backs. Together the two signs read *THE* and *END*, and they told the literal truth.

The Gemini launch team would soon be disbanded. And just hours after blastoff, wreckers began hacking the launch stand into scrap iron. It was time to make room for Apollo.

Gemini 12 rendezvoused with its Agena target rocket 3 hours and 45 minutes after launch. This was excellent time, considering the fact that about an hour after launch, radar reception got so bad that the on-board computers refused to accept the intermittent readings. Backup charts and alternative procedures worked out by Aldrin ("Dr. Rendezvous") at MIT, but never tested in space, made the rendezvous possible. It was a brain-wrought rendezvous, accomplished without a computer.

Besides additional all-important practice in docking and undocking with the Agena, the heart of the mission was Aldrin's EVA.

If there was any surprise in the overall Gemini program, it was that the difficulty of performing tasks during EVA had been grossly underestimated. Eugene Cernan had had his troubles during the EVA of *Gemini 9*, and Richard Gordon of *Gemini 11* had suffered from exhaustion and overheating. It was decided to forget the astronaut maneuvering unit on this mission and have Aldrin devote his time to a series of simple tasks that could be accurately measured in terms of work load.

Aldrin's months of practice in the underwater tank environment paid off, just as they would more than a decade later for shuttle astronauts who would practice building space station structures in space or repair satellites in orbit. His deliberate, careful motions helped him stay cool. The MIT astronaut pulled himself to the fantail of the Gemini adapter and performed the work tasks of the flight plan: plugging and unplugging connectors, screwing and unscrewing bolts, manipulating hooks and rings—about twenty assigned tasks in all.

Carefully monitored by the flight surgeon, Charles Berry, Aldrin displayed no unusual stress. He proved to mission planners that astronauts, if properly trained, could perform useful work outside their spacecraft and eventually gain skills that would be of utmost importance to humankind's future in space. Setting an EVA world record with a total of 5 hours and 30 minutes (which included one space walk and two stand-up, hatch-open periods), Aldrin became the world's first space worker.

After fifty-nine orbits, *Gemini 12* began its computer-controlled reentry on November 15, 1966, and the last Gemini spacecraft splashed

GEMINI MANNED SPACEFLIGHT SUMMARY

Spacecraft Name	Crew	Date	Flight Time (Hrs., Min., Sec.)	Revolutions	Remarks
Gemini 3	Virgil I. Grissom John W. Young	3/23/65	04:52:31	3	America's first two–man spaceflight.
Gemini 4	James A. McDivitt Edward H. White II	6/3–7/65	97:56:12	62	First walk in space by an American astronaut. First extensive maneuver of spacecraft by pilot.
Gemini 5	L. Gordon Cooper, Jr. Charles Conrad, Jr.	8/21–29/65	190:55:14	120	Eight–day flight proved man's capacity for sustained functioning in space environment.
Gemini 7	Frank Borman James A. Lovell, Jr.	12/4–18/65	330:35:01	206	World's longest manned orbital flight.
Gemini 6A	Walter M. Schirra, Jr. Thomas P. Stafford	12/15–16/65	25:51:24	16	World's first successful space rendezvous.
Gemini 8	Neil A. Armstrong David R. Scott	3/16/66	10:41:26	6.5	First docking of two vehicles in space.
Gemini 9A	Thomas P. Stafford Eugene A. Cernan	6/3–6/66	72:20:50	45	Three rendezvous of a spacecraft and a target vehicle. Extravehicular exercise—2 hours, 7 minutes.
Gemini 10	John W. Young Michael Collins	7/18–21/66	70:46:39	43	First use of target vehicle as source of propellant power after docking. New altitude record—458 miles.
Gemini 11	Charles Conrad, Jr. Richard F. Gordon, Jr.	9/12–15/66	71:17:08	44	First rendezvous and docking in initial orbit. First multiple docking in space. First formation flight of two space vehicles joined by a tether. Highest manned orbit—apogee about 853 miles.
Gemini 12	James A. Lovell, Jr. Edwin E. Aldrin, Jr.	11/11–15/66	94:34:31	59	Astronaut walked and worked outside orbiting spacecraft for more than five and a half hours—a record proving that a properly equipped and prepared man can function effectively outside his space vehicle. First photograph of a solar eclipse from space.

Courtesy NASA.

down less than 3 miles (4.8 kilometers) from the planned landing point.

The Gemini flag and pennant that had flown over the Houston Manned Spacecraft Center during all missions were lowered for the last time. President Lyndon Johnson summed up the Project Gemini years the day the last Gemini spacecraft fell into the sea:

"Ten times in this program of the last twenty months we have placed two men in orbit the earth in the world's most advanced spacecraft. Ten times we have brought them home.

"Today's flight was the culmination of a great team effort stretching back to 1961 and directly involving more than 25,000 people..."

Then the President looked toward the future: "Apollo will make America truly a spacefaring nation. The three-man Apollo is the certain forerunner of the multimanned spaceships of the not too distant future...ships that will bear the hopes of all men."

Taking the Moon's Measure

The moon beckoned during the early 1960s, the beginning of the Space Age, some 350 years after Galileo had observed it through his 30-power telescope. In 1610, the great Italian astronomer described the surface of the moon as "neither smooth nor uniform, not very accurately spherical . . . [but] uneven, rough, replete with cavities and packed with protruding eminences . . . ."

The United States had made a commitment before the world to land men and machines on the moon before 1970, but scientists in the early sixties were not much more knowledgeable about lunar surface characteristics than Galileo had been three and a half centuries earlier. Would the descending moon rocket, for example, sink deep into, or even bury itself in, a thick blanket of lunar dust? No one knew for sure, but both the United States and the Soviet Union decided to build and fly machines that would discover the answer.

The first, *Luna 1*, was launched by the Russians on January 2, 1959, and its 797-pound (362 kilogram) payload (named *Mechta* for "dream") was intended to crash-land on the moon's surface. But it missed the moon by 3,728 miles (5,998 kilometers) and went into orbit around the sun.

This was the first of twenty-four Russian probes in the Luna series (1959–1976), which would eventually: hit the moon; send back the first photographs of the moon's far side; orbit and reconnoiter the moon; test the lunar soil; return a sample of lunar soil; deliver *Lunokhod 1*, the robot moon rover; and more.

The Americans had four different spacecraft programs to snoop and probe the moon before men were sent. The earliest was the Pioneer program, which grew out of the International Geophysical Year and was inherited by NASA from the Department of Defense. *Pioneer 1, 2,* and *3,* launched in late 1958, failed to reach the moon but sent back useful data on the atmosphere and radiation belts. The most successful Pioneer probe was *Pioneer 4,* launched on March 3, 1959, which measured particles and fields during a moon flyby before it went into orbit around the sun. *Pioneer 4* was, however, some 37,000 miles (59,533 kilometers) wide of its moon mark.

After these initial moon probes, the Pioneer program was devoted to planetary missions, which included the first man-made objects to leave the solar system.

Then the Ranger program began in the summer of 1961, and it too had a high failure rate at first—expensive practice, but there was no other way. *Ranger 1* and *2* both failed to reach deep space; *Ranger 3* missed the moon by almost 23,000 miles (37,000 kilometers); *Ranger 4* landed on the far side of the moon in April 1962 (more than two years after the Russians hit the moon with *Luna 2*), but its timer failed and no data was returned to earth; it was a near miss with *Ranger 5,* which passed within 450 miles (724 kilometers) of the moon; aim improved when *Ranger 6* hit the moon, but its TV system failed. Finally, after almost three years of complete or partial failures, the last three Ranger spacecraft—*7, 8,* and *9*—delivered more than seventeen thousand detailed photos of the moon's surface. Even the smoothest-appearing mare areas of the moon, the photos revealed, were pockmarked with small craters. These photographs had a resolution that was some two thousand times better than the best earth-bound telescopic views. The quality of photographs taken of this region of *Mare Nubium* (Sea of Clouds) was so outstanding that the International Astronomical Union renamed this lunar area *Mare Cognitum* (Known Sea). On moon maps of today there are now two seas adjacent to one another.

As the Apollo testing and design programs consumed the skills and brains of hundreds of thousands of people, two more sophisticated moon probe programs, which flew concurrently, were begun: the Lunar Orbiter series of five spacecraft, which together made more than six thousand orbits of the moon, discovered the mascons (excess concentrations of mass under the maria) and photographed more than 99 percent of the lunar surface; and the Surveyor series of moon probes, soft-landing spacecraft that surveyed the moonscape with their TV cameras and analyzed the chemical composition of the lunar surface.

Five out of seven Surveyors flew successful missions, and their combined operating time on the moon was seventeen months. The Surveyors found out, among other things, that:

• The lunar surface was hard and could support spacecraft and astronauts.

• The surface-bearing strength of moon soil increased with depth.

• The chemical composition of the surface geology revealed a similarity to the basalt on earth.

• A surveyor's rocket engine could be remotely started and fired from earth. The short lift-off did not raise a cloud of dust and resulted only in shallow cratering.

• Laser beams from earth could be photographed from the moon—a significant communications test.

Surveyor 1, 3, 5, 6, and *7* sent back a total of more than 86,000 photographs, and the moon's

image was changed forever. As an added bonus, *Surveyor 3* bounced upon landing and photographed one of its own footprints. For decades science fiction and space artists had assumed that the moon had no erosion because it was airless and waterless, and they had depicted jagged peaks with other sharp-edge features. But the Surveyor robots sent back photos of craters with rounded rims. There was a mysterious erosion occurring on the moon; it was another reason for the Apollo ships to land and their astronauts to explore and return with pieces of the moon for scientists to study.

Early Apollo Days

Project Apollo was publicly announced at a news conference in July of 1960, almost a year before President Kennedy gave his famous to-the-moon speech to Congress and the American people. The original Apollo program, however, called for a manned circumlunar mission, not a manned landing on the moon's surface.

On May 25, 1961, when President Kennedy announced the national goal of landing men on the moon, the United States had only fifteen minutes of manned flight experience in space (Alan Shepard's suborbital *Freedom 7* flight on May 5, 1961) and a very general plan to have men circumnavigate the moon. There was no rocket available that could launch a moon expedition, nor was there even a consensus on what was the best way to reach the moon's surface safely by the end of the decade.

Hundreds of important decisions were made during the early days of the Apollo program, but certainly the major decision was which of the three basic modes should be used on a manned moon mission:

• Direct flight from earth to moon and return

• Earth orbit rendezvous

• Lunar orbit rendezvous

The direct flight to the moon required the building and testing of a gigantic Nova rocket, which would have had eight times the first-stage thrust of Saturn 1 (1.5 million pounds/668,000 kilograms). But Nova, which could place a payload equal to a locomotive (about 150,000 pounds, 68,000 kilograms) on the moon, could not be built in time to meet the deadline of Kennedy's goal, so a commitment was made to the Saturn 5, which had a first-stage thrust of 7.5 million pounds (3.4 million kilograms) and was able to lift 90,000 pounds (41,000 kilograms) of payload to escape velocity and go to the moon.

Because of Saturn 5's payload capacity, a direct flight from earth to moon was not possible. If Saturn 5 was to take men to the moon, a rendezvous mode—earth or moon—had to be chosen. It came down to a question of weight: Which mode, earth orbit rendezvous or lunar orbit rendezvous, could get the most weight to the moon safely and economically?

The earth orbit rendezvous plan called for orbiting a fuel tanker first and pumping its fuel into the Apollo third stage before it blasted off for the moon. The lunar orbit rendezvous called for a lunar excursion module, with two men aboard, to descend to the lunar surface from the Apollo command ship, which remained in orbit, and then to rendezvous with it after exploring the moon landing area.

Earth orbit rendezvous was favored during 1961–62 by several NASA groups, and John C. Houbolt, a leading proponent of the lunar rendezvous, felt that his ideas were not getting a fair hearing or being taken seriously. Houbolt persevered, however, and gradually individuals and groups endorsed the lunar orbit rendezvous. It certainly had weight and cost advantages. Early calculations put the advantage of the moon payload weight at 7,000 pounds (3,200 kilograms) versus 150,000 pounds (68,000 kilograms) for the direct moon landing payload. This along with the fact that earth orbit rendezvous required two Saturn 5 launches— one for the tanker and one for the Apollo spaceship—made the lunar orbit rendezvous attractive in terms of both payload weight and cost considerations. Finally, after a year-and-a-half battle waged by Houbolt and his supporters, Werhner von Braun and his rocket team were

won over to this mode, and a commitment to the lunar orbit rendezvous was made. It was publicly announced on July 11, 1962, after an estimated million man-hours of study by NASA, industry, and university scientists.

The lunar module concept, and its essential lunar orbit rendezvous with the command spacecraft, had seemed preposterous to many scientists at first. In the end, however, it enabled the United States to land on the moon several years earlier than otherwise possible—and at a savings of several billion dollars.

Early Saturns: 1964–65

The Apollo program required seven years of testing before astronauts could fly and test firsthand their rockets, modules, and space suits. And the final human feat of getting machines and men to land and walk in the ancient moon soil was attributable to the work of more than four hundred thousand people and hundreds of corporations and other organizations.

By 1965 the Apollo manager knew which tests were needed to qualify the spacecraft. As the last Gemini spacecraft flew in 1966 and gained invaluable experience in earth orbit, and as Surveyor and Orbiter robots photographed and analyzed the moon in 1966 and 1967, Project Apollo was quickly coming off the drawing boards and blasting skyward.

As early as 1964, the Saturn 1 rocket was being test-flown. On January 29, 1964, the second stage alone was tested. Then on May 28 and again on September 18, Saturn 1 rockets with second stages and boilerplate Apollo modules were test-fired successfully. The early Saturn 1 launch vehicle was also used and tested to launch three Pegasus satellites in 1965: *Pegasus 1* (February 16); *Pegasus 2* (May 25); and *Pegasus 3* (July 30). Named after the mythological winged horse, these huge satellites, with winged panels some 96 feet (30 meters) tip to tip holding some 2,300 square feet (214 square meters) of sensors, were sent into orbit to determine the rate of meteoroid penetrations—important knowledge to have for the safety of the future Apollo astronauts. With wings unfolded, the Pegasus satellites could be seen in the sky at night without a telescope or binoculars. The results of the Pegasus satellite experiments revealed that interplanetary dust particles were some ten thousand times less abundant than earlier measurement had indicated. This had influence on Apollo hardware design and cut down on overall spacecraft weight by about 1,000 pounds (454 kilograms).

The early version of Saturn 1 that launched the three Pegasus satellites also carried boilerplate Apollo command and service modules, and the Apollo launch escape system, which was tested during launch. All these Saturn 1 launches in the mid-sixties helped development of Saturn 1B and the great Saturn 5 that would carry manned expeditions to the moon.

Tests and Disasters: 1966–68

The unmanned flight tests of Apollo continued in 1966, 1967, and 1968. Three unmanned Apollo Saturn 1B missions were launched in 1966. All three of these flights tested the Saturn 1B rocket, and two of them were configured with Apollo command and service modules. A suborbital test launch of February 26, 1966, took the Apollo spacecraft up and then 5,500 miles (8,850 kilometers) downrange for recovery. On July 5, 1966, the second stage of Saturn 1B went into orbit, but no spacecraft was carried. Another suborbital test flight of a Saturn 1B was launched on August 25, 1966, and carried the command and service modules, both of which survived a high-speed reentry, similar to what an Apollo spacecraft and crew would have to experience on the return to earth.

Testing, evaluation, and redesign of some Apollo systems continued to ready the spacecraft for the first manned flight scheduled for February 21, 1967. The crew for this manned mission was announced on March 21, 1966: Virgil I "Gus" Grissom, Edward H. White, and Roger B. Chaffee.

As the three astronauts continued to train in the simulator, engineering changes were con-

stantly being made on the Apollo 204 spacecraft, and it was a real problem to keep the mission simulator current with the changes. At one time there were more than one hundred modifications outstanding. This upset Apollo commander Gus Grissom so much that, as an act of protest, he hung a lemon on the trainer.

A simulated countdown and launch were conducted on Friday, January 27, for the first Apollo crew. Grissom, White, and Chaffee slid into their Apollo couches and were sealed inside their spacecraft at 1:00 P.M., while a thousand men prepared for the ground test. On entering the cabin, Grissom reported a peculiar odor that reminded him of sour milk. The countdown was held while the cabin oxygen was sampled, but no impurities were found. The ground testing continued. Then at 6:32 P.M., a voice yelled. "Got a fire in the cockpit." Seconds later, after a few other broken phrases, the transmission ended with a cry of pain. Three astronauts were dead. Apollo 204 was a charred and useless wreck. And Apollo was eclipsed by tragedy and mourning. It would be more than a year and a half before the second manned Apollo mission flew. (For a more detailed account of the *Apollo 1* tragedy, read Chapter 5, "Space Disasters and Close Calls.")

Three more Apollo test flights were flown in 1967 and 1968 before *Apollo 7* took men into orbit for the first time:

Apollo 4, **(November 9, 1967).** The first Saturn 5 rocket launch. The command module reentered at moon-return velocity to test the heat shield. This first launch of the Saturn 5 first stage generated 7.5 million pounds of thrust at lift-off. The question was not whether the Saturn had risen, one witness remarked after launch, but whether Florida had sunk!

Apollo 5, **(January 22, 1968).** The first unmanned test of the lunar module in earth orbit, launched by a Saturn 1B. The LEM's ascent and descent engines fired twice in orbit.

Apollo 6, **(April 4, 1968).** The second Saturn 5 launch, which put unmanned Apollo hardware into orbit. Oscillations of the rocket during launch were recorded (called the "POGO" oscillations).

Apollo 7: *A "First-Class Spacemobile"*

During *Apollo 7*, the first manned Apollo mission, launched October 11, 1968, the Apollo command and service modules were tested in earth orbit. The *Apollo 7* crew was blasted into orbit by a Saturn 1B rocket and remained above the earth for almost eleven days—260.2 hours and 163 orbits. This was longer than what an actual moon mission required. The mission accomplished more tests than the flight plan had defined.

The service module rockets were fired eight times and performed flawlessly, and rendezvous practice between the command and service modules and the rocket's spent upper stage was successful. All in all, *Apollo 7*'s hardware proved itself ready for a lunar voyage.

The crew—commander Walter M. Schirra, command module pilot Donn F. Eisele, and lunar module pilot R. Walter Cunningham—were pleased with their spacecraft's performance. Schirra referred to it during the mission as a "first-class spacemobile," and Cunningham called it a "magnificent flying machine." Crew members were not so pleased when it came to their own body responses, however. Schirra came down with a head cold (there is logic to his selling Actifed® on TV), and Eisele and Cunningham came down with the first symptoms of upper respiratory infection. The crew orbited the earth with stopped-up ears and noses. Their flight surgeon, Dr. Charles Berry, prescribed the usual earthbound remedies: aspirin, decongestant, and plenty of water. Extra rest was not prescribed for obvious reasons.

Because of these physical problems, irritability was often showed by Schirra in his communications with the ground. He refused, for example, to turn on the TV camera for ground control and the major networks at the scheduled time because he was too busy: "I tell you this

flight TV will be delayed without further discussion," Schirra made clear, "until after the rendezvous."

Later, when the first live telecast was made from Apollo in orbit, the cast was ready. During their seven-minute program, two printed signs were held up for the earth audience to read: FROM THE LOVELY APOLLO ROOM HIGH ABOVE EVERYTHING and KEEP THOSE CARDS AND LETTERS COMING IN FOLKS. Public relations in orbit was born.

A safe splashdown occurred on October 22, 1968. *Apollo 7* had paved the way for the United States' first space spectacular—a Christmastime trip around the moon.

Apollo 8: *A Lunar Christmas*

The moon voyage of *Apollo 8*, second only in historical significance to the lunar landing of *Apollo 11*, has been referred to by some as NASA's greatest gamble. It was, in the end, a gamble that paid tremendous dividends to the United States. After this voyage around the moon, the United States was seen by the world as the unquestioned leader in the space race. It also gave the U.S. space community the confidence necessary to go all the way to the surface of the moon. The flight director at the Manned Spacecraft Center, Christopher Columbus Kraft, Jr. believed that *Apollo 8* was the pivot of the entire moon-landing program. "From there on," Kraft said, "we really knew what we were doing."

The *Apollo 8* mission was originally planned as an earth-orbit mission similar to *Apollo 7* to check out all operating systems. The main and important difference was that the spacecraft and crew would be launched by a Saturn 5, not the smaller 1B.

The decision to have *Apollo 8* circumnavigate the moon and make history, even though it was a test flight, was made by high-level NASA managers in the summer of 1968. At first President Johnson, who had the final word, was opposed to it, as was NASA's boss, James E. Webb, but they were soon persuaded by logical arguments. A moon voyage, even without a landing,

*You are cordially invited to attend
the departure of the
United States Spaceship Apollo VIII
on its voyage around the moon,
departing from launch Complex 39A, Kennedy Space Center,
with the launch window commencing at
seven a.m. on December 21, 1968*

r.s.v.p.   *The Apollo VIII Crew*

The flight of Apollo 8, *the world's first circumnavigation of the moon, was NASA's greatest gamble, but its success finally gave the United States the lead over Russia in the space race.* Courtesy NASA.

would be a U.S. coup and would prevent the Russians from winning another round. A second political consideration that no doubt came into play was President Johnson's problems with the Vietnam War and his last few months in office as an unpopular President. But even more important to justifying the trip was the potential wealth of knowledge to be gained, knowledge that would carry the Apollo program forward to the moon landing missions.

Frank Borman, James A. Lovell, and William A. Anders blasted off the earth atop the mighty Saturn 5 rocket on December 21, 1968, for humankind's first circumlunar flight. More than a quarter of a million people who had assembled to witness the launch watched the Saturn 5 climb slowly skyward on a column of fire as big as a naval destroyer. The tens of thousands of people watching the lift-off shook, just as Florida beneath their feet shook, when the giant moon rocket thundered skyward.

During the second earth orbit over Hawaii, at 2 hours, 50 minutes into the flight, the third stage ignited and burned for 5 minutes, 19 seconds, boosting their speed to escape velocity of 24,200 miles (38,938 kilometers) an hour. *Apollo 8* was on its way to rendezvous with the moon.

"You are go," said Michael Collins, capsule communicator at Mission Control. "You look good." And then commander Borman uttered one of the great cosmic understatements of the Space Age. He said—as he and his crew pushed off to the moon for the first time in human history, traveling faster than humans had ever traveled before—simply and courteously, "Thank you, Michael."

The mission of *Apollo 8* was filled with firsts:

• The first manned flight of the thirty-six story Saturn 5 rocket.

• The first spaceship and crew to pass out of the earth's gravitational control and come under the influence of the moon's gravity.

• The first human beings to travel at escape velocity from earth—some 440 times a car speed of 55 miles (88.5 kilometers) an hour.

• The first humans to see the entire earth. As *Apollo 8* sped away from the earth, James Lovell said, "I can see Gibraltar at the same time I'm looking at Florida."

• The first humans to see the mysterious back side of the moon and the front side up close, with their own eyes.

• The first spaceship to orbit the moon (ten orbits in twenty hours).

The day before Christmas 1968, after a sixty-six hour voyage to the moon, *Apollo 8* went into lunar orbit with a successful four-minute burn of the service module engine. With a smaller burn later, the spacecraft was in a near-circular orbit 60.7 by 59.7 miles (97.7 by 96.1 kilometers) above the moon.

The sixteen hours Borman, Lovell, and Anders spent orbiting above the moon's surface gave humanity a new view of their planet earth. "The vast loneliness is awe-inspiring," said Lovell, "and it makes you realize just what you have back there on earth." TV images sent back to earth were seen by a billion people in sixty-four countries. On this lunar Christmas Eve, the crew took turns reading the first ten verses of the Book of Genesis, and commander Borman closed with a "Good night, good luck, a Merry Christmas, and God bless all of you—all of you on the good earth." A miniature bottle of brandy was smuggled on board for some Christmas cheer.

It was, declared *The New York Times*, "the most fantastic voyage of all times." But that was before *Apollo 8* returned to earth. Once in moon orbit, the service propulsion system had to fire properly to bring them back to earth. There was no lunar module system on this flight that could be used as a backup, and this fact alone made the mission a very high risk. The lives of the crew depended on a 303-second (5-minute) burn of the SPS engine. But no one would know if it had ignited, because it happened on the far side of the moon, when communications were down. Finally Houston heard Lovell: "Please be informed there is a Santa Claus." *Apollo 8* was on its way home to earth after making history. After a fifty-seven hour flight (nine hours shorter than the outward journey), the earth brought *Apollo 8* home to its Pacific Ocean on the morning of December 27. The Space Age had sired the Moon Age.

Apollo 9: Gumdrop *and* Spider

It was an impossible act to follow, *Apollo 8's* Christmas visit to the space around the moon.

But Project Apollo was comprised of many missions, each building on those that flew before and each new mission pushing state-of-the-art technology to its limits. In many ways, those last 60 miles (96.5 kilometers) to the moon's surface from an orbit above would be the most difficult, and the craft to take men there, the lunar excursion module (LEM for short), had yet to get off the ground.

That was the main mission of *Apollo 9*, which flew from March 3 to March 13, 1969, with astronauts James A. McDivitt, David R. Scott, and Russell L. Schweickart at the controls. The ten-day, earth-orbiting flight was the first manned flight of all lunar hardware, including the all-important first flight of the lunar module. It was a complex flight, which included the first space walk on an Apollo mission by astronaut Schweickart.

Apollo 9 was also the first flight to use names for the command and service module (Gumdrop) and the lunar module (Spider). The names were chosen by the mission astronauts, and this continued through *Apollo 17*, the concluding flight in the Apollo program. These crew-chosen names were a refreshing change in an all-too-often space language full of bureaucratic acronyms.

Once in orbit, the crew separated their command and service module from the Saturn 5 third stage, maneuvered clear of it, turned their spacecraft around, and then docked with *Spider*, the lunar module.

During the first and second days in orbit, the crew concentrated on hundreds of systems checks, including multiple firings of the service propulsion system. Then on the third day, McDivitt and Schweickart entered *Spider*, powered it up, and fired its descent engine while in the docked position. *Spider's* crucial test finally came on the fifth day in orbit. McDivitt fired its descent engine and left *Gumdrop*, the command spacecraft. Eventually *Spider* was 115 miles (185 kilometers) behind *Gumdrop*, and McDivitt and Schweickart began their return rendezvous maneuvers after jettisoning their descent stage. *Spider* soon caught up with *Gumdrop*. As command module pilot Scott saw the returning lunar module, he exclaimed, "Oh, I see you out there, coming in the sunlight. You're the biggest, friendliest, funniest-looking spider I've ever seen." The two spacecraft docked about six hours after their initial separation. The lunar landing *Spider* was spaceworthy.

After 151 orbits around the earth, *Gumdrop* splashed into the Atlantic, ending an above-earth journey that cost $340 million, or about $1.4 million for every hour. But it was all three acts—the entire Apollo show—put on above earth. Everything had worked. It was time for a dress rehearsal above the moon.

Apollo 10: Charlie Brown *and* Snoopy

The decade was almost over, the 1960s had only a few months to run, and men had yet to walk on the moon. True, three astronauts had circled the moon ten times in *Apollo 8* and had come to within 70 miles (113 kilometers) of its surface, but President Kennedy's promise to the world called for "landing a man on the moon and returning him to earth" before the end of the decade. That had not been done. *Apollo 8* had not even flown carrying a lunar excursion module and could not have landed men on the moon.

Apollo 10 could have landed but didn't. The eight-day flight was a dress rehearsal for the finale. NASA managers wanted to thoroughly test the lunar excursion module in a lunar environment, and to carry out all the activities of the "Big Trip" with the exception of the final descent and landing. In particular, they wanted to see how the LEM's guidance and navigation system would behave in the moon's uneven gravity fields. No politics were involved in the flight of *Apollo 10*, as they were for *Apollo 8* at the end of President Johnson's term. The second lunar voyage had to clear the path for *Apollo 11*, and it did so with marks of excellence.

Commander Thomas P. Stafford and his crew, John W. Young and Eugene A. Cernan, left the earth on May 18, 1969, for the second lunar circumnavigation in human history. On board they carried a small, high-quality color

TV camera to send some cosmic shorts back to the home planet. They named their command and service module *Charlie Brown* and their lunar module *Snoopy.*

About 4,000 miles (6,436 kilometers) into the journey, they separated from the Saturn third stage, turned around, and docked with *Snoopy,* leaving the third stage behind.

The second morning, the crew issued a historic weather report from 100,000 miles (160,900 kilometers) out. Italy was clear south of Rome, but much of Europe was cloudy. Greece, Arabia, Israel, and Jordan were clear. South Africa was cloudy. Most of Russia north of the Black Sea was clear, but the rest was overcast. Early the morning of Wednesday, May 21, *Charlie Brown* and *Snoopy* passed into the moon's gravitational influence and went behind the moon. Late in the afternoon, their Apollo engine burned for almost six minutes, which put them into lunar orbit: "*Apollo 10* can tell the world that we have arrived." After two orbits of moonscape-watching and crater identification, the crew initiated another engine burn and circularized their orbit.

On May 22, Stafford and Cernan powered up the LEM *Snoopy* and undocked from *Charlie Brown.* After a visual inspection of *Snoopy* by Young in the command ship, Stafford fired up *Snoopy's* descent engines and flew down to within 50,000 feet (15,240 meters) of the moon's surface, where they photographed the selected Apollo lunar landing sites. At their lowest altitude over the moon, Stafford told Houston, "You can also tell Jack Schmitt (the astronaut-geologist who trained them) that there are enough boulders around here to fill Galveston Bay." *Snoopy's* crew had been well trained in identifying lunar features and called out such moonmarks as Diamondback Rille, which Stafford compared to a "dry desert out in New Mexico or Arizona," and Landing Site 2.

As the *Snoopy* crew began the important test called staging—separating from the descent stage, just as future crews would do on the moon's surface, and burning their ascent engine to begin rendezvous with *Charlie Brown*— something went wrong. "Son of a bitch," shouted Cernan. *Snoopy* was throwing a tantrum, gyrating wildly. Stafford soon took over manually and finally regained control. A switch was in the wrong position, Houston claimed. (For a more detailed account of this event, read Chapter 5, "space Disasters and close calls.")

Snoopy's descent to some 9 miles (14.5 kilometers) above the moon's surface and the rendezvous and docking had taken almost eight hours. The crew rested and did more moon reconnaissance before firing up on May 24 and beginning their fifty-four hour return to earth, which brought them safely down in the Pacific, 395 miles (636 kilometers) east of Pago Pago on May 26, 1969, after a flight lasting 192.1 hours. Up next: *Apollo 11.*

Apollo 11: Columbia *and* Eagle

A Dangerous Descent. On a Sunday afternoon in the late 1960s, the earth changed forever. It was Sunday, July 20, 1969. The time was 4:17:43 P.M. Eastern Daylight Time. After a dangerous descent, Neil Alden Armstrong and Edwin Eugene "Buzz" Aldrin, Jr. soft-landed their moonship, the *Eagle,* in the powdery, ash-like soil of the desolate and windless Sea of Tranquillity.

This moment of lunar landfall, and the moment of Neil Armstrong's first steps in the ancient moon soil some six and a half hours later, symbolize humankind's never-ending reach outward. When all else is forgotten in the twentieth century, *Apollo 11* and the other voyages of Apollo to the moon will be remembered. The year 1969 will be known to future generations as the year that humankind burst from its terrestrial bonds. After the *Eagle* descended onto the dry lunar sea, the human mind would never again be the same. The event created the twentieth-century pyramids: pyramids made not of stone, but of new ideas filled with human possibility, thrust inside humanity's head by the triumph of *Apollo 11.*

Every baby boomer remembers where he or she was on that summer afternoon in July 1969, when commander Armstrong's *Eagle* touched

FACING PAGE LEFT: *The historic blast-off of* Apollo 11, *which would take men to the surface of the moon for the first time in human history.* ABOVE: *The signed* Apollo 11 *insignia was reproduced on an invitation to the* Apollo 11 *prelaunch briefing at the Cocoa Beach Theatre.* FACING PAGE RIGHT: *Also shown is the most unusual customs declaration ever filed.* Courtesy NASA.

down on the moon, his heart pounding more than twice its normal beat—156 beats a minute—as he accomplished one giant leap for mankind. But many *Star Wars* kids were not even born; others were too young to remember. For them, and for those of us who will never forget, this is a recounting of the dramatic and dangerous descent of the moonship *Eagle* to the Sea of Tranquillity.

On Sunday morning, July 20, 1969, the lunar module *Eagle* undocked from *Columbia,* the command ship. "The *Eagle* has wings," Armstrong radioed to earth. The two spacecraft then flew in formation, and Michael Collins in *Columbia* visually inspected the *Eagle* to ensure it was not damaged and would function properly. Collins reported that everything appeared ready for descent. "I think you've got a fine looking flying machine down there, despite the fact that you're upside down," Collins quipped. Houston

take them to an altitude of 50,000 feet (15,240 meters) above the moon.

At five minutes into the burn, as the *Eagle* descended to 6,000 feet (1,829 meters), a yellow caution light came on.

"Program alarm!" Armstrong reported loudly. "It's a 1202!" Houston came back with a go. The 1202 was an executive overflow of the onboard computer, which meant that the computer was forced to postpone some things because it had too much to do at once. At about 3,000 feet (914 meters) above the surface, the yellow caution light flashed again, this time a 1201 program alarm, another overflow condition. Again the ground told the crew that descent was still go and to ignore the alarm.

Armstrong and Aldrin kept on responding to four more such alarms in about four minutes. Steve Bales, the computer flight controller back

gave *Eagle* a go for powered descent. An hour after undocking, on the far side of the moon, the *Eagle*'s descent engine was fired, making the first of two engine burns to reach the moon's surface, a descent that would last only 12 minutes and 34 seconds after millions of man-hours and billions of dollars had been spent to prepare for it. *Eagle* was still some 60 miles (95.5 kilometers) above the surface when Armstrong and Aldrin initiated the first engine burn, which would

in Houston, who made the go decision, was "believing" the landing radar. As it turned out he was right; it was the rendezvous radar, not the landing radar, causing the computer overload. Had Steve Bales thought otherwise, the *Eagle* would never have landed—Bales would have ordered a mission abort.

The alarms and instrument readings had taken the crew's full attention, and they were unable to look out the window from an altitude of 5,000 feet (1,524 meters) to determine their location. When they finally could look out, they were only 1,968 feet (600 meters) above the lunar surface and had only three minutes of fuel left. Immediately Armstrong saw that they were heading for a large boulder field that surrounded a crater; and the larger boulders were 16 feet (5 meters) in diameter, big enough to burst open the belly of the *Eagle*. It was here that Armstrong's heart rate soared to 156 beats a minute.

Aldrin continued to call out the descent and feet-per-second forward motion rates. They were now about 300 feet (90.5 meters) above the moon. It was then that Armstrong decided he must take manual control of the moonship and fly over the West Crater and boulder field in search of a smoother landing area. Ground control noticed that *Eagle*'s forward speed suddenly shot up to 80 feet (24 meters) per second— about 55 miles (88.5 kilometers) per hour. This was *not* according to the flight plan.

As he searched for a landing site, Armstrong asked Aldrin how much descent fuel was left, but Aldrin was too busy watching the computer; he didn't hear him. Armstrong then slowed the forward speed. They were only 100 feet (30.5 meters) above the moon. Finally he found a small smooth clearing about the size of a house lot and headed for it: On one side were craters; on the other side was a field of boulders. Then, abruptly, a red light flashed on the control panel, and a warning came on in Mission Control back on earth. Only 5 percent of their descent fuel remained. If they were not on the surface within ninety-four seconds, they would be forced to abort and fire *Eagle*'s ascent engine.

Only sixty seconds of fuel remained. Lunar dust kicked up by the descent engine obscured

Armstrong's view. It was like a ground fog, but it had movement and made it impossible for Armstrong to judge his altitude or forward motion. He tried to judge by picking out large rocks and watching them through the haze!

Thirty seconds of fuel remained! At 33 feet (10 meters) above the surface, *Eagle* started slipping to the left and moving backward. But there was no rear window. What dangerous obstacles were behind him? The rim of a crater?

Armstrong stopped the backward motion, but not the drift to the left. He didn't want to slow the descent any more; there were only seconds of fuel left. He was concentrating so hard that he did not feel the first touch on the moon's surface or hear Aldrin call out "contact light" when the footpad probes brushed the surface. The landing was gentle, at about 1 foot (0.3 meters) per second. They were down, with only twenty seconds of fuel left! *Eagle* was on the moon at a slight tilt—about 4.5 degrees from the vertical—some 4 miles (6.4 kilometers) beyond the programmed landing area. The first words from the moon were Aldrin's: "Okay, engine stop." And then seconds later, Neil Armstrong's famous "The *Eagle* has landed." The moon dust cleared. The blazing bright moonscape revealed itself. The earth had changed. Two men were on the moon.

Setting Up Tranquillity Base. Moonship *Eagle* remained on the moon's surface for 21.6 hours. About six and a half hours after landing the hatch was opened and Neil Armstrong backed out onto the lunar module's "front porch." It was 10:39 P.M. Eastern Daylight Time—prime TV time in the United States. Seventeen minutes later, Armstrong planted his left boot on the surface of the moon and spoke his "small step . . . giant leap," phrase that will never be edited out of the history books.

"The surface is fine and powdery . . . . I can pick it up loosely with my toe. It does adhere in fine layers like powdered charcoal to the sole and sides of my boots. I only go in a small fraction of an inch, maybe an eighth of an inch . . . . It's actually no trouble to walk around."

Aldrin followed at 11:11 P.M. When he planted his left boot on the moon's surface for the first time, his urine bag, strapped to his left foot, broke. The second human on the moon squished his way across its surface; he felt it each time he took a step.

The total EVA time on the moon was 2 hours, 36 minutes. The following list describes the astronauts' main activities on the moon:

Shortly after stepping onto the moon's surface, Armstrong collected a "grab sample" (contingency sample) of soil by scooping it into a Teflon bag. He tucked the bag into a pocket above his left knee.

Armstrong described the historic plaque attached to the *Eagle*'s forward landing gear strut: "Here Men from the Planet Earth first set foot upon the Moon, July 1969 A.D. We came in peace for all mankind."

Armstrong removed the TV camera from the equipment compartment beneath the LEM. He scanned the lunar horizon before setting it on a tripod about 82 feet (25 meters) away.

Aldrin removed the solar wind composition experiment and set it up. It was an aluminum sheet hung vertically from a pole and was retrieved before the EVA ended. It was exposed

for 1 hour and 17 minutes. The experiment was designed to trap particles from the solar wind (an ionized or electrified gas constantly streaming away from the sun at about 1 million miles, 1.6 million kilometers, an hour) so that Swiss scientists on earth could measure them. As many as one hundred trillion atomic solar particles were embedded in foil and weighed less than a billionth of an ounce.

Armstrong and Aldrin raised the American flag. The moon soil had great resistance downward but little sideways, and they had difficulty getting the flagstaff in far enough to hold the flag upright.

ABOVE: *This famous photo of Buzz Aldrin, the second man to walk on the moon, was taken by Neil Armstrong. Few people realized that Aldrin "squished" his way across the moon's surface because his urine bag, located in his boot, broke when he took his first step.* LEFT: *Here the* Eagle *moonship is seen at its landing site as Aldrin deploys the seismic experiment.* Courtesy NASA.

The first moon astronauts were called before the TV camera to take a relayed telephone call from President Nixon at the White House.

Armstrong collected "bulk," undocumented rock samples within about 98 feet (36 meters) of the mooncraft. Using a scoop, he filled up one special rock box.

At the same time, Aldrin photographed all sides of the *Eagle*.

Armstrong set up the first of several experiments to be peformed—the laser ranging retro-reflector (LRRR)—which bounced narrow light beams back to earth so that physicists could measure earth–moon distances with great accuracy. The reflecting array, made from cubes of fused silica, was aimed toward the earth. Earth stations such as McDonald Observatory, Fort Davis, Texas, beamed the laser to the experiment at Tranquillity Base to calculate round-trip travel times for the measurements.

Aldrin set up the passive seismic experiments package (PSEP) about 60 feet (18 meters) away from the LEM. It was a seismic unit up to one hundred times more sensitive than those used on earth, and it could record tremors one million times smaller than what a human could feel. The unit also contained nuclear heaters, filled with 34 grams of radioactive plutonium, to help the instrument withstand the frigid lunar nights. This was the first major use of nuclear energy on a manned spacecraft mission. The seismic unit picked up the astronauts' footfalls on the moon and sent the signals back to Houston.

Armstrong and Aldrin collected and identified the "documented" lunar samples. Aldrin obtained two core-tube specimens but could only pound the tube in 5 inches (12.7 centimeters). At the same time, Armstrong gathered with tongs twenty-five more rock specimens, which went with Aldrin's cores into a special sealed rock box.

With a pulley rig, the astronauts hauled up the sample rock boxes and cameras.

Both reentered the Eagle, closed the hatch, and repressurized the cockpit.

The first two men on the moon had collected 48.5 pounds (22 kilograms) of moon rock and soil. They did not yet know it, but they had a new mineral in stow, one of the most common on the moon, to be named after them and their fellow crew member, Collins. The mineral was an oxide composed of iron, titanium, and magnesium, and it was called armalcolite: *arm* for Armstrong, *al* for Aldrin, and *col* for Collins. Even the rock would carry their names into posterity.

Tranquillity Base: 2009? The third manned flight to the moon, *Apollo 11*, was also the longest up to that time (8 days, 3 hours, and 18 minutes).

The *Eagle* lifted off from Tranquillity Base at 1:54 P.M. on July 21, 1969, and successfully rendezvoused with Michael Collins and *Columbia* almost four hours later. The lunar launch, never before attempted, was quiet and smooth after the crew heard the pyrotechnics fire to separate *Eagle*'s ascent stage from its landing base and saw the debris flying. After that, Armstrong and Aldrin had difficulty sensing acceleration.

Before leaving the moon, the astronauts opened the hatch one more time and threw out equipment and other flight items that would not be needed for their return to earth. The seismometer picked up the tremors as the crews' life-support backpacks and urine bags hit the moon's surface.

The moon voyage and landing were the most widely known and shared adventure in human history. July 17, the second day of the historic Apollo mission, was also the fortieth anniversary of Robert Goddard's launching of the first instrumented rocket, which included a thermometer, barometer, and camera. Just four decades later, a gigantic rocket had carried men to and landed them on the moon.

Where would humankind find itself in another forty years of space exploration? Inhabiting a base on the planet Mars? And living on a moon base built before the first Mars mission was launched? Someday, perhaps by 2019, fifty years after *Apollo 11*'s landing, there may be a Tranquillity Base Museum, a structure built over

The First Space Litter

The following is a list of items the *Apollo 11* left behind at Tranquillity Base:

• *EAGLE'S DESCENT* stage, with the famous plaque on one leg.

• TWO LIFE-SUPPORT backpacks.

• A BAG of gear, including cameras, tools, lunar overboots, bags, containers, armrests, brackets, and other miscellaneous items.

• AN *APOLLO 1* shoulder patch, commemorating Gus Grissom, Ed White, and Roger Chaffee—the three astronauts who died in the pad fire on January 27, 1967.

• MEDALS HONORING Russians Yuri Gagarin and Vladimir Komarov, who lost their lives participating in their country's space program.

• A 1½ inch (3.4 centimeter) disk of silicon with messages of goodwill from leaders of seventy-three nations, reduced in size two hundred times and micro-etched onto the disk.

• A GOLD olive branch symbolizing peace.

• AN AMERICAN flag and staff (which fell over at the time of lunar blast-off).

• A MAST of the solar wind experiment.

• A TV camera and power cable.

• A SEISMIC unit.

• A LASER reflector unit.

the place where men first walked on the moon. Space tourists from earth may comfortably travel 240,000 miles (386,000 kilometers) to the moon for recreation and a visit to the famous and historic moon site.

Apollo 12: Yankee Clipper *and* Intrepid

Four months after *Columbia* splashed down to worldwide acclaim, on November 14, 1969,

Apollo 12 blasted moonward. The command spaceship, *Yankee Clipper*, would stay in lunar orbit almost thirty hours longer and for fifteen more orbits than *Apollo 11*. On November 19, two of its crew members, Charles "Pete" Conrad, Jr., and Alan L. Bean, piloted the lunar module, *Intrepid*, to a landing site on the moon's dry Ocean of Storms (*Oceanus Procellarum*) not far away from the landing site of the earlier unmanned spacecraft *Surveyor 3*. Conrad and Bean remained on the lunar surface for 31.5 hours— 10 hours longer than Armstrong and Aldrin— and were the first men to stay a full day on the moon. They set up a more sophisticated set of surface experiments, gathered 75 pounds (34 kilograms) of lunar rock and soil, and retrieved parts of the unmanned *Surveyor 3* spacecraft, which had landed on the moon two and a half years earlier. The third crew member, Richard F. Gordon, Jr., remained in orbit in command of the *Yankee Clipper* spacecraft as his buddies explored the moon.

Highlights of the *Apollo 12* mission to the moon:

*T*hirty seconds after *Apollo 12* blasted off, it was struck by lightning, which caused the main circuit breakers of the spacecraft to go out. "We had everything in the world drop out," Conrad said. Power was restored by quick action of the crew and launch control.

*I*ntrepid *landed on the* Ocean of Storms, 250 miles (402 kilometers) from the great Copernicus Crater and 1,300 miles (2,092 kilometers) west of where *Apollo 11* landed. The surface of the moon was believed covered by debris thrown out when Copernicus underwent its cosmic excavation. *Intrepid* made a pinpoint landing, only 600 feet (183 meters) from *Surveyor 3*'s landing site.

*P*ete Conrad and Al Bean made two EVAs while on the lunar surface, each lasting just under four hours. They described the strange behavior of the lunar dust, such as the little dust clouds that kicked up around the astronauts' boots each time they put their feet down.

On the first EVA, a contingency sample was taken and the TV camera and S-band erectable antenna were set up. The antenna would transmit better quality color TV images. Within a few minutes, however, the camera was pointed at the sun by mistake, which burned out the tube and made TV coverage impossible from the moon's surface.

The astronauts then set up the Apollo lunar surface experiments package (ALSEP), which included a solar wind collector; a passive seismic unit; a lunar surface magnetometer, the bar antenna to transmit data to earth, and the mininuclear power station to power the experiments.

The seismometer in this scientific package was designed to last a year, but it exceeded all design expectations by operating until September 1977, when it was shut down for budgetary reasons and diminishing scientific value of data. It registered almost 2,300 moonquakes and meteorite impacts during that period.

After a rest period, Conrad and Bean began their second EVA on the moon, which included lunar rock and soil collecting and photography of the specific sites for documentation. Some green rocks were found.

The 75 pounds (34 kilograms) of rock samples collected by *Apollo 12* were so different from those gathered by *Apollo 11* that experts knew the moon had a very complex geological history. The samples were "a veritable feast," said the director of the U.S. Geological Survey, W. T. Pecora, compared to the "geological hors d'oeuvres" of *Apollo 11*.

The last major activity on this second EVA was to visit the *Surveyor 3* landing site. The robot lander had been on the moon for thirty-one months, and it was covered with a coating of fine dust that had been kicked up during the *Intrepid*'s descent. After photographing the spacecraft, they cut off samples of aluminum tubing, some electrical cables, and the soil sampling

Apollo 12's Intrepid *landed near the robot* Surveyor 3 *spacecraft, portions of which were brought back to earth. To the amazement of scientists, a terrestrial bacterium was found on a piece of foam inside the robot's TV camera. It had survived the hostile conditions for two and a half years.* Courtesy NASA.

scoop. They also broke off a piece of glass from a thermal glass and unbolted and removed the TV camera before returning to the *Intrepid*.

To the amazement of scientists, a terrestrial bacterium (alpha hemolytic *Streptococcus mitis*) was found on a piece of foam inside *Surveyor 3*'s television camera when it was examined after the mission. This primitive life had survived two and a half years of extreme heat, vacuum, and dry conditions. Because the bacterium is a benign inhabitant of the human respiratory tract and is dispensed during normal talking or during a sneeze or cough, experts concluded that it may have found its way into the equipment during manufacture or assembly on earth years before.

After ascent, rendezvous, docking, and transfer of the rock samples and *Surveyor* parts to the *Yankee Clipper*, the crew sent the lunar module *Intrepid* crashing into the moon at 5,000 miles (8,045 kilometers) an hour, producing the first artificial moonquake. On earth such an impact would have created a slight two-minute tremor, but on the moon it set the seismometer vibrating for two hours. The moon sang like a bell, said one scientist.

After 10 days, 4 hours, and 36 minutes, *Apollo 12*—the second manned mission to land on the moon—reentered the earth's atmosphere and splashed down safely. It was November 24, 1969. This year the crew's Thanksgiving Day would be spent in quarantine.

Apollo 13: Odyssey *and* Aquarius

The landing mission of *Apollo 13* (April 11 to April 17, 1970) was aborted when the number 1 oxygen tank in the service module exploded. The spacecraft was more than halfway to the moon when the explosion occurred. There was no turning back. The crippled spacecraft and its crew—James A. Lovell, John L. Swigert, and Fred W. Haise—had to circumnavigate the moon before returning to earth. But it became a mission of survival. The explosion of one oxygen tank damaged the second tank, and it began to leak badly. The crew lost their main source of

power because their fuel-cell batteries produced electricity by an oxygen-hydrogen reaction, and they lost most of their breathing oxygen. To survive, they were forced into the lunar module, *Aquarius*, and had to use its supplies.

As a rescue and survival mission, *Apollo 13* succeeded: Lovell, Swigert, and Haise returned to earth alive. Had the same accident occurred on *Apollo 8*, which flew without a lunar module, its three astronauts would have died in space.

For a more thorough account of the harrowing *Apollo 13* rescue mission, read Chapter 5, "Space Disasters and Close Calls."

Apollo 14: Kitty Hawk *and* Antares

America's first man in space, Alan B. Shepard, Jr., commanded the *Apollo 14* mission to the moon (January 31 to February 9, 1971) and became the fifth astronaut to walk on the lunar surface.

"It's been a long way, but we're here," Shepard commented as he stepped from the lunar module *Antares* onto the lunar soil for the first time. He was referring to more than their moon journey in their spacecraft *Kitty Hawk*. Shepard no doubt was also thinking of his long personal journey to reach the moon.

Almost ten years had passed since he made his suborbital flight in *Freedom 7* on May 5, 1961, to become America's first space hero. But soon thereafter he was grounded for several years because of a serious inner ear ailment—Ménière's disease—which caused unpredictable attacks of vertigo. Shepard won his wings again after a delicate operation corrected this chronic disorder of balance and hearing. At forty-seven years of age, Shepard became the oldest American to fly in space and the oldest to land and walk on the moon.

Accompanying Shepard on the *Apollo 14* mission were Edgar D. Mitchell, the lunar module *Antares* pilot, who also descended to the lunar surface, and Stuart A. Roosa, pilot of the command spacecraft *Kitty Hawk*, who made thirty-four circuits of the moon, with a total of sixty-seven hours in lunar orbit.

Commander Alan Shepard assembles hand tools on his equipment transporter. He would later use one of them as a golf club and become the moon's first golfer, sending golf balls hundreds of yards into the thin atmosphere.

The landing site of *Apollo 14* was changed and its launch date delayed after the *Apollo 13* mission was aborted. The mission, after spacecraft design changes, was targeted to the site where *Apollo 13* was intended to land before its oxygen tank exploded—Fra Mauro, in the lunar highlands. This area was named for a fifteenth-century Italian monk and cartographer who drew a famous, highly accurate (for the time) map of Africa and Asia in 1457 as well as some maps of the moon.

The flight of *Kitty Hawk* and *Antares* began on the afternoon of January 31, 1971; some four and a half days later the lunar module *Antares* landed on the moon. The mission set records and accomplished many firsts:

Two lunar EVAs totaled 9 hours and 25 minutes on the lunar surface, more than any previous Apollo mission.

The first lunar EVA lasted 4 hours and 50 minutes and was devoted to unstowing and setting up scientific equipment, most of which would remain on the moon, as well as to conducting some scientific experiments. The Apollo lunar scientific experiments package (ALSEP) was considerably more sophisticated than those of the two previous Apollo landing missions. While some of the experiments were similar to those deployed on earlier missions (the solar wind collector, the passive seismometer, and the laser reflector), new experiments included ionosphere and atmosphere detectors, and an active seismic experiment.

In the active seismometer experiment, astronaut Mitchell laid out vibration detectors (geophones) every 150 feet (46 meters) on a cable he had run across the lunar surface. He then took a tool containing explosive devices, called a thumper, which resembled a thin walking stick with a plate on the bottom, placed it against the moon's surface, and fired it. He would then move along the planned survey line and fire it again. Each firing (thirteen out of twenty one charges went off) created an impact of known force. The geophone sensors recorded the seismic waves and sent them to earth for scientists to study in order to learn how these miniquakes traveled through and below the lunar surface and to determine what was beneath the moon's surface.

The second part of the active seismic experiment was not activated until after *Apollo 14* had returned to earth. Mitchell set up a grenade launcher with four grenades. Later, they were remotely fired from earth, each impacting at a certain distance, so that experts could decipher the moon's innards.

A moon equipment cart was used for the first time on the second EVA. NASA named it the modularized equipment transporter (MET). The astronauts referred to it as their lunar ricksha.

The second EVA took Shepard and Mitchell farther afield than the previous Apollo missions had—about 3 miles (4.8 kilometers). It was the first time that doctors on earth ordered the astronauts to stop and rest. Shepard's heart rate reached 150 beats a minute at one point as he and Mitchell climbed toward the rim of Cone Crater, passing 12-foot (3.7 meter) high boulders on the way. Ground control noticed irregular heartbeats as well, and doctors were concerned—thus the order to rest. Because the astronauts were somewhat disoriented, they had difficulty recognizing their whereabouts in relation to Cone Crater. In the most disappointing moment of the mission, Shepard and Mitchell turned back to *Antares*, not realizing they were only about 75 feet (23 meters) from the crater's rim. Still, they had traveled much farther than the first four astronauts on the moon.

At the end of the second EVA, Shepard did something that would win him a place in the trivia book craze of the 1980s. He became the moon's first golfer. Just before he entered *Antares* and left the moon's surface for the last time, he removed golf balls from a pocket and with a makeshift club—a detached handle from one of the geological implements—sent them soaring through the lunar vacuum. His golfer's form was unusual because his bulky spacesuit forced him to do one-arm swings. "There it goes," said Shepard, "miles and miles and miles." Later on he estimated that the first ball went 200 yards (183 meters) and the second went 400 yards (366 meters).

Edgar Mitchell also attempted a from-the-moon first. During his personal time and rest periods on the mission, he carried out an experiment in mental telepathy with four people back in the United States. Using a deck of twenty five cards, each with its own symbol, Mitchell attempted to transmit impressions of the cards to the people back on earth. Did the telepathy experiment work? The best score was fifty-one symbols out of two hundred—not very impressive, considering the fact that random guessing can score forty out of two hundred. Mental telepathy from the moon apparently did not succeed in this experiment.

Some 94 pounds (42.6 kilograms) of lunar material were collected and returned to earth, much of it precisely documented as to location and geological content. The lunar ricksha was used to transport the rocks.

*A*pollo 14 *was the first* moon mission to accomplish a rendezvous between *Antares*, the lunar module, and *Kitty Hawk*, the command spacecraft, during the first orbit—about two hours after lift-off rather than the usual four or five hours. To accomplish this, the ascent engine was burned a second time.

*A*pollo 14 *created two artificial* moonquakes so that scientists could study the seismic measurements recorded by the instruments Apollo flights had set up on the moon. Such measurements would help experts learn about the moon's interior structure.

First, the Saturn-5's third stage was sent on a collision course with the moon. The impact force equaled the explosive power of eleven tons of TNT, and the moon reacted like a bell, vibrating for up to three hours. The vibrations traveled to a depth of 22 to 25 miles (35 to 40 kilometers).

Second, once Shepard and Mitchell were back in the *Kitty Hawk* after their moon expedition, Mission Control sent *Antares* crashing to the moon. Although its impact force was not as powerful as the third-stage crash (equivalent to 1,600 pounds, 725 kilograms, of TNT), it was a scientific milestone because it was recorded by two moon seismic stations—those of *Apollo 12* and *Apollo 14*—which gave more accurate information about the moon's structure. The resulting tremors lasted about ninety minutes.

The third manned moon mission was the most complex lunar voyage yet flown, and it was successful. It gave the oldest astronaut and the first American in space, Alan Shepard, some new records to hold. His space days were over, but he had some lunar memories to cherish after *Apollo 14*'s safe landing in the Pacific south of Samoa on February 9, 1971.

Apollo 15: Endeavor *and* Falcon

The last three Apollo flights to the moon incorporated major design changes in the lunar module, which allowed the astronauts to more than double their exploration time to the moon. *Apol-*

lo 15's lunar module, the *Falcon*, was the first of these spacecraft to have enlarged propellant tanks, an additional battery, more life-support supplies, and provisions for the first lunar terrain vehicle, which NASA named the lunar roving vehicle (LRV) and the crew, thankfully, simplified to Rover.

On July 26, 1971, the *Apollo 15* moon mission blasted off from earth toward a landing site in the Hadley-Apennine region near the Apennine Mountains. David R. Scott, who had flown with Neil Armstrong on *Gemini 8* and who was the command module pilot on the above-earth mission of *Apollo 9*, commanded the fourth lunar landing mission. James B. Irwin was the pilot of the lunar module *Falcon* and Alfred M. Worden was the command module pilot for *Endeavor*. Worden too would come home with his own world's record.

Apollo 15 was the longest lunar mission, lasting 12 days, 7 hours from launch on July 26 to splashdown on August 7, 1971. *Endeavor*, the command spacecraft, orbited the moon 74 times in 145 hours. Scott and Irwin had three EVAs outside the moonship, for a total lunar surface time of 18 hours, 36 minutes. The *Falcon* remained on the moon for more than 3 days.

Apollo 15 mission highlights:

The lunar module Falcon, named after the U.S. Air Force mascot (all crew members were Air Force pilots), landed on the Swamp of Decay at 6:15 P.M., July 30, 1971. It came to rest with one leg in a depression, which caused it to tilt 11 degrees.

At the start of their first EVA, Scott and Irwin removed and unfolded the first moon car, the lightweight lunar roving vehicle, an electric-powered, four-wheel-drive car that would take them a total of 17 miles (27.4 kilometers) on the lunar surface. The Rover had its own navigation system that always knew where the *Falcon* was located.

Each wheel had its own quarter-horsepower motor. The top speed of the Rover on level moonscape was 7 miles (11 kilometers) per hour, and its battery contained enough power to travel 55 miles (88.5 kilometers). The actual distance from the *Falcon* that the astronauts could travel was restricted to 6 miles (9.7 kilometers) by mission rules. This would enable Scott and Irwin to walk back to their spacecraft if the Rover ever broke down. Even with this restriction, the Rover allowed the astronauts to explore an area ten times that covered by the *Apollo 14* astronauts on foot.

The *New York Daily News* during coverage of the mission referred to the Rover as "the merry moon-mobile." It weighed only 76 pounds (34.5 kilograms) on the moon and 455 pounds (206 kilograms) on earth, but it could carry two and a half times its own weight across the lunar surface.

For the first time, the astronauts wore redesigned, improved space suits and backpacks, which provided the men with increased mobility and EVA time on the moon.

"Okay, Jim, here we go," said Scott, as he started driving the Rover from the left side driver's seat. "Whew," replied Irwin, "hang on!"

The first EVA Rover drive, referred to as Traverse 1, took the two men 6.28 miles (10.1 kilometers), to the rim of Hadley Rille, a deep, canyonlike feature with an average depth of 1,000 feet (305 meters), an average width of one mile (1.6 kilometers), and an average length of 70 miles (113 kilometers). "That's beautiful...spectacular!" exclaimed Jim Irwin as he stood at the edge of this great lunar canyon.

Arriving back at the *Falcon*, the astronauts set up the Apollo lunar surface experiments package (ALSEP), in many ways similar to those used during the *Apollo 12* and *Apollo 14* missions. Scott and Irwin then climbed back aboard the *Falcon* ending their first EVA after 6 hours, 34 minutes, for some much deserved rest.

August 1, a new month of the moon: Scott and Irwin began their second EVA at 7:49 A.M. They drove their lunar dune buggy almost 8 miles (12.9 kilometers), to the base of Mount Hadley Delta, which rose 11,700 feet (3,566 meters) above the plain where *Falcon* landed. There, where the plain met the mountain flanks, Scott and Irwin collected rock and dust samples,

TOP: Apollo 15's Falcon *carried in its belly the first lunar rover moon car, which allowed the astronauts to explore more of the moon's surface.* BOTTOM: *Astronaut David Scott stands by the Rover at the edge of Hadley Rille, the floor of which is 1,000 feet (305 meters) below.* Courtesy NASA.

dug sample trenches, and documented verbally and photographically many of their activities. It was at this location that they found green rock—green glass beads in a grayish matrix. Their lucky EVA continued. "Guess what we just found!" Scott exclaimed. It was a "genesis" rock, one of the oldest found on the moon—more than four billion years old, older than any rock on earth. It became the most famous rock of the 169 pounds (76.6 kilograms) of rock samples their mission returned to earth.

The third EVA began on their third day on the moon and took them 7.8 miles (12.5 kilometers) along the rim of Hadley Rille, where they saw the floor 1,000 feet (305 meters) below and described the outcrops and debris at the different levels of the canyon wall opposite them. After more rock collecting, they returned to their *Falcon* base and pulled out drilling cores, then prepared for launch off the moon.

Before entering the Falcon for the last time, Scott drove the Rover to its final parking place and adjusted the camera to view the launch. Then, in a small depression 20 feet (6 meters) north of the Rover, the commander of *Apollo 15* placed a white plaque, bordered in black, in the lunar soil. Printed on the plaque were the names of fourteen American and Soviet astronauts who died during the pioneering years of the Space Age. In front of the plaque, Scott placed a small human figure lying flat. This aluminum sculpture, designed by Paul Van Hoeydonck, was called "The Fallen Astronaut" and symbolized the men who had died. Said Scott, "Many people have contributed to this pinnacle we've reached, and we know of fourteen individuals who contributed all they had."

The launch of Falcon off the moon was televised by the color TV that was left on the Rover some 300 feet (91.5 meters) to the east of the moonship. The camera, controlled from earth, showed *Falcon*'s ascent stage break away from its landing base in a shower of colorful sparks and debris. Then the surprise: Scott and Irwin played a tape during ascent for the entire world to hear: "Off we go into the wild blue yonder."

The Air Force crew had finally got their chance to outdo Army and Navy.

After rendezvous and docking with *Endeavor* (named after Captain James Cook's scientific sailing ship) and its pilot, Alfred M. Worden, the *Falcon* was cast adrift to crash on the moon and produce seismic data. Worden, who had been alone for over three days, had kept extremely busy with several new in-moon-orbit experiments.

The next day, in lunar orbit, a scientific satellite was launched from a bay in *Endeavor*'s service module. This was the first satellite ever launched from a manned spacecraft. It provided data for about a year on lunar gravity, mascons, solar flares, the earth's magnetic field, and the moon's weak magnetic field.

On August 5, 1971, after *Endeavor* had left lunar orbit and begun its return to earth, Alfred M. Worden became the world's first astronaut to take a space walk in deep space, when the spacecraft was some 197,000 miles (317,000 kilometers) from earth. Worden was outside the ship for sixteen minutes, just long enough for him to retrieve two cassettes of film that were taken of the moon from orbit. This first-of-a-kind space activity has now become commonplace in the age of the space shuttle, even if it is done closer to the home planet.

Splashdown was in the mid-Pacific, north of Hawaii, on August 7, 1971, after a total flight time of 295 hours and 12 minutes (more than 12 days)—the longest Apollo voyage to the moon.

Apollo 16: Casper *and* Orion

A Saturn 5, carrying the largest payload ever lifted into moon orbit, thundered off the pad at Kennedy Space Center on April 16, 1972. It was the flight of *Apollo 16*—its command spacecraft, *Casper,* and its lunar module, *Orion,* which held in its belly the second Rover moon car.

Commanding the mission was John W. Young. No rookie to spaceflight, he had flown on *Gemini 3* and *Gemini 10* and was command module pilot for *Apollo 10*, the dress rehearsal flight for the first moon landing. *Apollo 16* would take Young those last 9 miles (14.5 kilometers) to the lunar surface. Other crew members were Thomas K. Mattingly II, pilot of the command spacecraft, *Casper*, and Charles M. Duke, Jr., pilot of the lunar module *Orion*, which was named for the famous constellation.

As the mission entered moon orbit on the afternoon of Wednesday, April 19, the complete spacecraft, crew and equipment, weighed 76,109 pounds (34,523 kilograms). The next afternoon, *Orion* separated from *Casper* and began descent procedures to the lunar surface. This fifth manned expedition to the moon would bring the total of man-hours spent on the surface to more than eighteen days.

Here are the mission highlights:

A serious problem occurred in *Casper's* backup steering system shortly after *Orion* and *Casper* undocked. *Orion* was ordered to halt its descent while the vibrations of the main rocket engine were thoroughly evaluated by several groups on earth. (For a more complete account, read Chapter 5, "Space Disasters and Close Calls.") After a six-hour delay of the descent and lunar landing, Young and Duke were finally given the go to land *Orion* in the lunar highlands—the rugged, rock-strewn Cayley Plains of the Descartes region. It was the southernmost landing site of the Apollo missions and was at a higher elevation than other touchdown points—about 8,000 feet (2,400 meters) higher than *Apollo 11*'s Tranquillity Base, which was about 150 miles (240 kilometers) to the northeast.

Young and Duke left the *Orion* three times to set up experiments, explore on the Rover, observe, and collect a total of 213 pounds (96.6 kilograms) of rock and soil samples. These EVAs had a total time of 20 hours and 15 minutes on the moon's surface. The cameras on the surface and in orbit were kept busy: a total of 10,830 frames of film were exposed.

John Young, who commanded Apollo 16 *to the moon, walks on its surface during a Rover EVA. A decade later, Young would fly the first mission of the space shuttle.* Courtesy NASA.

EVA 1 (7 hours, 11 minutes) began with a scientific first: setting up and using an astronomical observatory on another planetary body. Called the far ultraviolet camera/spectroscope, it photographed ultraviolet emissions from clouds of hydrogen and other gases enveloping the earth and other celestial objects. This is impossible to do from earth because of the atmosphere.

The other scientific equipment was set up, including a cosmic ray detector that scientists hoped would answer many questions—including the composition, origin, and age of these elusive particles.

Because of the time needed for the scientific experiments, Young and Duke made their shortest trek in the Rover, heading west to explore two small craters, Flag and Spook. They took rock and soil samples about halfway into Flag Crater and also obtained a football-size rock there. This became the largest moon rock of the Apollo program, and it represented about 50 percent of the entire lunar sample weight returned by the first landing mission, *Apollo 11* (48.5 pounds, 22 kilograms). Dubbed "Big Muley" and officially designated sample number 61016, it weighted 25.89 pounds (11.7 kilograms) on earth. On their return to *Orion* base, Young put the Rover through some rigorous paces for the benefit of the engineering teams who designed and built it. During this so-called Grand Prix, Young drove the Rover over small craters at 7 miles (11 kilometers) an hour, causing both front wheels to jump off the moon's surface. He made sharp turns, skids, and short stops, kicking up a rooster tail of moon dust all the way. Duke, who filmed the Grand Prix, told his earth audience, "Man, I'll tell you, Indy's never seen a driver like this."

EVA 2 (7 hours, 23 minutes) took place on Saturday, April 22, and covered 7.1 miles (11.4 kilometers). This time they headed south to Stubby Crater, climbing the 20-degree slope of Stone Mountain in their Rover until they were 700 feet (213 meters) above the plain where *Orion* had landed—almost halfway up the 1,600 foot (488-meter) high mountain. "What a

The ascent stage of Apollo 16's Orion, *just before docking with the command ship,* Casper, *after returning from the moon, on April 23, 1972. Courtesy NASA.*

view!" exclaimed Duke as he looked down to the plain.

On their way down Stone Mountain, Young and Duke set the moon speed record for four-wheel lunar vehicles, with a top speed of 11 miles (17.7 kilometers) per hour. It has stood an easy test of time because *Apollo 17* did not beat it and that was the only other Apollo mission to fly to the moon.

EVA 3, on Sunday, April 23, was the shortest time outside *Orion* (5 hours, 42 minutes). They drove the Rover to the rim of North Ray Crater. With a diameter of 3,000 feet (914 meters), North Ray Crater was the largest crater directly explored by astronauts in the entire Apollo program. Near the crater's rim was a huge boulder the astronauts named "House Rock" because it was the size of a house—about 66 feet high by 98 feet long (20 meters high by 30 meters long). On the slopes of North Ray Crater, the astronauts measured the highest reading of magnetism taken on the moon. This surprised scientists, some of whom suspected that North Ray Crater may have been formed by a comet strike millions of years ago.

As spectacular as North Ray Crater was, the allotted time for EVA 3 was running short; the six-hour landing delay had shortened the lunar stay. After making additional short stops to take geological samples on the return trip to *Orion*, Young and Duke arrived back at the landing site having covered 7.1 miles (11.4 kilometers) in the Rover. After completing some science experiment chores, the two astronauts squeezed through the hatch for the last time and prepared for a TV-covered moon launch that would take them up to the friendly *Casper* in orbit.

Orion docked with *Casper* on the evening of Sunday, April 23, and then, a day later, *Casper* cast off its moon mooring and fired its engines for a return to earth on Thursday, April 27.

There were now only two more Apollo astronauts who would walk on the moon. The last mission of the twentieth century was up next.

Apollo 17: America *and* Challenger

The last manned voyage to the moon in the twentieth century, *Apollo 17*, began its journey in the early morning Florida darkness on December 7, 1972, at 12:33 A.M. The first night launch of an Apollo moon rocket turned night into day, peaceful quiet into thundering, physical jolts of sound.

Eugene A. Cernan commanded the mission to the Taurus-Littrow Valley of the moon, which was surrounded by the Taurus Mountains on one side and the Littrow Crater on the other. The landing area was located at the southeast corner of *Mare Serenitatis*, which is on the northeast corner of the moon.

Commander Cernan was a veteran astronaut: co-pilot on *Gemini 9* and lunar module pilot for *Apollo 10*, the dress rehearsal before *Apollo 11* landed. The two other crew members were both rookies—Ronald E. Evans, a U.S. Navy captain, who was the command module pilot of *America*; and Harrison H. Schmitt, the Apollo program's first bona fide Ph.D. scientist-geologist, who was the pilot of the lunar module *Challenger*.

The last Apollo moon mission would set records that would remain unchallenged well into the twenty-first century.

• The longest Apollo flight: 301 hours, 51 minutes, 59 seconds (12 days, 13 hours, 52 minutes).

• The longest time spent on the surface of the moon, 75 hours, with the most hours spent outside the LEM for EVAs, of which there were three, totaling 22 hours and 6 minutes.

• The most lunar rock and soil samples ever returned by an Apollo mission, with a total weight of 243 pounds (110 kilograms).

• The longest distance traveled, during three EVAs, by the Rover moon car: 21 miles (34 kilometers).

As with *Apollo 15* and *Apollo 16*, the larger payload capacity allowed *Apollo 17* more sophistication in terms of scientific experiments and the ability to explore on the moon's surface. Rover number three was packed in *Challenger*'s hold.

Nor was such sophistication limited to the lunar surface. A complex scientific instrument package was in orbit in the service module of *America*, constantly operated and tended by Evans. These orbital science experiments were very successful and included five orbits of data by a laser altimeter and by panoramic and mapping cameras, which used several miles of film to record the lunar surface and produce some of the best maps yet obtained for the moon. The lunar sounder returned more than five hundred million soundings to help scientists probe the depths of the moon, and the infrared scanning radiometer made about one hundred million independent temperature measurements to produce thermal moon maps.

The twenty-seventh U.S. manned spaceflight and the sixth manned Apollo flight to the moon entered lunar orbit on the afternoon of December 10, 1972. Not quite a day later, lunar module *Challenger* descended to the moon. Both Cernan and Schmitt recognized moonmarks they had studied during flight training: "There's Camelot, right on target," Cernan said, recognizing a crater during final descent.

The *Challenger* landed with a vertical speed of just 2.5 miles (4 kilometers) per hour, and the

crew noticed a final jolt as the mooncraft settled back on its rear leg.

Five hours later, commander Cernan stepped out onto the plains of Taurus-Littrow and dedicated his first moon step to those tens of thousands of people who built *Apollo 17* and made its voyage possible.

Mission highlights included:

The rover moon buggy, which the crew spent an hour deploying on their first EVA, got the toughest workout on this last moon expedition; it put down over 21 miles (34 kilometers) of tracks on the moon. The first lunar fender bender occurred when Cernan caught a fender with his hammer and knocked it off his Rover as they unloaded the ALSEP scientific package. Eventually, with some help from Young and company back on earth, the rear fender was repaired at the start of EVA 2, by patching it with unused plastic-coated map sheets.

And then, on the third EVA of the mission, in the foothills of the Sculptured Hills, commander Cernan reported "a couple of dented tires."

"What's a dented tire?" Houston asked.

"A little golf-ball size, or smaller, indentation in the mesh," Cernan answered.

"Sounds like a dented tire," earth responded.

The sixth automated research station, more sophisticated than the earlier ones, was set up on the first EVA. The heat-flow experiment was again set up, as on *Apollo 15* and *Apollo 16*, as were the passive and active seismic experi-

ABOVE: *Scientist-astronaut Harrison H. Schmitt collects lunar rake samples during an EVA at the Taurus-Littrow landing site. This* Apollo 17 *moon landing was the last voyage to the moon in the twentieth century. Schmitt is now working toward the manned Mars mission.* FACING PAGE TOP: *Astronaut Ron Evans retrieves a film canister during his deep space walk on* Apollo 17's *return to earth.* FACING PAGE BOTTOM: *The tip of the great Saturn 5 moon rocket, the command module, returned to earth on December 19, 1972. This splashdown ended the Apollo moon program; twelve men landed and walked on the moon, and then returned safely to earth.* Courtesy NASA.

APOLLO MANNED SPACEFLIGHT SUMMARY

Spacecraft Name	Crew	Date	Flight Time (Hrs., Min., Sec.)	Revolutions	Remarks
Apollo 7	Walter M. Schirra, Jr. Donn F. Eisele R. Walter Cunningham	10/11–22/68	260:8:45	163	First manned Apollo flight demonstrated the spacecraft crew, and support elements. All performed as required.
Apollo 8	Frank Borman James A. Lovell, Jr. William A. Anders	12/21–27/68	147:00:41	10 rev. of moon	History's first manned flight to the vicinity of another celestial body.
Apollo 9	James A. McDivitt David R. Scott Russell L. Schweickart	3/3–13/69	241:00:53	151	First all-up manned Apollo flight (with Saturn 5 and command and service, and lunar modules) First Apollo EVA. First docking of CSM with LEM.
Apollo 10	Thomas P. Stafford John W. Young Eugene A. Cernan	5/18–26/69	192:03:23	31 rev. of moon	Apollo LEM descended to within 9 miles of moon and later rejoined CSM. First rehearsal in lunar environment.
Apollo 11	Neil A. Armstrong Michael Collins Edwin E. Aldrin, Jr.	7/16–24/69	195:18:35	30 rev. of moon	First landing of men on the moon. Total stay time: 21 hrs., 36 min.
Apollo 12	Charles Conrad, Jr. Richard F. Gordon Jr. Alan L. Bean	11/14–24/69	244:36:25	45 rev. of moon	Second manned exploration of the moon. Total stay time: 31 hrs., 31 min.

ments. New instruments included: the lunar-surface gravimeter; the lunar atmospheric composition experiment; and the surface electrical properties transmitter, which would help scientists understand the electrical nature of the material beneath the moon's surface.

Just short of a day, twenty-two hours, was spent on the lunar surface during three EVAs, which resulted in a wealth of geological samples, documentation, and a few surprises.

During the second EVA, on a southern traverse of 12 miles (19 kilometers), Cernan and Schmitt had stopped the Rover at station 4 on their preplanned route, near Shorty Crater,

named after a character in Richard Brautigan's novel, *Trout Fishing in America*. Schmitt noticed something unusual.

"Hey!" he shouted. "There is orange soil."

Everyone became excited, the last two men on the moon and the last Apollo flight team in Houston.

"Hey, it is. I can see it from here!" Cernan confirmed.

What they thought might be evidence of volcanic activity—a real find for lunar research—was later determined to be a color produced by chemistry alone. Still, it was one of the most dramatic and unexpected geological finds of the entire Apollo program.

APOLLO MANNED SPACEFLIGHT SUMMARY *(cont'd.)*

Spacecraft Name	Crew	Date	Flight Time (Hrs., Min., Sec.)	Revolutions	Remarks
Apollo 13	James A. Lovell Jr. John L. Swigert, Jr. Fred W. Haise, Jr.	4/11–17/70	142:54:41	—	Mission aborted because of service mod-module oxygen tank failure
Apollo 14	Alan B. Shepard Jr. Stuart A. Roosa Edgar D. Mitchell	1/31–2/9/71	216:01:59	34 rev. of moon	First manned landing in and exploration of lunar highlands. Total stay time: 33 hrs., 31 min.
Apollo 15	David R. Scott Alfred M. Worden James B. Irwin	6/26–7/7/71	295:11:53	74 rev. of moon	First use of lunar roving vehicle. Total stay time: 66 hrs., 55 min.
Apollo 16	John W. Young Thomas K. Mattingly II Charles M. Duke, Jr.	4/16–27/72	265:51:05	64 rev. of moon	First use of remote-controlled television camera to record lift-off of the LEM ascent stage from the lunar surface. Total stay time: 71 hrs., 2 min.
Apollo 17	Eugene A. Cernan Ronald E. Evans Harrison H. Schmitt	12/7–19/72	301:51:59	75 rev. of moon	Last manned lunar landing and exploration of the moon in the Apollo program returned 243 pounds of lunar samples to earth. Total stay time: 75 hrs.

Courtesy NASA

*O*n the third and last EVA at Taurus-Littrow (on the afternoon of Wednesday, December 13, 1972), Cernan and Schmitt drove the Rover to the northeast end of the valley, to the gentle slopes of North Massif Mountain. On the way they found a huge split boulder—about 60 feet (18 meters) long, 33 feet (10 meters) wide, and 20 feet (6 meters) high—which they sampled.

At the last stop of EVA 3, the astronauts took two core samples on the rim of Van Serg Crater. There was concern on earth about their time safety margins, and Houston ordered the astronauts back to the *Challenger* base, where they completed experiment tasks and parked the Rover and its TV camera in a position to cover humankind's good-bye launch from the moon.

Shortly before the two men stepped off the lunar surface for the last time, Cernan uncovered a plaque on *Challenger*'s landing leg and read it for his earth audience: "Here man completed his first exploration of the Moon December 1972 A.D. May the spirit of peace in which we came be reflected in the lives of all mankind."

Challenger's ascent stage engine ignited at 5:55 P.M., December 14, 1972. This last man-made fire on the moon thrust the astronauts off the surface toward Apollo's final lunar rendezvous and docking.

LEAVING THE MOON BEHIND (FOR NOW)

Man-on-the-Moon-Speed

Twelve astronauts and six lunar excursion modules spent about three hundred hours on the moon's surface during the Apollo landings (July 1969 through December 1972). About one-fourth of those (79.4) were spent walking and driving around on the lunar surface, covering a total distance of 54.5 miles (87.7 kilometers). The average speed of man on the moon was, therefore, 0.686 miles (1.1 kilometers) per hour. If the astronauts had to make the return trip to earth at this speed, it would take them almost forty years.

Better than Gold

The total quantity of lunar rock and soil brought back on the Apollo missions was 841.6 pounds (381.75 kilograms), which averages out to about 70 pounds (31.75 kilograms) for each of the twelve astronauts who walked on the moon's surface. The material was divided among approximately 550 scientists for analysis once it was returned to earth, which amounts to about 1.5 pounds each. During analysis, the original number of samples, 2,196, was broken up into 50,000 pieces. One estimate puts a value of $28,500 per pound on this material, but this is not a market value and is grossly underestimated, being just over five times the 1984 average price for a pound of gold.

Paper Moon

Just two years after *Apollo 17*'s voyage to the moon in 1972, there were over thirty thousand pages of scientific printed material on the results of lunar rock and soil analysis from the mis-

sions. This would make a stack of material almost 14 feet (4.75 meters) high. If all these pages were placed end to end, they would add up to about 27,000 feet (8,230 meters)—the estimated rim-to-floor depth of the moon's deepest crater, Newton.

The Round-Trip Moon Rock

Almost 2,200 individual samples of moon rock and soil were brought back to the earth during the Apollo years, and of that number only one piece of rock—sample 12002, a piece of basalt brought back by the *Apollo 12* crew—was returned to the moon. Why? Well, it wasn't the fact that it was contaminated or otherwise dangerous; it was rather to test the amount of magnetism that a moon rock could develop by exposure to the Apollo spacecraft and the earth's magnetic field. One of the big surprises coming from the Apollo program was that the moon had possessed a strong magnetic field at one time in its past (a key to internal structure and history). This fact was detected in the analysis of the lunar samples. Lunar sample 12002 thus became a kind of control rock. It was completely demagnetized and sent back to the moon on the *Apollo 16* mission, to become the only moon rock to make a round trip.

Mysterious Rusty Rock

There is no rust on the moon. Moon rocks in all their billions of years have never been exposed to water or free oxygen because neither exists in the lunar environment. Elaborate precautions were initiated to protect the moon rock samples from the earth's atmosphere, which could instantly rust them. The rock samples were stored in an atmosphere of dry nitrogen, and exposure to the laboratory atmosphere was limited as much as possible without restricting the experiments. With such foresight and planning, it was a surprise to scientists when they found "Rusty Rock" 66095, which contained the only water found in any of the lunar material brought back to earth. No final explanation for Rusty Rock has been agreed on by everyone. The three most credible explanations for the water are: earth contamination, a comet strike on the moon, or lunar volcanic gas. Some lunar mysteries remain even after Apollo.

The Moon Gems

The most beautiful moon materials returned by the Apollo voyages were the glasses found in lunar soil and rock. These glasses came in a whole range of colors: pale yellows, browns, reds, blacks, greens, and oranges. Most of the lunar glasses were created by the tremendous force and heat of meteorite impacts.

In fact, *Apollo 11* astronauts Armstrong and Aldrin saw the impact-created glasses as patches splashed onto rocks and craters. The two most famous moon glasses, however, did not originate in the usual way and were in the form of tiny glass droplets. *Apollo 15* astronauts found the "green clods," which were rich in emerald-green glass spheres, at Spur Crater. These green lunar gems made up 20 percent of the soil by volume. They were dated at 3.3 billion years and were therefore related to the ancient lava flows from the *Palus Putredinis* (Swamp of Decay). The biggest surprise for the *Apollo 17* crew was to find orange soil at Taurus-Littrow—soil that stood out dramatically in the usual bleak gray moonscape. The orange beads found in the soil were 3.8 billion years old and were also related to the earliest great lava floods on the moon.

These green and orange glass beads are believed to have come from within the moon, perhaps as deep as 200 miles (322 kilometers down). Volcanic in origin, they probably formed as droplets in the volcanic fire fountains and, in fact, bear a strong resemblance to glasses from the Hawaiian lava fountains.

Moon Nostalgia

The fifteenth anniversary of *Apollo 11*'s moon landing was celebrated on July 20, 1984, and the three-man crew, Neil Armstrong, Buzz Aldrin,

and Michael Collins, gathered together and made some public appearances, including a "Today Show" interview on NBC television. The famous Apollo crew also had one of their rare reunions five years earlier, in July 1979, when the tenth anniversary of the historic landing was celebrated.

"Many of the individual parts of the flight I recall as if it were yesterday," Armstrong admitted. When asked if the first step on the moon was his ultimate high for the journey, he replied that it wasn't. "If there was an emotional high point, it was the point after touchdown when Buzz and I shook hands without saying a word. That still in my mind is the high point."

Apollo's greatest achievements may be seen as the basic spacefaring skills taught, the complex machines and facilities built, and the vast scale of successful organization created to extend the limits of human activities beyond planet Earth in less than a decade.

"We know now that human beings are not chained to this planet," Armstrong said. He believes that a permanent presence in space is the next step, which will come when the United States has its space station operational sometime in the 1990s.

Michael Collins, who was command pilot of *Columbia* in orbit above the moon when Armstrong and Aldrin were on the surface at the Tranquillity Base, sees an even larger picture. "When the history of this galaxy is written, and I think it may already have been, if the planet earth is mentioned at all, it won't be because its inhabitants visited their own moon. That first step, like a newborn's first cry, would be automatically assumed. What would be worth recording is what kind of civilization we earthlings created and whether or not we ventured out to other parts of the galaxy. Were we wanderers? Human history so far indicates we are indeed."

"I certainly believe we will return to the moon," Armstrong told his audience in 1984. No one disputes this. The only question is: When? Perhaps there will be a permanent lunar base whose inhabitants will celebrate the fiftieth anniversary of *Apollo 11*'s landing on the moon in 2019.

2

The Shuttle Ships

A Spaceship Named Shuttle

Before January 28, 1986, most of us thought the space shuttle was a technological wonder. This wide-winged spaceship, created by the minds and hands and machines of men and women on planet Earth, represented a maturing Space-Age technology that would take ever-increasing numbers of people, even ordinary citizens, into space to experience a new reality and do the important work that would build our future. Twenty-four times it blasted its crew and cargo into orbit; twenty-four times it returned safely to the earth's surface. Before the disaster, about the worst that could be said of the space shuttle was that it had a dull and mundane name.

Here was the most complex machine ever built on earth, and it was working. Whole new technologies were forged during the decade in which it was built and made ready to fire astronauts into orbit, where they would live and work within the hostile space environment. *Hostile environment?* Yes. Space was hostile, and the way of getting there was dangerous. But who was talking about these negative realities? No one. The ever-present dangers of spaceflight were played down, and the reliability and safety of modern-day spacefaring were emphasized to the public. Before the *Challenger* disintegrated in midair, we saw the shuttle crews—the commanders, the pilots, the mission specialists, the payload specialists—return safely from space month after month with smiles on their faces. We did not dwell on the fact that without their engineered cocoons—their pressurized spaceship quarters or their self-propelled space suits—they would be dead in seconds, asphyxiated, frozen, or burned. Not many of us thought of the inhuman energies required to blast these shuttle ships into space as the crews, one after the other, returned to earth.

Before the *Challenger* exploded in less than a second during the most dangerous phase of all

shuttle flights, we believed what NASA told us: Shuttle spaceflights were routine and safe. We had forgotten, focusing on NASA's goals, how dangerous space adventures still were. No one wants to know how fragile life is when a shuttle blasts its crew into orbit, balanced on a great and powerful fire that human ingenuity does everything to control. Yes, most of us had forgotten how hostile space is to human life, and how dangerous it is to ride a rocket to orbit. Then the instant destruction of *Challenger* and the seven dead crew members changed all that. Even though astronauts and politicians alike professed that men and women would soon continue to fly the shuttle into orbit, manned spaceflight would not be the same for the rest of the twentieth century. Routine spaceflight was dead; danger had returned. The shuttle would fly again, and more people would be actively interested in the space adventures. Regretfully, some of them would be drawn to the danger, not the accomplishments and promise, of spaceflight.

Tarnished Magnificence

Before the last flight of *Challenger*, the space shuttles were considered the greatest flying machines of the Space Age. Even John Young, the first commander of a space shuttle mission, called them that. These spaceships that thundered and crackled through the atmosphere on their way to orbit month after month, that allowed their spacefaring crews to accomplish thousands of hours of important work, that were taking the United States and its allies into the future, had been stuck with the unfortunate name space shuttle, which undercut their exciting reality and impressive accomplishments. This fleet of four spaceships was taking the Space Age into the twenty-first century. Then the *Challenger* exploded and scattered into the Atlantic below. One of the four white spaceships fell to earth; a technological pyramid had disintegrated before our eyes. For two years, the U.S. manned space program stopped flying. The name "space shuttle" suddenly seemed more appropriate; it implied that it was ordinary and could sometimes break down like any other flying machine. What the astronauts had called their magnificent flying machines had been badly tarnished.

The Ho-Hum Lull Before Tragedy

At least the individual ships in the space shuttle fleet had names worthy of their accomplishments and their promise: *Columbia*, *Challenger*, *Discovery*, and *Atlantis*. But the general public never greeted the shuttle flights with the enthusiasm it had shown during the early Apollo missions to the moon. Why was there such a ho-hum attitude from millions of taxpayers in the United States before the 1986 tragedy?

For one, NASA successfully promoted its idea of routine spaceflight to the public, even though each shuttle mission was anything but routine. Twenty-four successful flights, with less and less media coverage as the flights continued, also enforced the illusion that spaceflights to orbit were easy and safe. And in the opinion of many, the space shuttle also lost the Space-Age beauty contest to the Saturn 5 moon rocket, at least for those old enough to remember; the shuttle just isn't as impressive-looking as the late, great moonship. Even some of its pre-disaster nicknames, such as "The Flying Brickyard" (it drops like a brick when it reenters the atmosphere) or "Polish Bomber" (its cargo bay doors open upward), spoken with some degree of affection by its makers, demonstrated a degree of ambivalence about its physical appearance. Perhaps the shuttle design clashed with the early fantasy designs of space art and literature. But appearances are deceiving in spaceships. Think of the dozens of beautiful spaceships from artists in this century that never got out of the studio and onto an engineer's drafting board. And given the choice, who wouldn't sacrifice aesthetic design for engineering reliability? If it worked and did what it was built to do, then it was a good spaceship.

In the aftermath of the *Challenger* loss, it was difficult not to think of the space shuttle as a

ABOVE: *The space shuttle* Challenger *heads toward its second earth-orbital mision on June 18, 1963.* PRECEDING PAGES: *The first EVA of the space shuttle program on the maiden flight of space shuttle* Challenger *in April 1983. Astronauts Story Musgrave (left) and Donald Peterson (right) move about the cargo bay, tethered to safety slide wires.* Courtesy NASA.

failed compromise—a compromise of America's superior technological ability forced upon NASA by Congress during the early 1970s because of budgetary pressures resulting from the Vietnam War.

Still, as the investigations conclude, as the decision-making and technological fixes are made, and as the astronauts prepare to fly again, the space shuttles enter the last decade of the twentieth century as better and safer spaceships than they were before the *Challenger* exploded, killing seven of America's best. But the word "routine" is no longer used in press releases, and everyone is painfully aware of what they should have known before: Spaceflight, like all of human existence, is precarious and fraught with danger.

By the year 2000, the redesigned shuttle spaceships will have transported thousands of people to jobs in space and a new way of life that may someday be the norm for our species. Tens of thousands of human hours in space will accumulate as the pace of space activities steadily increases and confidence is restored.

What is all this activity above earth about? The thousands of specific tasks the space shuttles allows us to perform in space, be they commercial, scientific, or defense-related add up to nothing less than the foundation for our future. When the new space shuttle is built, perhaps NASA should consult with a presidential speechwriter or a brilliant Hollywood producer to find a deserving name for the spaceship that is carrying future generations toward some mysterious destiny beyond the earth.

The Ugly Duckling of the Space Age

When the Saturn 5 moon rockets roared off the earth and shook the Florida coast, they were going somewhere special; toward the moon. When space shuttles fly, they crackle through the sky and head into low earth orbit, just a few hundred miles or kilometers above our heads. The adventure of stepping onto a mysterious world is no longer there, and perhaps this is why many people, before the disaster, considered the regular launches into orbit nothing to get excited about.

Actually, ever since President Nixon announced on January 5, 1972, that the United States would develop the space shuttle, NASA promoted this new space program by emphasizing that it was a reusable spaceship, with a lifetime of one hundred missions. Each spaceship built would routinely fly to orbit, accomplish its mission goals, and return safely to *terra firma*. But airplanes did almost that thousands of times each day. Many people forgot that the space shuttle could fly fifty times higher than an airplane, in a vacuum beyond the earth's atmosphere. In theory the space shuttle would eventually rent its services to customers around the world and pay for itself. It all sounded so easy to the public. After all, we had just put twelve men

FACING PAGE: *Space shuttle* Discovery *leaves the launch pad at Johnson Space Center in Houston, Texas on April 12, 1985.* BELOW: *Space Shuttle* Atlantis *clears the launch tower at Kennedy Space Center on October 3, 1985. Courtesy NASA.*

on the moon. No one really cared if 99 percent of the moon rocket that took them there was abandoned along the way. If the space agency wanted to give the impression that its new spaceship would make spaceflight to and from orbit routine, it succeeded—until January 28, 1986.

But where was the excitement of the early Apollo missions to the moon? Certainly not in the name space shuttle or the Space Transportation System, generic labels that sounded too much like the busy air service between cities or the mass transportation systems within them. To appreciate this spaceship, a person had to appreciate what it could do, not where it was going. Not just what it could do on a single mission, either, but what the entire fleet could accomplish in hundreds of missions over twenty years or more. This was not news of a dramatic touchdown or lift-off from the moon's surface, but rather a slow accumulation of experience and knowledge gained over many years. Even the destroyed *Challenger* tragically served this end. As the shuttles flew for almost five years before they were grounded, space enthusiasts were excited, but many people felt that the Space Age was starting to get old. Most kids waited for the next special effects yarn from Hollywood and occasionally got interested in some zero-g antics on the TV news.

Perhaps this feeling was unconsciously reinforced as millions of TV viewers tuned in to watch the first few test-flight launches of *Columbia* in the early 1980s. There was the space shuttle on the launchpad, appearing squat and fat compared to the tall, sleek Saturn 5 moon rocket. And when *Columbia* landed in the California dessert like an airplane, it appeared deceptively conventional. None of the high-tech breakthroughs that allowed the space shuttle to fly into space and return through the metal-melting heat of reentry could actually be seen by anyone. No one appreciated the tremendous engineering that went into creating the world's first reusable rocket engine, or the more than thirty thousand heat-shield tiles that would prevent the space shuttle from burning up during dozens of blazing reentries.

Much of the general public had these and other false impressions about the space shuttle. In name and appearance, this new spaceship was the ugly duckling of the Space Age, and the *Challenger* explosion reinforced this. Few had the foresight and knowledge to know that it would grow up to be, in the aftermath of the disaster, the swan in our local cosmic pond.

Shuttle's Deceptive Guise

By viewing two same-scale drawings or photos of the space shuttle and the Saturn 5 moon rocket on their launchpads, it is impossible to know that one is by far the other's technological superior. The space shuttle is smarter and more versatile; the mighty Saturn 5 is more powerful and much taller. One does well to remember that the failure of one solid rocket booster on the shuttle was the failure of old technology that went back decades. While the space shuttle can perform hundreds of tasks above earth, including satellite and spacecraft launches, the Saturn 5 was designed to accomplish only one major goal: to land men on the moon and bring them back to earth. The mighty moon rocket was 365 feet (111 meters) tall, its three propulsion stages stacked one atop the other, with the Apollo spacecraft on the top. The space shuttle at launch, from the bottoms of its solid rocket boosters to the nose tip of its huge external fuel tank, stands 184 feet (56 meters) high, about half the length of Saturn 5. But wait. The squat look of the shuttle is deceptive because its major components—the two solid rockets, the external fuel tank, and the piggybacked orbiter spaceship—are clustered together, not stacked as they are on Saturn 5. If the space shuttle's orbiter and solid rockets were somehow detached and moved to the top of the external fuel tank, these major shuttle components would stand 425 feet (129.5 meters) high and dwarf the Saturn moon-ship. The space shuttle keeps all its power lower to the ground. The width of its gigantic external tank with the two solid rockets attached to either side comes to almost 52 feet

(15.8 meters), almost 20 feet (6 meters) wider than the diameter of Saturn 5's first stage. But a spaceship's size and shape does not necessarily indicate its power. At launch, the stocky shuttle spaceship delivers some 83 percent of the power that the great Saturn 5 rocket did, and it's only going 1/1,600th the distance of an Apollo moonship.

The Smart Spaceship

No one disputes Saturn 5's superior power, which was needed to put 90,000 pounds (41,000 kilograms) into orbit around the moon. Comparing their thrust powers at launch, the five first-stage Saturn F-1 engines blasted about 7.5 million pounds (3.5 million kilograms) of thrust against the launchpad; the space shuttle's solid boosters and three advanced main engines put out some 6.4 million pounds (2.9 million kilograms) of thrust. Of course, the Saturn 5 had a second and third propulsion stage to drive 280,000 pounds (127,008 kilograms) into earth's orbit before going to the moon; today's space shuttle is considered a one-and-a-half-stage rocket, with its main engines and the solid rockets, and is capable of delivering 65,000 pounds (29,484 kilograms) of payload into low earth orbit. Add to that the orbiter's own weight of 140,000 pounds (63,504 kilograms), and these advanced rockets can deliver a total of 205,000 pounds (92,988 kilograms) into the high frontier. These first few hundred miles (kilometers) into earth orbit represent the toughest part of any spaceflight: getting out of the earth's gravity well, which in terms of energy is 4,000 miles (6,436 kilometers) deep.

But in the Space Age, what goes up does not necessarily have to come down. The Saturn moon rocket weighed 6.1 million pounds (2.77 million kilograms) before it thundered off earth. But only the Apollo command module, weighing about 11,000 pounds (4,990 kilograms) and measuring about 11 feet (3.4 meters) high and 13 feet (4 meters) in diameter splashed back to earth with its three-man crew. The entire space shuttle orbiter, 122 feet (37 meters) in length and with an empty weight of 165,000 pounds (74,844 kilograms), lands on solid ground to fly again another day. Also, its solid rocket engines, fuel expended, are recovered in the Atlantic or the Pacific to be refurbished and used again. Unlike the Saturn 5 rocket, which sacrifices most of itself during its moon journey, the shuttle spaceship was designed to survive its voyage and return to earth, sometimes even laden with satellites and spacecraft it picks up in orbit.

Without doubt the space shuttle is a much smarter spaceship. Its intelligence can be seen by comparing what happens at launch. Compared to the Saturn 5 moon rocket, which hesitated on the pad in the midst of all its thundering flame and smoke before ever-so-slowly lifting itself off the earth, the space shuttle lifts off smartly, with steady ease, leaving no doubt that it knows exactly where it's going. Even the *Challenger* did that as it lifted off for the last time.

A fish-eye lens captured this view of the space shuttle Columbia's flight deck. In the center are the three CRT display units for the computers. The commanders and pilots can ask more than one thousand questions and get their answers displayed immediately. The original computer hardware, since upgraded, was forty times faster than that of the Saturn 5 moon rocket. Courtesy NASA.

Those Magnificent Flying Computers

The shuttle astronauts ride atop great columns of fire, several times longer than the spaceship itself, each time they blast into orbit. We know their lives depend on the rockets and their flows of hydrogen and oxygen. But their lives also depend on the microchips and electrical microcurrents that travel through the complex on-board and launch and ground-control computers, which automatically control the spaceship during its critical mission phases. Without millions of software instructions communicating throughout the computer system at the speed of light, the space shuttle would never get off the ground. Not surprisingly, the space shuttle is a computer dependent spaceship.

Back in February 1974, during the early development stages of the space shuttle, the director of the Johnson Space Center, Christopher C. Kraft, went to Washington and reviewed the flight quality of the space shuttle for the Subcommittee on NASA Oversight of the House Committee on Science and Astronautics. Wind tunnel tests showed some flight instabilities at supersonic velocities during reentry, and some congressmen and astronauts were concerned.

"A lot of our astronauts have been very unhappy and outspoken about the lack of good flying qualities," Kraft told the subcommittee members. He went on to tell them that it would be the computer-controlled small gas thrusters that would compensate for these instabilities rather than any movable ventral fin, which the design team thought would create heating and control problems during the spaceship's critical glide back to planet Earth. What Kraft told the congressmen back in 1974 was that if the United States wanted the space shuttle in the decade of the 1970s (it wound up being two years late), NASA was "not going to build an aerodynamically stable vehicle. That's an impossibility. We're building a fly-by-wire vehicle at present that's going to get its stability from black boxes [the computer system]. And that's the name of the game in the shuttle."

The space shuttle, in other words, would be run by a committee of computers. They would comprise the most complex autopilot ever built and could act thousands of times faster than a human pilot could. Each space shuttle in the fleet has five computers that perform guidance, navigation, and controlling tasks. One of the five computers is a backup during the critical launch and reentry phases of each mission. The four computers are capable of doing 325,000 operations each second. And before launch, during the last nine minutes, the computers process about 88,000 different sets of instructions. What follows is a list of other computer features and capabilities.

• Space shuttle astronauts can ask the computer more than one thousand questions about the flight and the condition of their spaceship and get answers and data immediately in several different formats on any of three CRT display units on the flight control panels.

• The computer complex of each space shuttle orbiter has thirty-eight subsystems, with an additional four on the solid rocket boosters.

• The data processing rate for the original computer system was 787,000 thirty-two-bit words a second; it is over one million for the recent upgraded system.

• The original processor for the CRT display units was forty-two times faster than the one in the Saturn 5 rocket computer; the upgraded version is more than sixty times faster than the Apollo system.

• The computer programs are the most sophisticated spacecraft programs ever developed and are more complex than most earthbound systems. Their reliability is exceptionally high because during critical flight times, all four computers back one another up and cross-check one another.

• The space shuttle's operational programs that enable the computer to accomplish its operational functions are written in a programming language called HAL/S. Is this where the infamous computer HAL of the film *2001* got his name?

• The five application programs that control the shuttle's data processing for trajectory, telemetry, command, network communications, and control during a mission form one of the largest and most complex systems ever produced. These five programs contain over 600,000 lines of programming, and the trajectory programs alone contain 220,000 lines.

• During launch, the flight controllers are given new trajectory data every two seconds on their CRT screens; the upgraded system is even faster.

The space shuttles have more computer power than any previous spacecraft. The original IBM hardware on the shuttle could process data forty times faster than the Saturn 5 moon rocket computer, and the memory capacity for the shuttle's original hardware was five times greater. The hours of testing before launch involve more than forty thousand associated test parameters. Before a Saturn 5 rocket blasted off to the moon, 250 engineers had to work for six months to test the vehicle. With the shuttle, all the testing has to be done within two weeks, so an entirely new computer system was developed by IBM.

Shuttle's Second-Generation Computer System

Through 1985 the space shuttle flew an extremely reliable computer system (even considering the failure of two computers on *Discovery*'s Spacelab 1 flight in November 1983 because of microscopic bits of solder floating inside the chips), but its design was well over a decade old. In 1987 the space shuttle fleet was scheduled to receive new upgraded computer systems.

The differences between the original computer system and the second-generation computer system (the system future mission will be flying with) generally follow the advances in the entire computer industry—faster, lighter, less expensive, and able to do more with less power. But what is most amazing about the new IBM system is that the new, state-of-the-art hardware can run on the millions of lines of original software faster and with less power than the hardware used during the first decade of shuttle flights. The development costs and time of creating all new software would have been prohibitive, and so the integration of new hardware with old software is a definite breakthrough.

The first-generation computer system ran close to its limits of memory capacity during the critical phases of launch and reentry. The new computer system has twice the memory capacity—more than 262,000 words. The larger memory will give space shuttle flights a greater safety margin and generally increase operational capabilities and options. The new hardware is also about three times faster than the first-generation system. System reliability (operational hours between any failure) has also been substantially increased, to about twice its present level, to upwards of ten thousand operational hours. All these advances are accomplished with equipment that is about one-half the size of the original equipment and with about 20 percent less power. And even the taxpayers are getting more for their money because the new IBM system is less expensive than the first-generation hardware.

While the shuttle's computer nervous system has grown and matured, so has the number of tasks it must do in the next two decades. The second generation of space shuttle computers will certainly not be the last.

Technological Edges and Shuttle Snafus

Real technological frontiers had to be crossed before the space shuttle would ever fly, not as many as the Apollo program faced, perhaps, but still technological challenges that had to be conquered. In the early 1970s, everyone associated with the shuttle knew that the heat-shield system was essential to the success of the entire space shuttle program. Without a heat shield that could protect the spacecraft time and time again, during dozens of blazing reentries, there

could be no reusable space shuttle. Everyone knew that this was the most formidable and challenging technology that had to be created from scratch. And the aerospace community recognized that the development of the shuttle's rocket engines, the first in the world ever to be fired more than once, while not as formidable an engineering task, would also require dogged engineering persistence. As it turned out, these two technical breakthroughs proved considerably more difficult than anyone had anticipated.

The Heat Shield Tiles. To protect the space shuttle and its crew during the fiery reentry, when areas of the spaceship's exterior experience temperatures of more than 2,700 degrees F. (1,428 degrees C.), some 31,000 protective tiles are custom fitted on the ship's underbelly, beneath the wings, and on other heat-bearing surfaces.

The tiles are made of silicate fibers, which are derived from sand stiffened with clay. They come in two different basic sizes, 6 inch (15.24 centimeters), and 8 inch (20.32 centimeters), and vary in thickness from 0.5 to 5 inches (1.27 centimeters to 12.7 centimeters), depending on their location on the space shuttle and the amount of reentry heat to which they are subjected. Each one is individually machined to fit the contours of the fuselage and wings, and specifically measured gaps separate the tiles, depending on the structural stress of the fuselage skin beneath. These gaps range from .00004 to .0000133 inch (.0001016 to .0000337 centimeter). No two tiles in the 31,000-piece jigsaw are alike. The black-coated tiles are on those areas that will receive the highest heat (the shuttle's underbelly and under the wings), and the white tiles on areas that receive the lower temperatures. The insulating tiles are extremely light and cannot weigh more than 1.7 pounds (0.77 kilograms) for every square foot of surface. When you hold a tile in your hand, it feels as light as a feather; the mind finds it hard to accept that this seemingly insubstantial material can protect a spaceship from burning up in the earth's atmosphere. For the Apollo insulation that peeled off and could not be reused, the

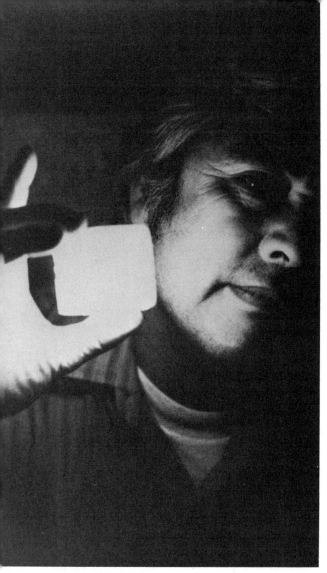

Some 31,000 custom-fitted protective tiles, no two alike, absorb the tremendous heat during reentry, as high as 2,700 degrees (1,482 degrees C°). Here a Lockheed technician holds the white hot material in his bare hand without injury. This demonstrates how fast the unique silica tile material casts off heat. Courtesy Lockheed Missiles & Space Company, Inc.

weight limit was more than twice that amount, almost 4 pounds (1.8 kilograms) for the same square area.

The insulating qualities of these new tiles were astounding and delivered everything they promised. They could throw off heat so quickly that a white-hot tile directly out of an oven, with a temperature of 2,300 degrees F. (1,260 degrees C.), could be held in a bare hand without burning or causing other injury. The new technology worked. The serious problem turned out to be how to apply the tiles to the shuttle and keep

them on. On one airborne flight, with the shuttle perched atop the specifically adapted Boeing 747, one hundred tiles were ripped off. The tiles were not sticking. If this happened on an actual shuttle mission, it could spell disaster. The new technology was worthless unless the tiles stayed on. And the tile problem was one of the major reasons the maiden flight of the space shuttle *Columbia* took place about two years later than was originally scheduled.

Installing the heat-shield system was much more complex than anyone in the program had expected. It turned out to be more than a bonding problem. The main oversight was that the thermal heat shield was not tested for mechanical integrity before most of the tiles were installed on *Columbia*. When stress-tested, thousands of the tiles failed the tests; many of them split.

The decision was finally made to make the base of each tile more dense to avoid the splitting. Their density would be increased by treating them with ludox, a silica-boron compound. But the process took a great amount of time. It has been estimated that it took 670,000 hours to properly install the tiles on the space shuttle *Columbia*—about 335 man-years! That worked out to about 1.7 installed tiles per worker per week. The space shuttle's major snafu was finally, for the most part, solved. *Columbia* would blast off to orbit late, but at least it would get off the ground and return without burning up.

Space Shuttle Main Engines (SSMEs). At the same time that the thermal tile problem was being resolved, a second major snafu was developing, and it came to a head in 1979—the second goal year that was set for the maiden test flight. The shuttle's main engines were experiencing serious problems during tests; and there was no way to fudge on main engine reliability. Between March 1977 and November 1979 there were fourteen engine-test failures. In the first five years of engine development, some 895 modifications to the engines were made.

The space shuttle main engines, three mounted on each spaceship in the fleet, were the most powerful hydrogen-oxygen engines ever built.

The combined thrust of these three high-pressure engines would be more than 1.1 million pounds (499,000 kilograms) at lift-off, not counting the thrust power of the two solid rocket boosters that would develop several million more. And they were the world's first rocket engines designed to fly more than a single mission, in their case fifty missions.

What were the problems? Many. Valve breakdowns, hydrogen line ruptures, faulty seals, and cracked turbine blades were examples. Early optimism about the known technology of main engine development was groundless. The development of the SSMEs involved major technological advances, but NASA had told Congress during the approval process that the engines would require only some innovation. Everyone underestimated the technological challenges involved in creating the world's first reusable engines with the most powerful thrust possible for their weight and size. The high-pressure fuel pumps in the engines, for example, were so powerful that they were capable of emptying an Olympic-sized swimming pool in just twenty-five seconds. Money and time were both in short supply, but major technological advances require ample supplies of both. Engine-test failures damaged or destroyed the engines and their testbeds; this meant that NASA had a short supply and had to order more. Costs rocketed and made original estimates appear ridiculous.

Slow progress was made. By the end of 1978, there was an accumulation of almost 35,000 seconds (9.7 hours) of test firings. This was approaching the halfway mark of NASA's requirement of 80,000 seconds (22 hours) before an orbital flight could be launched. But the problems were far from over; more failures occurred. The series of failures was traced to improper welding wire. All the engines were shipped back to Rocketdyne in California for rewelding and other modifications. Finally the engines began passing their tests in late 1980. On January 17, 1981, the final engine test of the program was conducted. It ran for a total of 629 seconds (10.48 minutes), which was about 2 minutes longer than the 8 minutes and 30 seconds that the maiden *Columbia* needed to get into orbit. And finally, after more than two years of delays, spaceship *Columbia* fired up its main engines on April 12, 1981, and began the space shuttle era.

The Versatile Shuttle

The space shuttle has been called a space truck, and no one can argue with the fact that its most important single purpose is to deliver satellites and spacecraft to low earth orbit on a regular schedule. Whatever can fit inside its roomy cargo bay (large enough to accommodate five African bull elephants) and is within certain weight limits can be taken into orbit or picked up and brought back down to earth. Hundreds of satellites for communications, military reconnaissance, weather, navigation, search and rescue, and scientific observations of the earth's surface or the faraway universe will be deployed in orbit by the shuttle fleet over the next two decades. Dozens of multipurpose and dedicated satellites have already been successfully deployed for governments, scientific organizations, and corporations around the world. The first twenty-four shuttle missions, broke out close to the following percentages, based on the customer:

NASA and U.S. scientific missions	36 percent
Commercial and foreign	22 percent
Department of Defense	35 percent
Reflights for failed missions	6 percent

Of course, the shuttle spaceships provide much more than just a delivery and pick-up service. They can be launchpads for interplanetary spacecraft such as *Galileo* to the planet Jupiter and *Ulysses* to an unusual polar orbit around the sun (see Chapter 8, "Robots from Earth"), or for state-of-the-art instruments such as the Hubble Space Telescope (Chapter 9, "Commerce and Science in Orbit"). When they fly with the Spacelab modules in their hold, they become dedicated research laboratories for government or industry (Chapter 9). On certain missions, they become military bases for testing top-secret technology. They become construction headquarters for testing and building large space structures as well as astronomical observatories for special cosmic events or for scanning the distant universe with sophisticated high-tech "eyes" in all wavelengths.

Each time one of the shuttle spaceships flies successfully, it also becomes a life science laboratory, collecting important information on how the crew members respond to life in zero gravity. And as the shuttle ships cover their tens of millions of miles in orbit around the planet Earth, dozens of sensors collect data on spacecraft performance, creating a base of knowledge from which their future replacement ships will be designed and built.

The shuttle ships do so much that even the men and women who fly them are hard-pressed to learn the complexities in their area of responsibility. Joe H. Engle, commander of the second test flight of *Columbia*, summed it up for the novelist James Michener during an interview.

"Look at this pile of manuals we have to know by heart," he told the author. "Hydraulics, propulsion, communications, digital processing, life-support systems, environmental control, orbital maneuvering, reaction controls, navigation guidance, mechanical functioning, big-arm manipulation, reentry data, glide control." Getting to know the world's most complex machine does not come easily.

Shuttle Limits

The question should be: What *can't* the space shuttle do when it comes to space activities? First, it is a low-earth-orbit spaceship and cannot fly to geosynchronous orbit to deploy certain types of satellites such as the military's early warning systems or the global communications systems. These satellites are deployed with special launch systems attached, such as the payload assist modules (PAMs) or upper stage rockets, to transport them to high orbit locations.

Mission time is also limited to about ten days, although varying certain mission factors, such as the number of crew members, can extend

mission duration for a few days. And future plans call for longer missions that can be achieved in several ways, such as putting supplies into storage satellites and launching an auxiliary power station the shuttles can tap into for extra electricity.

And then there is the very obvious fact that this most sophisticated of all flying machines cannot even fly from Vandenberg Air Force Base in California to Cape Kennedy in Florida under its own power, but has to return to the Cape piggyback fashion on a Boeing 747. Without its strap-on rocket power, the shuttle spaceship is all but helpless in our planet's atmosphere—except, of course, for its gliding ability. Almost all the power the shuttle uses to reenter the atmosphere and return to earth comes from the orbital velocity of more than 17,000 miles (27,353 kilometers) per hour that the original blast-off rockets gave it. Its weight, delta wings and other aerodynamic characteristics, and computers are what make possible the almost perfect deadstick landings.

This generation of space shuttle will never make quick hops between Los Angeles and New York, New York and London, or London and Melbourne. The jet engines that were originally envisioned to give the shuttle its own power through the atmosphere were victims of program budget cuts in the early 1970s. So while the shuttle spaceships will carry at least some of the space station components into orbit, deploy hundreds of satellites that are changing our planet, make possible breakthroughs in zero-gravity research, and launch our most sophisticated robots to explore distant worlds, these good ships will never make a suborbital hop from Orlando to Houston.

The Maiden Flights

Each space shuttle in the original fleet, the four that have flown missions to space and the test vehicle, *Enterprise*, which was never intended to get into orbit, proved themselves in the late 1970s and the first half of the 1980s. What follows is a summary of their maiden voyages.

Enterprise: **August 12, 1977.** The first space shuttle to be rolled out of Rockwell International's assembly plant, the *Enterprise* (also known as Orbiter 101) was used for five approach-and-landing tests in 1977 to see if the predictions of the wind tunnel tests were correct. Could this untested, 150,000 pound (68,040 kilogram) spaceship glide safely through the atmosphere to a landing strip at Edwards Air Force Base in the Mojave Desert? If the *Enterprise* landed safely after the Boeing 747 jumbo jet carried it to an altitude of 22,000 feet (6,706 meters) and released it, then the space shuttle testing program was off to a good, although delayed, start. It would also set a world's record as the heaviest glider ever to fly.

After test pilot Fred Haise (remember *Apollo 13*?) pushed the release button, the test shuttle was gliding toward a desert touchdown. The *Enterprise* was safely down in just under six minutes, having fallen, as predicted, more like a brick than a spaceship. It was, officials said, a "beautiful" flight. One space shuttle at least had finally gotten off the ground.

The *Enterprise* never saw space service, but it was used to test preflight and launch facilities at both Cape Kennedy and Vandenberg Air Force Base, and it was asked to give up some of its parts for use as spares. That this shuttle never flew disappointed many of the space enthusiasts and science fiction buffs who had lobbied to have President Ford christen it the *Enterprise* after the spaceship in the "Star Trek" TV series.

Columbia: **April 12, 1981.** Finally, after years of delay, the space shuttle's moment of truth had arrived. It would be tested from launchpad to orbit, through reentry to landing, for the first time. Although the shuttle had never been flight-tested before, except for its gliding performance in the lower atmosphere, there would be live astronauts aboard, commander John Young and pilot Robert Crippen. Only those people deeply involved knew what a dangerous mission this could turn out to be. There were so many unknowns; only ground-test data suggested success.

Spaceship *Columbia* left the surface of the earth at 7:00 A.M. Eastern Standard Time on April 12, 1981, gushing out huge columns of orange and yellow flame and rising above great billows of steam on the launchpad. The 6.5 million pounds (2.9 million kilograms) of thrust at launch was more than ninety-nine times more powerful than the Redstone 3 rocket that Alan Shepard rode on his historic flight of May 5, 1961. At launch *Columbia* weighed slightly under 4.5 million pounds (2 million kilograms).

This was the first orbital test flight of the space shuttle. It was also the first time a U.S. manned spacecraft flew an orbital mission without prior tests. The *Columbia* carried no payload in its cargo bay on this test mission, but it was still the heaviest spacecraft ever flown at 214,000 pounds (97,070 kilograms). *Columbia* was also the largest manned spacecraft ever flown, with a length of 122 feet (37 meters). This was about 6 feet (1.8 meters) longer than Skylab. For the first time in a United States spacecraft, the crew quarters contained a normal earthlike atmosphere.

President Reagan and the First Lady meet astronauts Henry Hartsfield and Thomas Mattingly on July 4, 1982, after the fourth and last orbital test flight of the space shuttle Columbia. *Courtesy NASA.*

Young and Crippen put their spaceship through its tests, including the all-important orbital maneuvers. Data results surprised the Houston team because of their accuracy. The huge payload bay doors were opened and closed several times ("Doors all opened hunky dory," Bob Crippen told the ground), all systems were checked, and minor problems such as a few missing heat-shield tiles and a leaking oxygen regulator valve were reported, discussed, and corrected if possible.

When *Columbia* was over Australia during its sixteenth orbit, the crew played a tape of "Waltzing Matilda" for the men on night watch at the Orroral Valley ground station.

After ten long years of hard work, everyone was delighted at how well this maiden voyage was going. "The spaceship is just performing beautifully," John Young said before they began preparations for reentry. "The dream is alive again," he said, summing up the space shuttle's first orbital mission.

Soon *Columbia* entered the fringes of the earth's atmosphere to begin its 4,400-mile (7,080-kilometer) descent path through the atmosphere, during which the temperatures on its heat shield rose to some 2,750 degrees F. (1,510 degrees C.) on the leading edge of the delta wings. During the blazing heat of reentry, there was the expected sixteen-minute radio blackout as the spaceship was surrounded by an envelope of intense heat as it glided under computer control to the California coast. Never had such a long reentry glide, from the east coast of Australia to California, been attempted in a spacecraft.

Radio contact was established again when *Columbia* was at about 180,000 feet (54,864 meters) and descending at more than ten times the speed of sound. All was well, ground control told Young and Crippen: "perfect energy," and "perfect ground track." Two sonic booms were heard as the first space shuttle made its final turn for approach. The crowds on hand saw the shuttle as it appeared over the dry lake bed. Touchdown came minutes later, and *Columbia* rolled some 9,000 feet (2,743 meters). It was 10:20 A.M. Pacific Standard Time on April 14, 1981, when the first space shuttle returned safe-ly to earth after a flight of 36 orbits lasting 54 hours, 21 minutes. *Columbia*, the first space shuttle, had flown for the first time.

"This is the world's greatest flying machine, I'll tell you that," said John Young after landing. This veteran of the Gemini and Apollo spaceflights tells no lies.

Challenger: **April 4, 1983.** When this spaceship made its maiden voyage, the Space Transportation System was fully operational. The first four test flights were completed, and one operational flight of *Columbia* had flown.

This mission had an important job to do. Its cargo bay carried the heaviest payload to date: 46,615 pounds (21,144 kilograms). This included the first tracking and data relay satellite (TDRSS A) and its upper stage booster to take it to geosynchronous orbit. This important satellite was an advanced space-to-ground communications facility, the first of three that would be in place to support Spacelab missions and the tremendous quantity of data that had to be

BELOW: *Judith Resnik on her first and last flight into space before the* Challenger *disaster. This was taken just after a TV broadcast to earth during which she held up the sign HI DAD to greet her father. The mission was the maiden flight for space shuttle* Discovery, *in late August and early September of 1984.* ABOVE, LEFT AND RIGHT: *As a mission specialist, Resnik was responsible for deploying the huge solar array that rose more than 100 feet (30.5 meters) above the cargo bay. This tested techniques that would be used on the future space station.* Courtesy NASA.

transmitted to the ground from all the experiments. While the *Challenger* and its four-man crew had a great mission, the satellite's booster failed during its ascent to higher orbit. A complex set of maneuvers controlled by the ground over a period of some two months saved this important communications satellite.

The mission included the first U.S. space walk in nine years, when crew members practiced repair techniques and use of tools in the *Challenger*'s open cargo bay.

As the second space shuttle of the fleet, *Challenger* had several design changes over its sister ship. It had an extra 4,000 pounds (1,814 kilograms) of thrust. More than 20,000 pounds (9,072 kilograms) were designed out of the huge external fuel tank and the solid rocket boosters, and this weight reduction gave extra payload capacity.

Challenger completed its five-day maiden voyage on April 9, 1983, with a perfect landing at Edwards Air Force Base. The mission had only twenty-two anomalies in the flight plan, compared with the eighty-two on *Columbia*'s first flight. There were now two spaceships in the shuttle fleet.

Discovery: **August 30, 1984.** The inaugural flight of this third space shuttle orbiter lasted six days, and ninety-seven full orbits around the earth were flown. There were six crew members, including the second U.S. female astronaut, Judith Resnik, who deployed and tested a huge solar array panel. This same solar energy technology will be used in the future on large space structures, including the U.S. space station. Another crew member, Charles Walker, employed by McDonnell Douglas, became the first commercial payload specialist to fly on the space shuttle. His job was to operate electrophoresis equipment that was processing experimental drugs in the microgravity environment.

Three heavy communications satellites were launched from the payload bay: the AT&T *Telstar 3C*, the Satellite Business Systems *SBS-4*, and the Navy/Hughes Leasat communications satellite (also known as *Syncom IV*). A new "Frisbee" deployment technique was used for the first time to launch the 7.5-ton Leasat from *Discovery*'s cargo bay.

"We're three for three," mission specialist Dick Mullane told ground control after the third satellite was successfully deployed. Their assist rockets took them to geosynchronous orbit.

This twelfth space shuttle flight ended with a landing on September 5, 1984. Although it was the maiden voyage for *Discovery*, the mission goals were complex and varied, and their success showed that shuttle spaceflights had reached a new maturity. The shuttle fleet now had three flying spaceships.

Atlantis: **October 3, 1985.** This fourth space shuttle orbiter, and the last of the original fleet to be built, flew a classified military mission for the United States Air Force on its maiden flight. The four-day dedicated military flight, with a five-man crew aboard, was the twenty-first time a space shuttle went into orbit. In carrying out its secret work, reported by many sources to be the deployment of two defense communications satellites that were hardened against damage from nuclear radiation, *Atlantis* set a new altitude record for the space shuttle program of 320 miles (515 kilometers). This was about half the stated maximum altitude limit for the current space shuttle, which is set by capabilities of the orbital maneuvering system and other system factors.

Because of the flight's security classification, there were no press interviews set up by NASA,

The first operational flight of space shuttle Columbia *successfully deployed commercial satellites in late 1982. Here is the proud and happy crew: clockwise from the top left, Bill Lenoir, Bob Overmyer, Joe Allen, and the commander, Vance Brand, holding the sign.* Courtesy NASA.

and the media were not able to monitor space-to-ground communications as they can with nonmilitary flights. As with all secret missions, encryption devices which scrambled all flight communications were used extensively.

Air Force colonel Karol Bobko was the commander of *Atlantis* on its first flight. After landing on October 7, 1985, at Edwards Air Force Base, Bobko did not shed much light on the mission, except to say that the $1.2 billion spaceship "performed superbly." We can only take his word for it.

Space Shuttle Firsts

Every space shuttle mission until the last flight of *Challenger* set new records and accomplished significant firsts in manned space flight. If all of them were listed, including the hundreds of unique experiments that have been flown since the space shuttle became operational, an entire book would be needed to present them. What follows is a selected list.

The Shuttle Astronauts

As the spaceships change, so do the men and women who fly them. The space shuttle's versatility is reflected in its crews, and the increased diversity of the astronaut's backgrounds is a direct result of the dramatic increase in the kinds of space activities going on above earth today. Women have flown in space for the first time in the U.S. space program, and their accomplishments are significant; Sally Ride was first, in June 1983, and Kathryn Sullivan was the first American female space walker in October 1984. The first four women astronauts were selected for the program in 1978, but their ranks now number more than twelve.

Doctors, U.S. senators and congressmen, and dozens of scientists from disciplines ranging from astronomy to plasma physics to zoology also flew before flights were halted. Foreign astronauts who have flown on the shuttle include payload specialists from Canada, France, Ger-

many, Indonesia, Italy, Japan, Mexico, and Saudi Arabia for starters, and even more countries will be represented in the years to come, foreshadowing the international character of the space station in the 1990s.

What kind of people are shuttle astronauts? They are all people who want to get into space and are willing to work hard and make sacrifices along the way to reach their above-earth goal. They are intelligent and well educated, the majority with advanced degrees in the sciences. The commanders and pilots, of course, generally have aviation and military backgrounds. One of the minimum qualifications to apply for the pilot astronaut program (as opposed to the mission specialist astronaut program) is one thousand hours as the pilot in command flying jet aircraft.

Astronauts who prove themselves as pilots on shuttle missions usually advance to mission commanders. What follows is some background information on a few selected astronauts who are commanders, pilots, mission specialists, payload specialists, and spaceflight participants. Because there are close to one hundred astronauts in the space program at any given time, this is only a somewhat arbitrary sampling, but other astronauts are mentioned throughout this chapter in the context of mission activities.

Commanders. As of 1986, all the commanders save one come to NASA directly from military service. The exception is Vance D. Brand, who joined NASA as a civilian but did serve as an aviator with the U.S. Marine Corps. Almost two-thirds of all astronauts on active flight status hail from the military services, demonstrating that flight experience and discipline are still important prerequisites for the majority of people who enter the space program.

The oldest space shuttle commander is John W. Young, a twenty-year veteran of the space program. Born in 1930, Young holds the U.S. record for the most space flights, with seven to his credit. He became an astronaut in September 1962 and was among the second group chosen after the famous original seven Mercury astronauts were selected in April 1959. His first flight

SELECTED SPACE SHUTTLE FIRSTS

Launch Date	Space Shuttle and Mission	Mission Firsts
8/12/77	*Enterprise* approach and landing test	Heaviest glider ever flown (150,00 pounds, 68,040 kilograms).
4/12/81	*Columbia* (STS-1)	First orbital flight test of space shuttle orbiter; first U.S. manned spacecraft to fly without prior flight tests; heaviest spacecraft.
11/12/81	*Columbia* (STS-2)	First operation of remote manipulator arm built in Canada; first shuttle science payload.
3/22/82	*Columbia* (STS-3)	First maneuvering of a payload by robotic arm; electrophoresis test.
6/27/82	*Columbia* (STS-4)	Final of four orbital test flights; first Department of Defense cargo; first commercial electrophoresis test for McDonnell Douglas; first Getaway Special flown.
11/11/82	*Columbia* (STS-5)	First operational flight; first deployment of commercial satellites.
4/4/83	*Challenger* (STS-6)	Maiden flight of *Challenger*; first shuttle space walk (EVA) by Peterson and Musgrave.
6/18/83	*Challenger* (STS-7)	First American woman in space, Sally K. Ride; first flight of five-person crew; first use of robotic arm to deploy and retrieve satellite, *Shuttle Pallet Satellite* (*SPAS*); first M.D. in space, Norman E. Thagard.
8/30/83	*Challenger* (STS-8)	First shuttle night launch and landing; first black American in space, Guion S. Bluford.
11/28/83	*Columbia* (STS-9)	First flight of Spacelab (seventy-one scientific investigations); first flight of six-person crew; first flight of non-American on shuttle, Ulf Merbold from Germany; first in-flight computer failures.
2/3/84	*Challenger* (41-B)	First untethered space walk and first use for free flight of Manned Maneuvering Unit backpack by Bruce McCandless; first landing at Kennedy Space Center.
4/6/84	*Challenger* (41-C)	First repair of satellite in orbit, *Solar Maximum* satellite.
8/30/84	*Discovery* (41-D)	Maiden flight of *Discovery* orbiter; first commercial payload specialist, Charles D. Walker; deployment and testing of huge solar array panel by Judith Resnik; heaviest payload to date, 47,516 pounds (21,553 kilograms).

was on *Gemini 3* in March 1965, with the late Vigil I. Grissom. Young will serve as commander of the *Atlantis* flight that will deliver the Hubble Space Telescope into orbit in the late 1980s, and he also commanded the *Columbia* Spacelab 1 mission in November 1983. He went to the moon twice, once in orbit on *Apollo 10* and once on the lunar surface as commander of *Apollo 16*. If Chuck Yeager is the legendary test pilot, then John Young deserves equal legendary billing as the most experienced veteran astronaut. Young is currently chief of the astronaut office at the Johnson Space Center, and he was the spokesman for all astronauts in the aftermath of the *Challenger* disaster.

Shuttle missions during the first few years of the program were commanded by other veterans, most of them coming from selection groups 5 and 7, which go back to 1966 and 1969. The youngest shuttle commanders come from selection group 8, which was chosen in August 1979. As of 1986, the youngest shuttle commander is Robert L. Gibson, who was born October 30, 1946.

Robert L. Crippen, who accompanied John Young on *Columbia*'s maiden voyage, holds the record for flying the most shuttle flights, a total of four. After his pilot's mission with John Young on the very first shuttle mission, he commanded three others: STS-7 (June 1983); 41-C

	SELECTED SPACE SHUTTLE FIRSTS (cont'd.)	
Launch Date	Space Shuttle and Mission	Mission Firsts
10/5/84	Challenger (41-G)	First seven-person crew; first American woman to walk in space, Kathryn D. Sullivan; first Canadian astronaut, Marc Garneau.
11/8/84	Discovery (51-A)	First retrieval of satellites in orbit, Palapa B-2 and Westar VI, and their return to earth; Joseph P. Allen held the Westar satellite up over his head for one trip around earth.
1/24/85	Discovery (51-C)	First dedicated Department of Defense mission.
4/12/85	Discovery (51-D)	First U.S. senator, E. J. "Jake" Garn, to fly on shuttle.
4/29/85	Challenger (51-B)	Spacelab 3 provided 250 billion bits of data; first "flying zoo" mission, with two monkeys and twenty-four rats aboard.
6/17/85	Discovery (51-G)	First time four satellites were launched; first laser test; first French astronaut, Patrick Baudry; first Saudi Arabian astronaut, Sultan Salman Al-Saud; one hundredth American in space, Steven R. Nagel.
7/29/85	Challenger (51-F)	Spacelab 2 mission; first flight for professional astronomer, Karl Henize; first early engine cutoff during flight.
8/27/85	Discovery (51-I)	Record space walk time of over seven hours to repair satellite; first human launch of satellite by James "Ox" van Hoften.
10/3/85	Altantis (51-J)	Maiden flight of Atlantis, the last orbiter in shuttle fleet; secret Department of Defense mission to deploy defense satellites.
10/30/85	Challenger (61-A)	First "rented" mission to foreign country, West Germany; Spacelab D-1 flown and main payload control from Germany; three foreign astronauts aboard.
11/27/85	Atlantis (61-B)	First EVA building of space structures to test techniques for building the space station; beam tower was 45 feet (14 meters) high.
1/12/86	Columbia (61-C)	Record number (seven) of on-pad scrubs for space shuttle launch before it finally flew almost a month late.
1/28/86	Challenger (51-L)	The world's worst manned space disaster, killing all seven crew members; the U.S. manned space program is grounded.

(April 1984); and 41-G (October 1984). Crippen was born on September 11, 1937, in Beaumont, Texas. His degree is in aerospace engineering. He came to the space program as a captain in the U.S. Navy. In 1985 Crippen became deputy director of flight operations at the Johnson Space Center.

Pilots. Some of the pilot astronaut candidates who have recently entered the program or who will do so in the near future will be the commanders on the shuttle missions that build the space station in the 1990s.

The majority of astronaut pilots have commanded or will command their own missions in the future. Gordon Fullerton, for example, who was the pilot on the third space shuttle mission in March 1982, had to wait until July 1985, sixteen flights later, before he commanded his own mission—the Spacelab 2 mission, 51-F. As a colonel in the U.S. Air Force, Fullerton was transferred from the canceled manned orbital laboratory program to the astronaut group in 1969. He flew in the shuttle approach and landing tests in 1977. Fullerton was born in Rochester, New York, in 1936 and earned a master's in electrical engineering from the California Institute of Technology in 1958. He proves that patience is definitely a virtue for shuttle pilots.

Charles F. Bolden had the good fortune to be

assigned pilot next to commander John Young when space shuttle *Atlantis* delivers the Hubble Space Telescope to orbit in the late 1980s. Bolden was in the ninth group of astronauts, who were selected in May 1980. The space telescope mission will be his second shuttle space flight, more than twenty years after Young took his first space flight in the *Molly Brown* Gemini spacecraft back in 1965. A graduate of the U.S. Naval Academy and the University of Southern California, Bolden came to the astronaut program as a major in the the U.S. Marine Corps, having flown combat missions in Vietnam. One of the younger pilots, he was born in 1946 in Columbia, South Carolina.

Mission Specialists. Because there are so many types of missions but only one type of spaceship to fly them, there is more variety in the educational and professional backgrounds of the astronauts trained as mission specialists than of those trained as pilots. Using as an example the future *Atlantis* mission 61-J that will take the Hubble Space Telescope into orbit, there will be three highly trained mission specialists to deploy the world's most sophisticated telescope and make certain it will operate as planned to begin unlocking the secrets of the universe.

Who are these astronauts who will tend the launching of the most important astronomical instrument since Galileo used his small telescope in 1610 to observe the moon and the moons of Jupiter? They are Steven Hawley, Bruce McCandless, and Kathryn Sullivan. Here are brief biographies of these two men and one woman who will blast into space with the Hubble Space Telescope.

Steven Hawley is the astronomer-in-residence for the Hubble Space Telescope mission, and he will be the youngest astronaut on the flight, born on December 12, 1951, in Ottawa, Kansas. He flew on the maiden voyage of space shuttle *Discovery* (flight 41-D) in August–September 1984. The husband of Sally Ride, Hawley earned his doctorate in astronomy from the University of California at Santa Cruz in 1977 and was in the eighth group of astronauts selected in 1978 (so was his wife).

Kathryn Sullivan (on the right, over the robot arm) became the first American woman to walk in space during the Challenger *flight in October 1984. Her astronaut partner in the cargo bay is David Leestma.* Courtesy NASA.

Bruce McCandless became the world's first untethered spacewalker during the flight of *Challenger* in February 1984. His EVA skills, including repair in space, are important for the Hubble Space Telescope mission. Before the flight, he and his astronaut colleague, Kathy Sullivan, will have trained for in-space repair for hundreds of hours in a special NASA swimming pool that imitates weightlessness. The pool contains a full-scale model of the Hubble Space Telescope. Said McCandless, "It's like trying to repair an alarm clock while wearing welder's gloves."

A captain in the U.S. Navy, McCandless graduated from the U.S. Naval Academy and then went on to get his master's in electrical engineering. He will be the oldest of the three mission specialists on the Space Telescope flight, having been born in Boston, Massachusetts, on June 8, 1937.

Kathryn Sullivan, like Hawley, is another of the young astronauts who came to the space program in the late 1970s. She was born October 3, 1951, in Paterson, New Jersey. In 1978, just before entering the astronaut program, she earned her doctorate in geology from Dalhousie University, Halifax, Nova Scotia.

Sullivan became the first American woman to

walk in space during the *Challenger* shuttle flight 41-G in October 1984. She too was chosen for the EVA skills that were thought necessary for the Hubble Space Telescope mission.

So this important mission has as its mission specialists an astronomer, a Navy man trained as an electrical engineer, and a geologist. Two of them know how to take weightless strolls in space and make repairs on the space telescope, and one of them knows as much or more about the Hubble Space Telescope than anyone else on planet Earth.

Payload Specialists. This crew assignment category includes all foreign astronauts, such as Ulf Merbold of Germany, the first foreign payload specialist to fly the shuttle, on the ninth flight at the end of 1983, and the Canadian, Dutch, French, Indian, Indonesian, Japanese, Mexican, and Saudi Arabian astronauts who flew on subsequent flights.

All the astronaut specialists from industry are also assigned to flights as payloads specialists. Charles D. Walker, chief test engineer at Mc-Donnell Douglas, became the first commercial astronaut on *Columbia*'s flight 41-D in August–September 1984 and was responsible for the electrophoresis equipment and experiments that hold so much promise for future medical treatment. As a young man Walker was influenced by one of the original seven astronauts, Virgil "Gus" Grissom, who grew up just a few miles away from Walker's home in Indiana. He received a science degree in aeronautical engineering from Purdue University and had other industrial experience before joining McDonnell Douglas in 1977. Walker was followed by commercial specialists from Lockheed, RCA, and other companies. One, Gregory Jarvis from Hughes, died aboard *Challenger*.

Just about anyone can be assigned a payload specialist slot on a shuttle flight, including senators and congressmen, although the rules may change. Senator E. J. "Jake" Garn of Utah became the first politician to orbit the earth in the space shuttle. His main contribution to the flight may have been the medical data he provided, data on how his blood circulation changed dur-ing weightlessness and how his digestion processes changed. He was the world's first senatorial guinea pig for space medicine, and he did not avoid getting space sick during the first few days. The senator did have some memorable reflections on his return. "As I looked at our earth in the black velvet of space . . . and saw the vastness of the universe extending beyond imagination. . . . I did not question that there are other worlds such as ours, where other children of God are living and working to fulfill the measure of their creation."

Senator Garn was followed by U.S. congressman Bill Nelson of Florida, in January 1986, on a flight of *Columbia*. These two members of the U.S. Congress serve on committees that have much to say about the U.S. space program.

Space Flight Participant. Christa McAuliffe became the first astronaut in this assigned category when she won the national competition for the teacher-in-space program. Because of the media's interest in her, she was the best known of the crew members when shuttle *Challenger* pointed to an orbit it would never reach in January 1986. A social studies teacher at Concord High School, Concord, New Hampshire, McAuliffe took a temporary leave of absence from her teaching duties so that she could fly in space and tell the world about it in spoken and written words. McAuliffe was scheduled to teach two lessons that were to be broadcast on live TV to schools across the United States. She planned to wear a T-shirt in space with the words, "I touch the future, I teach."

The next profession that was scheduled to be represented in the space flight participant category was journalism, but this program went on hold when the shuttle fleet was grounded. After that? Anybody's guess.

The Free-Flying Astronauts and Their MMUs

Some of the most breathtaking photographs so far produced on space shuttle missions are those taken of and by the free-flying astronauts who

Bruce McCandless takes the world's first untethered space walk on February 7, 1984, using the manned maneuvering unit for the first time. During this historic EVA, he also photographed the Challenger from above. Courtesy NASA.

are strapped into their space backpacks: Bruce McCandless floating against the blackness of space, with one of his booted feet touching the curvature of the earth; Dale Gardner at the bottom of a large communications satellite, with a "stinger" tool attached, retrieving the satellite so that it can be returned to earth; George Nelson working between the solar wings of the Solar Maximum Mission satellite that was retrieved, repaired in the shuttle cargo bay, and then released again into orbit, this time fully operational.

Just a flick of the wrist or finger on their arm-rest hand controls, and they can move freely to and from or above and below their spacecraft with a degree of freedom and independent mobility never before possible in manned space-flight. They become human satellites, separately orbiting the planet Earth at more than 17,000 miles (27,353 kilometers) an hour. But they do not have as much time as they would like for above-earth sightseeing and the spectacular views of the earth below them; they are out there to deploy, retrieve, or repair satellites, or to accomplish other tasks that are out of reach or too complex for the shuttle's robotic arm.

Appearing like legless space chairs, the manned maneuvering units (MMUs for short) have so far performed flawlessly in their varied tasks, and they receive raves from the astronauts who fly them. Of course, who would expect less from a piece of high-tech equipment that costs about $10 million to build, even though much of the hardware was already available from other space projects? Some left-over Viking Mars lander hardware, for example, was utilized in building the manned maneuvering units. The nitrogen fuel tanks, which hold the propulsion fuel, were off-the-shelf tanks used in helicopters. Even the electronic control assembly, although designed specifically for the MMUs, was made from available, proven components. And the entire unit had to be kept within dimensions that would allow it to fit through the hatch between the cargo bay and the crew quarters if it had to be brought into the pressurized quarters for repair.

Each maneuvering unit weighs 338 pounds (153 kilograms) on earth, and this includes two full tanks of nitrogen propellant. Its dimensions are: height, 4.1 feet (1.25 meters); width, 2.7 feet (0.8 meters); and depth (with the hand controller arms fully extended), just under 4 feet (1.2 meters). The control arms are slanted downward at a 30-degree angle to allow the astronauts to get closer to whatever object their particular task requires. There are twenty-four fixed nitrogen jet thrusters, each of which produces 1.7 pounds (.77 kilograms) of thrust by releasing cold nitrogen gas. Heaters protect the propulsion and electronic components from the -150 degrees F. (-101 degrees C.) of space, and two silver-zinc batteries provide the power.

The right- and left-hand controls on the unit's arms have separate functions. The right-hand controller commands attitude changes in pitch, roll, and yaw, while the left-hand controller commands the thrusters to fire for point to point motions such as forward, backward, left and right, up and down. There is no instrumentation to tell an astronaut of his or her velocity or attitude; the pilot's own visual cues are all he has to determine these factors. Once you are an astronaut (and that is the hard part), only eighteen hours of training will put you in the MMU driver's seat.

These space backpacks and their future modified versions will be used increasingly in future projects. In the 1990s, space-working astronauts will be flying among the beam trusswork of the space station, moving the modular structural parts into position and assembling them. As the numbers of space people increase during the last decade of this century, it is highly likely that someone will get into trouble in the hostile environment of space. If an astronaut gets into a life-threatening situation, it will be the MMU that comes to the rescue. Martin Marietta, the aerospace firm that manufactured the space backpack, has done intensive simulations of such space rescues that could retrieve stranded astronauts, unconscious or conscious, up to a distance of 450 feet (137 meters). Bruce McCandless, who donned his MMU and became the world's first untethered space walker on February 7, 1984 (flight 41-B), jetted out some 300 feet (91.5 meters) from the space shuttle *Challenger*. His life was wholly dependent on the equipment he carried on his back. He was no doubt reassured to know that most of the hardware and electronic components he depended on to survive had proved themselves over the years in other equipment.

The Shuttle's Left Arm

A giant robotic arm, 50 feet (15.2 meters) long and weighing 900 pounds (408 kilograms), is the space shuttle's powerful flexible appendage that can reach out and pluck a satellite from space, put one into orbit, act as a foothold for space-working astronauts, and do countless other chores from its base in the cargo bay.

Built in Canada, the carbon-composite arm of lightweight tubing is formally known as the remote manipulator system (RMS). Using the human arms as the basic design reference, engineers gave the RMS shoulder, elbow, and wrist joints as well as a "hand" (what NASA calls the "end effector") with snare wires that can hold on to a grapple fixture on the satellite or other payload it is carrying to or from the spacecraft's cargo bay. The motions of its joints are driven by DC electric motors, and they are controlled by the mission specialists from a control panel in the aft section of the flight deck. TV cameras mounted on the wrist and elbow provide the operators with visual cues for maneuvering the arm to its target. Direct visual observations through windows looking into the cargo bay are also used by the mission specialists. Several different operating modes are possible; these operating modes range from full manual to computer-controlled operation, and they are accomplished with hand controls and a keyboard at the control station.

When not in use, the shuttle's robot arm is latched into three cradle pedestals along the left side of the cargo bay. If a mission requires another arm, a right robotic arm can be installed on the right side of the cargo bay, but both arms cannot be operated at once. So far, shuttle missions have been able to get along with just the left arm.

RMS was first put through its paces on the second shuttle flight, STS-2, which flew in November 1981. Flexing its muscles and turning its joints for the first time, the RMS was deemed spaceworthy by astronauts Joe Engle and Richard Truly. In the more than thirty flights since, the arm has proved its versatility again and

again. Beyond its basic satellite or other payload deployment and retrieval functions, it has been used as a "cherry picker" to hoist astronauts to nearby satellites for retrieval, as it did for Joe Allen when he and fellow astronaut Dale Gardner brought the ailing *Westar VI* satellite back to the cargo bay; it has been used as a working platform in the cargo bay, for example, when James van Hoften and George Nelson repaired the *Solar Max* satellite; and it has been used as a space tool when an improvised "flyswatter" device was attached to it and a lever was snared on *Leasat 3* in an attempt to reactivate the disabled satellite. It is a first-class robot arm for work in space, and there will be several of them going through their motions when the space station is assembled in orbit in the 1990s.

While the RMS arm has not yet experienced a mechanical or electronic sprain, it could happen. What will be done if the drive mechanisms jam and the arm cannot be moved to its stowed, deactivated position in the cargo bay? Something would have to be done because the cargo bay doors must be closed before the shuttle descends and makes its fiery passage through the atmosphere. NASA does have a plan. If the arm was locked frozen and space walks and troubleshooting did not work, the arm would be amputated with an explosive charge!

In-Orbit Repair and Salvage

Several hundred million dollars' worth of communication and scientific satellites have been repaired or salvaged in orbit by space shuttle crews. With their maneuvering jetpacks on their backs and the agile robotic arm reaching and grasping, space shuttle astronauts have proved that satellite repair and salvage will be an increasingly important activity above earth in the decades to come. The space station will have its own repair facility, and remote-control repair vehicles, equipped with their own repair kits, are also planned for the first decade of space station operations. What follows is a summary of some of the world's first satellite repair and salvage operations in space.

The world's first satellite repair mission took place during a Challenger *flight in April 1984. After capturing the satellite in the cargo bay, astronauts George Nelson and James van Hoften repaired the $235 million satellite and released it back into orbit. Courtesy NASA.*

The *Solar Max* Mission. The Solar Maximum Mission (*Solar Max*) satellite, a sophisticated scientific satellite whose function was to learn more about the energy output of our sun, had been drifting uselessly in space for three years because of three blown fuses in its attitude control box. Then, in April 1984, space shuttle *Challenger* came to the rescue. The mission plan was to retrieve the satellite, anchor it to a special support in the cargo bay, replace its defective modular components with new ones, and send it back to orbit, repaired and functioning.

The plan did not unfold as written. Astronaut George Nelson jetted over to the satellite and attempted to latch onto the satellite with a special backpack device. By firing his thrusters in a proper sequence, he was to stop the satellite from spinning and bring it closer to *Challenger* in order for the robot arm to grab it and bring it into the cargo bay for repair. There was only one problem. After three attempts, Nelson had failed to latch on, and his attempts had set the satellite spinning even more.

After the mission, NASA and Nelson knew why all the careful planning and training had failed. In an interview, the astronaut said, "The satellite was built a little bit after the drawings were made, and so the device we had built to attach to the satellite was built to a different set of drawings. It turned out that the backpack device didn't fit on to the satellite. There was no way it would grab hold of the side of the satellite." This oversight on drawing specifications almost cost the success of the entire mission to salvage the $235 million satellite.

Nelson had to retreat to the *Challenger*. Commander Robert Crippen made several attempts to snag the satellite with the manipulator arm, but failed. The shuttle backed off station to plan new attempts for the next day.

The entire mission plan was rewritten, and with some help from the ground, the spinning satellite was eventually stabilized so that the space shuttle could try again. Finally, early on the fifth day of the mission, the shuttle *Challenger* successfully grabbed on to the *Solar Max* satellite and hauled it into the cargo bay, and preparations were made for the repair.

The next day astronauts Nelson and Van Hoften worked on the satellite in the cargo bay, replacing the attitude control box and making other repairs, which took about two hours. Tests were run on the repaired satellite's performance, and the robot arm swung it back into orbit the following day.

The "Ace Satellite Repair Company," as the crew good-humoredly referred to themselves, had pulled off the complicated mission that at first looked like it would fail. It was an important success for science. The world's first in-orbit satellite repair and servicing call had finally succeeded.

A New Lease for *Leasat*. The second attempt to repair a satellite in orbit was not part of the mission plan, as the *Solar Max* repair had been.

Space shuttle *Discovery* was launched on April 12, 1985, and one of its main mission objectives was to place two communications satellites into orbit. The deployment of both satellites went well, but one of them, the Hughes-built *Leasat 3*, also known as *Syncom IV-3*, was dead in space hours after deployment because of unknown design or engineering problems that occurred on the ground. Its built-in rocket motor, which was to fire up and take the satellite into its high geosynchronous orbit, failed. The satellite was all but worthless in a lower orbit.

Hughes engineers thought they knew what the problem was—a lever on the side of the satellite that was to activate a timer device during deployment had not been moved far enough. The *Discovery* crew got creative and improvised a flyswatter-type device that would attach to the

FACING PAGE AND BELOW: *Some spectacular photographs taken during the record EVA of more than seven hours to repair the* Leasat *communications satellite during the* Discovery *mission in September 1985. Astronaut James van Hoften accomplishes the first manual launching of a satellite from the cargo bay after* Leasat *was repaired. Then the two astronauts took photos of one another. Astronaut William Fisher is standing on the foot restraint on the edge of the cargo bay, and van Hoften is riding the robot arm after his satellite launch by hand.* Courtesy NASA.

end of the robot arm. With it they could move the lever and activate the timer, which in turn would fire the satellite's engine. None of this was in the flight plan.

The device worked—at least it moved the lever several times as it was intended to do. But the advice from the ground was obviously wrong. *Leasat 3* was left in its useless orbit to await another service call by the same space shuttle *Discovery* in August 1985.

Discovery once again rose amid great clouds of steam with crackling thunder from Cape Kennedy on August 27, 1985. After successfully launching three satellites, the crew set out to save the ailing $85 million *Leasat* communications satellite, which the Navy leases from Hughes. During the intervening months, Hughes engineers had determined that the problem was an inoperative satellite sequencer. A parts and repair kit for the *Leasat* flew to orbit with the crew.

On August 31, the fifth day of the mission, *Discovery* rendezvoused with the huge 7.5 ton, 14 by 14 foot (4.3 by 4.3 meter) *Leasat* satellite. Soon thereafter astronauts William F. Fisher and James D. van Hoften donned their space suits and MMU jetpacks and prepared to retrieve the unwieldy satellite and attach it to the robot arm so that it could be repaired. Securing the satellite to the arm proved to be difficult because of the complex relative motions of the space shuttle and the satellite. "We are having trouble matching the rates," astronaut John Lounge told van Hoften. Maneuvers included flying the shuttle around the satellite while astronaut van Hoften stood atop the robot arm on a special work platform.

Finally van Hoften and Fisher secured the giant satellite to the arm and brought it into the cargo bay for repair. Nearly three hours into the first of two EVAs, a space walk that would set an American duration record, the two astronauts began the repair procedures. Two hours later, when the shuttle was passing over the South Pacific, astronaut Fisher had managed to bypass the defective sequencer so that the *Leasat* satellite could process commands directly from the ground. After an omni antenna was deployed,

the first EVA ended and set a record of 7 hours, 8 minutes.

On their second EVA the next day, Fisher and van Hoften manually struggled again to get the unwieldy satellite into an appropriate position relative to the spacecraft for a cast off back into orbit. The satellite kept moving out of the desired alignment.

"Some other force is acting on it, something is much different now," Fisher said. "I'm trying to keep it from hitting the spaceship." It was soon discovered by ground control that the shuttle's small thrusters were automatically firing to hold altitude and this was causing the problem. Commander Engle then put the spaceship into free drift mode, and the problem was corrected. The arm rose from the cargo bay with van Hoften standing on its work platform holding the bulky *Leasat*. With several pushes of a bar, van Hoften gave the satellite its needed spin of three revolutions per minute. The repaired *Leasat* spun away from the *Discovery* and ground control took over to prepare to fire it into high orbit. This was the world's first hand-launched satellite and the first commercial satellite to be repaired in space.

Satellite Salvage and Return to Earth. Two other communications satellites, Western Union's *Westar VI* and Indonesia's *Palapa B-2*, were salvaged in space by the shuttle *Discovery* mission 51-A in November 1984. It was the first time that satellites were returned to earth from orbit, once again demonstrating the versatility of the space shuttle and its crew, once again proving that the minds and arms and hands of astronauts often succeeded when hardware failed. The satellites, launched from the shuttle *Challenger* mission in February 1984, were in uselessly low orbits when their built-in rocket motors malfunctioned shortly after deployment. Had it not been for this world's first space salvage operation, two sophisticated satellites, with a combined worth of about $70 million dollars, would have had to be written off, and commercial space operations would have taken a serious step backward. Because the insurance underwriters had to pay out $180 million to the

The world's first space salvage operation occurred during a Challenger *mission in November 1984, when two satellites were retrieved, secured in the cargo bay, and brought back to earth. Here the* Westar VI *is retrieved by astronaut Dale A. Gardner with the special "stinger" tool.* Courtesy NASA.

satellite owners (this cost included loss of anticipated revenue), they decided that the salvage operation was worth another $10.5 million payout gamble; if the space salvage operation was a success, the satellites could be refurbished, resold, and relaunched after their return to earth, thus substantially cutting the large insurance payout costs. If the gamble was not taken, satellite insurance was close to becoming unobtainable or unaffordable for commercial satellite companies.

Astronauts Joseph P. Allen and Dale A. Gardner did the EVA retrieval work, first securing the *Palapa B-2* and later in the mission the *Westar VI.* For the retrieval, a specially designed tool, nicknamed the "stinger," was used by the astronauts to grasp the bulky satellites and maneuver them in closer to the spaceship with their backpack thrusters so that the robot arm could hook onto them and pull them into the cargo bay, where they would be secured for reentry. But on the *Palapa* satellite, an unanticipated protrusion made the use of a securing bracket impossible,

so the astronauts improvised. Joe Allen stood on the end of the remote arm, his feet in foot restraints, and held the satellite with his arms while Gardner clamped it down in the shuttle cargo bay. This procedure took an entire orbit, as astronaut Allen held the satellite above his head for an entire trip around the world.

Two days later Allen and Gardner retrieved the *Westar* and secured it in the shuttle's hold, and it was two for two for the world's first space salvage operation. "We have two satellites latched in the cargo bay," mission commander Fred Hauck announced to the world a short time later. There were a lot of smiling insurance men on the ground. NASA's "can do" management philosophy and dedicated and superbly trained astronauts had done it again. This world's first space salvage had important implications for future space activities. Space workers who, along with robots, would build the space station would have to perform many of the same EVA activities when future space shuttles served as construction bases during the early construction phases.

Those Democratic Getaways

In addition to the dozens of large communications and scientific satellites that have been delivered to orbit by space shuttle missions, besides the Hubble Space Telescope and the interplanetary spacecraft such as *Galileo* and *Ulysses* that will be launched into the distant solar system, about fifty small, self-contained payloads, NASA's so-called Getaway Specials, have flown in the cargo bays of the space shuttles on a space-available basis through 1985.

These are truly democratic payloads because there are no restrictions as to who can fly them. If the payload and its scientific goals fall within NASA guidelines, it will eventually fly. Students, scientists, engineers, and private industry or government institutions can fly their Getaway Specials. Three payload categories cue up on a rotational basis for available space in the shuttle cargo bay: educational, U.S. government, and everything else, which includes private industry and foreign governments.

Even after the Challenger *disaster, the space shuttle remains the mainstay of U.S. space activities for two more decades. Without the shuttle fleet, no space station can be built and serviced. In November 1985, an Atlantis mission proved that astronauts could build large portions of the space station by hand. Here astronauts Sherwood Spring and Jerry Ross prove that structural parts and building techniques designed on earth work well in space and can be used to build the space station.* Courtesy NASA.

The first Getaway Special was flown on the fourth space shuttle mission in 1982. Utah State University students flew several experiments that studied the effects of weightlessness on the growth of brine shrimp, algae, duckweed, and fruit flies.

Since that first Getaway Special payload, dozens more have flown. A few examples: The Japanese have experimented with forming artificial snowflakes in space using water from Mount Fuji; high school students from Camden, New Jersey, flew an ant colony to see how ants would adapt to, and behave in, space; the George W. Park Seed Company sent a variety of vegetable seeds into orbit to see how they would germinate and grow; movie producer Steven Spielberg sent a plant experiment above earth to see

how the plants would respond to the lack of gravity and how they would absorb liquids; the U.S. Air Force Academy sent up an experiment to measure the dynamics of a vibrating beam; the U.S. Department of Agriculture used their canister for a gypsy moth egg experiment; and in 1985 Utah State University had a special canister launch a mini-satellite called *Nusat*, which was used to calibrate FAA radar.

Before the *Challenger* was lost, NASA planned to launch forty to sixty Getaway Special payloads each year. Although that number will now be less, their cumulative benefit to global research and development still will be tremendous. The fact that anyone or any organization with a good payload and a few thousand dollars can put their experiments in space again underscores NASA's policy of an open space program that potentially can benefit all the people on planet Earth. Future Getaway Specials, on the next generation of spaceships, will eventually carry large numbers of people, not just experiments, to vacations and adventures in orbit.

Future Shuttle

The redesigned space shuttle that has resulted from the *Challenger* disaster will be the mainstay of U.S. space activities for two more decades. Babies born in 1985 will leave their teenage years and enter their twenties in 2005, perhaps the last year that the shuttle will fly. The original fleet of four shuttles was expected to fly at least four hundred missions. But now that the *Challenger* is gone, and its replacement is years away, the number of missions that will be flown by the year 2000 is difficult to predict. No doubt there are hundreds of future missions to fly, and the 1990s will see the space shuttle deliver the components of the space station and the astronaut workers and robots who will assemble it. Even in the wake of the *Challenger* loss, it is still possible that the other shuttle spaceships will turn out to be the DC-3s of the Space Age; they may prove to have much longer flight lives than expected. While it is still too early in the shuttle program to tell, the space shuttles may continue to fly even after their follow-on spaceships are flying regular schedules.

Will the space shuttle drastically change over the next two decades? Besides the redesign of the solid rocket booster seals and other modifications resulting from the investigations after the *Challenger* exploded in flight, the spaceships will remain basically the same. Refinements will continue to be made, and some of those planned before the disaster will no doubt be carried through. But caution will rule in the light of investigation findings. Design changes intended to increase payload capabilities will be carefully scrutinized and will be much more difficult to get approved. Additional payload weight will be very important when the space station components are hauled into orbit. A second redesign

of the huge external fuel tank (a lighter external tank had been flying since the maiden voyage of *Challenger* in April 1983) was planned before the disaster. This would have given the space shuttles an extra 4,000 pounds (1,814 kilograms) of thrust. Whether these lighter external tanks ever fly depends on detailed investigation findings, and to what degree unmanned expendable rockets are now utilized.

A redesign of the engine casings, manufacturing them with a filament-wound technique of fiberglass-epoxy, was also planned. This too would result in the ability to get more weight into orbit. If approved, these and other modest design refinements could eventually translate in getting 5,000 to 8,000 additional pounds (2,268 to 3,629 kilograms) into space. This could actually save a complete mission in terms of getting the space station modules and structural members into orbit in the early 1990s.

Some modifications to the upper flight deck and the mid-deck are underway. A so-called heads-up instrument panel will allow the commanders and pilots to see their instruments on the windshield straight ahead. And on the mid-deck, some of the locker space in the aft section will be converted to space for experiments overseen by crew members.

A private U.S. company is also proposing the manufacture of an add-on module to expand the work area for commercial and scientific users. Called the Spacehab, this module would fit in the forward cargo bay right next to the crew quarters and would be connected to the cabin mid-deck by the same tunnel adapter that was developed for Europe's much larger Spacelab scientific module. The proposed Spacehab would add 1,000 cubic feet (28 cubic meters) of working and storage space, and it could carry up to 6,000 pounds (2,722 kilograms) of payload. This could mean more than seven astronauts flying on future missions. If the Spacehab gains final approval by NASA, it could be built by the Seattle, Washington, firm Spacehab, Inc. and could begin flying on shuttle missions by the late 1980s.

In the aftermath of the *Challenger* disaster, NASA canceled the Centaur shuttle upper stage

SPACE SHUTTLE MISSION SUMMARY

Flight	Crew	Duration Days/Hrs./Mins. Hrs. Elapsed Orbits/Miles	Date/Time Launch Landing	Comments
STS-1 *Columbia*	John W. Young, Cdr. Robert L. Crippen, Plt.	02/06:20:53.1 54:20:53.1 36/933,757	04/12/81 06:00 A.M. CST 04/14/81 12:20 P.M. CST Edwards AFB, CA.	
STS-2 *Columbia*	Joe H. Engle, Cdr. Richard H. Truly, Plt.	02/06/:13:12 54:13:12 36/933,757	11/12/81 9:10 A.M. CST 11/14/81 3:23 P.M. CST Edwards AFB, CA	First operation of remote manipulator arm; mission cut to two days due to fuel cell failure.
STS-3 *Columbia*	Jack Lousma, Cdr. C. Gordon Fullerton, Plt.	08/00:04:49 192:04:49 129/3.9 million	03/22/82 10:00 A.M. CST 03/30 82 10:04 A.M. CST Northrup Strip U.S. Army Sands Missile Range, N.M.	Bad weather at Northrup forced one-day landing delay.
STS-4 *Columbia*	Thomas K. Mattingly, Cdr. Henry W. Hartsfield, Plt.	07/01:11:11 169:11:11 112/2.9 million	06/27/82 09:59:59 A.M. CDT 07/04/82 11:11:11 A.M. CDT Edwards AFB, CA.	Landing on Edwards main runway 22. Final orbital test flight
STS-5 *Columbia*	Vance D. Brand Cdr. Robert F. Overmeyer, Plt. Dr. Joe Allen, M.S. Dr. Bill Lenoir, M.S.	05/02:14:27 122:14:27 81/1.85 million	11/11/82 06:19:00 A.M. CST 11/16/82 08:33:27 A.M. CST Edwards AFB, CA.	First operational flight
STS-6 *Challenger*	Paul J. Weitz, Cdr. Karol J. Bobko, Plt. Donald H. Peterson, M.S. F. Story Musgrave, M.S.	05/00:23:42 120:23:42 80/1.82 million	04/04/83 01:30:00 P.M. CST 04/09/83 08:33:27 A.M. CST Edwards AFB, CA.	TDRS failed to reach geosynchronous orbit due to inertial upper stage guidance error; thruster boost plan in work to reach geosynchronous. First flight of *Challenger.*
STS-7 *Challenger*	Robert L. Crippen, Cdr. Frederick H. Hauck, Plt. Sally K. Ride, M.S. John M. Fabian, M.S. Norman E. Thagard, M.S.	06/02:23:59 146:23:59 98/2.22 million	06/18/83 06:33:00 A.M. CDT 06/24/83 08:56:59 A.M. CDT Edwards AFB, CA.	Edwards AFB landing after KSC weather waveoff.
STS-8 *Challenger*	Richard H. Truly, Cdr. Daniel C. Brandenstein, Plt. Dale A. Gardner, M.S. Guion S. Bluford, M.S. William E. Thornton, M.S.	06/01:08:40 145:08:40 98/2.22 million	08/30/83 01:32:00 A.M. CDT 09/05/83 02:40:40 A.M. CDT Edwards AFB, CA.	First shuttle night launch and landing.

Mission / Orbiter	Duration / Orbits / Cost	Dates / Location	Remarks
STS-9 *Columbia*	09/07:47:24 223:47:24 148/3.33 million	11/28/83 10:00:00 A.M. CST 12/08/83 05:47:24 P.M. CST Edwards AFB, CA.	Two shifts, round-the-clock science operations; first non-U.S. crewman	John W. Young, Cdr. Brewster H. Shaw, Jr., Plt. Owen K. Garriott, M.S. Robert A Parker, M.S. Byron Lichtenberg, P.S. Ulf Merbold, P.S. (West Germany)
STS-10				DOD flight canceled.
STS-11 41-B *Challenger*	07/23:15:55 191:15:55 127/2.87 million	02/03/84 07:00:00 A.M. CST 02/11/84 06:15:55 A.M. CST KSC, FL.	First KSC landing; satellites failed to reach geo-synchronous orbit. EVA/MMU.	Vance D. Brand, Cdr. Robert L. Gibson, Plt. Bruce McCandless II, M.S. Robert L. Stewart, M.S. Ronald E. McNair, M.S.
STS-13 41-C *Challenger*	06/23:40:05 167:40:05 107/2.88 million	04/06/84 07:58:00 A.M. CST 04/13/84 07:38:05 A.M. CST Edwards AFB, CA.	Highest STS operating latitude (269 nm) to date.	Robert L. Crippen, Cdr. Francis R. Scobee, Plt. George D. Nelson M.S. James D. van Hoften, M.S. Terry J. Hart, M.S.
41-D *Discovery*	06/00:56:04 144:56:04 97/2.21 million	08/30/84 07:41:53 A.M. CDT 09/05/84 08:37:57 A.M. CDT Edwards AFB, CA.	First "Frisbee" satellite deployment; first commercial payload specialist.	Henry W. Hartsfield, Cdr. Michael L. Coats, Plt. Judith A. Resnik, M.S. Steven A. Hawley M.S. Richard M. Mullane, M.S. Charles D. Walker, P.S (McDonnell Douglas)
41-F				Canceled.
41-G *Challenger*	08/05/:23/37 197:23:37 133/3.4 million	10/05/84 06:03:00 A.M. CDT 10/13/84 11:26:37 A.M. CDT KSC, FL.	First American female EVA; first seven-person crew; first American orbital fuel transfer; first Canadian.	Robert L. Crippen, Cdr. Jon A. McBride, Plt. Kathryn D. Sullivan, M.S. Sally K. Ride, M.S. David D. Leestma, M.S. Marc Garneau P.S. (Canada) Paul D. Scully-Power, P.S.
51-A *Discovery*	07/23:44:56 191:44:56 127/2.87 million	11/08/84 06:15:00 A.M. CST 11/16/84 05:59:56 A.M. CST KSC, FL.	Retrieved *Palapa B-2* and *Westar VI* from STS-11 launch.	Frederick H. Hauck, Cdr. David M. Walker, Plt. Anna L. Fisher, M.S. Dale A. Gardner, M.S. Joseph P. Allen, M.S.

Revised from Johnson Space Center Data. Courtesy NASA.

SPACE SHUTTLE MISSION SUMMARY *(cont'd.)*

Flight	Crew	Duration Days/Hrs./Mins. Hrs. Elapsed Orbits/Miles	Date/Time Launch Landing	Comments
51-C *Discovery*	Thomas K. Mattingly, Cdr. Loren J. Shriver, Plt. Ellison S. Onizuka, M.S. James F. Buchli, M.S. Gary E. Payton, P.S. (DOD)	03/01:33:27 73:33:27 48 orbits	01/24/85 01:50:00 P.M. CST 01/27/85 03:23:27 P.M. CST KSC, Fl.	
51-E *Challenger*	Karol J. Bobko, Cdr. Donald E. Williams, Plt. M. Rhea Seddon, M.S. Jeffrey A. Hoffman, M.S. S. David Griggs, M.S. Patrick Baudry, P.S. (France)	Canceled.		
51-D *Discovery*	Karol J. Bobko, Cdr. Donald E. Williams, Plt. M. Rhea Seddon, M.S. Jeffrey A. Hoffman, M.S. S. David Griggs, M.S. Charles D, Walker, P.S. (McDonnell Douglas) E.J. "Jake" Garn, P.S. (U.S. senator)	06/23:55:19 167:55:19 108/2.5 million	04/12/85 07:59:05 A.M. CST 04/19/85 07:54:24 A.M. CST KSC, Fl.	*Syncom* failed to activate after deployment. "Flyswatters" strapped to RMS used in attempt to trip switch.
51-B SL-3 *Challenger*	Robert F. Overmyer, Cdr. Frederick D. Gregory, Plt. Don L. Lind, M.S. Norman E. Thagard, M.S. William E. Thornton, M.S. Lodewijk van den Berg, P.S. Taylor G. Wang, P.S.	07/00:08:47 168:08:47 110/2.9 million	04/29/85 11:02:18 CDT 05/06/85 11:11:05 CDT Edwards AFB, CA.	Two monkeys, twenty-four rats.
51-G *Discovery*	Daniel C. Brandenstein, Cdr. John. O Creighton, Plt. Shannon W. Lucid, M.S. John M. Fabian, M.S. Steven R. Nagel, M.S. Patrick Baudry, P.S. (France) Sultan Salman Abdelazize Al Saud, P.S. (ARABSAT)	07/01:39:00 169:39:00 112/2.5 million nm	06/17/85 06:33:00 CDT 06/24/85 08:12:00 CDT Edwards AFB, CA.	

Mission	Crew	Duration / Orbits	Dates	Remarks
51-F SL-2 *Challenger*	C. Gordon Fullerton, Cdr. Roy D. Bridges, Jr., Plt. F. Story Musgrave, M.S. Anthony W. England, M.S. Karl G. Henize, M.S. Loren W. Acton, P.S. (Lockheed) John-David Bartoe, P.S. (U.S. Navy civ.)	07/22:45:26 190:45:26 126/2.85 million	07/29/85 04:00:00 P.M. CDT 08/06/85 02:45:26 P.M. CDT Edwards AFB, CA.	
51-I *Discovery*	Joe H. Engle, Cdr. Richard O. Covey, Plt. James D. van Hoften, M.S. John M. Lounge, M.S. William F. Fisher, M.S.	07/02:17:42 170:17:42 111/2.5 million nm	08/27/85 05:58:01 A.M. CDT 09/03/85 08:15:43 A.M. CDT Edwards AFB, CA.	Launch three satellites; capture, salvage, repair *Leasat (Syncom IV-3)*.
51-J DOD *Atlantis*	Karol J. Bobko, Cdr. Ronald J. Grabe, Plt. David C. Hilmers, M.S. Robert C. Stewart, M.S. Maj. William A. Pailes, P.S. (Air Force)	04/01:44:38 97:44:38	10/03/85 10:15:30 A.M. CDT 10/07/85 12:00:08 P.M. CDT Edwards AFB, CA.	
61-A SL D-1 *Challenger*	Henry W. Hartsfield, Cdr. Steven R. Nagel, Plt. James F. Buchli, M.S. Guion S. Bluford, M.S. Bonnie J. Dunbar, M.S. Reinhard Furrer, P.S. (West German Space Agency) Ernst Messerschmid, P.S. (West German Space Agency) Wubbo Ockels, P.S. (European Space Agency)	07/00:44:51 168:44:51 110/2,501,290	10/30/85 11:00:00 A.M. CST 11/06/85 11:44:51 A.M. CST Edwards AFB, CA.	First foreign dedicated Spacelab. First eight person crew.
61-B *Atlantis*	Brewster H. Shaw, Cdr. Bryan D. O'Connor, Plt. Mary L. Cleave, M.S. Sherwood C. Spring, M.S. Jerry L. Ross, M.S. Rodolfo Neri Vela, P.S. Charles D. Walker, P.S. (McDonnell Douglas)	06/21:04:50 165:04:50 108/2,466,956	11/26/85 06:29:00 P.M. CDT 12/3/85 03:33:50 p.m CDT Edwards AFB, CA.	Two EVA's for EASE/ACCESS (Experimental Assembly of Structure in EVA/Assembly Concept for Construction of Erectable Space Structures) Construction.

Revised from Johnson Space Center Data. Courtesy NASA.

SPACE SHUTTLE MISSION SUMMARY *(cont'd.)*

Flight	Crew	Duration *Days/Hrs./Mins.* *Hrs. Elapsed* *Orbits/Miles*	Date/Time *Launch* *Landing*		Comments
61-C *Columbia*	Robert L. Gibson, Cdr. Charles F. Bolden, Plt. Franklin Chang-Diaz, M.S. Steven A. Hawley, M.S. George D. Nelson, M.S. Robert J. Cenker, P.S. (RCA) Bill Nelson, P.S. (U.S. congressman)	06/02:04:09 146:04:09 96/2,197,305	01/12/86 01/18/86	05:55:00 A.M. CST 07:59:09 A.M. CST	Edwards AFB landing after KSC weather wave-off.
51-L *Challenger*	Francis R. Scobee, Cdr. Michael J. Smith, Plt. Judith A. Resnik, M.S. Ellison S. Onizuka, M.S. Ronald E. McNair, M.S. Gregory Jarvis, P.S. Sharon C. McAuliffe, S.F.P	00/00:01:14.534	01/28/86	10:38:00 A.M. CST	World's worst space disaster in history of Space Age. Seven new members killed, five men and two women. First time Americans were killed during flight.

Revised from Johnson Space Center Data. Courtesy NASA.

booster rocket, which was liquid-fueled, for safety reasons. This rocket was designed to be carried to orbit in the shuttle cargo bay, where it would be deployed and then launch interplanetary probes such as *Galileo* to Jupiter, *Ulysses* to a polar orbit around the sun, and *Magellan* to Venus. In mid-1986, however, NASA recommended that it be replaced with the Boeing Intertial Upper Stage (IUS). This upper stage will also expand the space shuttle's capability of getting larger and heavier satellites into geosynchronous orbit. Such increased delivery power will eventually put future communication platforms into high orbit and will cut down on the crowding of these prime space locations. The IUS is not reusable, but reusable orbital transfer vehicles are scheduled for service in 1992 or 1993. A few years later, perhaps by 1995, orbital maneuvering vehicles may be in service to perform remote-control satellite repair, structure assembly, and tug operations. Such new spacecraft, designed to fit in the shuttle cargo bay, will enhance space shuttle operations during the last decade of this century.

Beyond Shuttle: A Twenty-First-Century Spaceship

What will the second-generation space shuttle be like? A joint research program between NASA and the Department of Defense is already actively pursuing the answer to this question, and the United States will probably know by 1990 what will be the basic design and capabilities of its after-shuttle spaceship.

While all options are open during the research phase, recent breakthroughs in what the aerospace industry calls ramjet/scramjet technology (air-breathing engines that ingest air for much of their ascent through the atmosphere) suggest that this technology will be incorporated into such a twenty-first-century spaceplane. It could have happened even sooner if the huge sums needed to replace *Challenger* were put into the next generation of aerospace technology. Such a

spacecraft could take off horizontally from conventional runways and then accelerate directly to orbit (single-stage-to-orbit concept), first using a hydrogen engine system that would consume the air it is flying through for its ascent to the upper reaches of the atmosphere and a top speed of sixteen times the speed of sound (almost 12,200 miles, 19,630 kilometers, per hour), and then getting its final power kick into orbit with rocket thrust. Unlike the space shuttle, which drops like a heavy, powerless glider to earth, it could have complete control over its descent through the atmosphere and could land at almost all conventional airports.

In theory, this concept would eliminate the need to carry large amounts of fuel for the launch phase (as the shuttle does with its giant external tank) as well as the need for large ground-support crews. This could dramatically bring the cost of to-orbit transport down to about 10 percent of the expensive per-pound shuttle payload cost—about $1,500 a pound, depending on how the cost is reckoned. Once payload cost can be reduced to $150 a pound and less, space will become more affordable and more useful to a larger number of commercial and scientific enterprises (read Chapter 9, "Commerce and Science in Orbit").

When will this twenty-first-century spaceship fly? Officials project that a research test vehicle, capable of exploring the hypersonic velocities through the atmosphere of more than eight times the speed of sound, can be flying in the 1990s. This would be followed by a full-scale test-flying of the advanced spaceplane near the turn of the century. These sleek, powerful spaceships will be able to climb out of the earth's steep gravity well all on their own, but they will owe their existence to the workhorses of the Space Age, the space shuttle ships that built the space station to which they will so often fly. What will become of *Columbia*, *Discovery*, and *Atlantis* and *Challenger*'s replacement? They will be explored and walked through by thousands of people each year as they enjoy a well-earned rest and retirement at aviation and space museums around the world.

3
---
*Space
Stations &
Beyond*

The Space Station Waltz

It was Stanley Kubrick's 1968 film, *2001: A Space Odyssey*, that first brought space station images before the general public, a year before the United States landed men on the moon. After the movie's opening evolution-on-earth sequence, we visually follow a twirling bone, thrown into the air by a dominant leader ape, which takes us into the future as Kubrick's film magic creates a time jump.

The sound track becomes a waltz. It is "The Blue Danube," the most popular waltz of the four hundred Viennese waltzes of Johann Strauss, Jr., and the audience is high above earth, watching space machines spin and dance above the blue planet. A quick sunrise is seen from orbit, and then a large, double-wheeled space station comes into view, slowly turning to Strauss's "Blue Danube" waltz. One wheel of the space station, the central hub, and the connecting spokes are fully constructed; the other wheel is still under construction. It is clear that an elegant machine is evolving in orbit above earth, and it is connected to earth by routine, scheduled space flights. The shuttle ramjet *Orion* ascends from earth, approaches the space station, and prepares to land in the central bay in the station's hub. The waltz plays on, and the space station spins on it as it circles the planet.

Less than thirty years after Kubrick and Arthur C. Clarke's film classic, during the last decade of this century and five hundred years after a Genoese navigator named Columbus set sail for the New World, there will be a real space station in orbit above earth giving humankind a permanent presence in space and another stepping stone on our species's evolutionary road.

It will look nothing like the space station wheel so often envisioned by writers, filmmakers, scientists, and engineers of the twentieth century. We know much more now than we did

PRECEDING PAGES: *In the artist's rendering of the space station's modular cluster, an orbital maneuvering vehicle that can repair and refuel satellites is anchored to a robot arm between missions.* Boeing Aerospace Company.

in the early days of the Space Age. The form of the real space station will have none of the elegance of the classic wheel-shaped structure; its form will follow its function. And it will neither spin nor have the artificial gravity that such motion produces because its inhabitants will live and work in it for periods of three months, not years, and the human body can remain healthy in weightlessness for that amount of time. Also, weightlessness—what exacting scientists prefer to call microgravity—is one of the most valuable resources in orbit, and the absence of gravity will be essential for much of the scientific and manufacturing work to be accomplished there.

What will the real space station of the 1990s look like? Rather like an unwieldy structure built by a child whose project was assembled by reaching for the large pieces of a toy construction set. Yes, it will look more like a giant Erector® set or a Lego® construction in orbit. It will bear little resemblance to the designs conveyed in the science fact and fiction literature of the past few decades. Dozens of various-shaped parts, large and small, will be attached to the skeleton of crisscrossed trusswork beams. The logic of the structure is there, but it cannot be seen by the eye alone; the eye needs the mind's help to know the hidden functions of the parts. The space station will have dozens of functions and dozens of structural parts to accomplish them. It will be an open-ended design, capable of evolving over time as new users and functions are defined and new hardware is attached. It will be the largest spacecraft ever put into orbit above earth, a machine of many modular branches, each one able to grow or change direction and perform new tasks. On one such branch, there will be a nest for birds—rockets that will fly to higher orbits, to the moon, or to Mars and the other planets of the solar system. It is only the beginning of the dance; the space station waltz plays on.

Space Stations of the Mind

Before the aerospace companies began their space station studies, when the Space Age was already rushing to the moon, the famous *Col-*

lier's magazine series "Man Will Conquer Space Soon" depicted in 1952 the spinning wheel that was the dominant conceptual design of the twentieth century. Wernher von Braun's contribution to the series, an article titled "Crossing the Last Frontier," described in words and pictures a wheel-shaped space station with a 250-foot (76-meter) diameter that contained three interior decks. In a polar orbit some 1,000 miles (1,609 kilometers) above earth, the wheel rotated three times each minute to produce artificial gravity. Von Braun suggested that, besides serving as a jumping off station to the moon, such a space station would make an ideal observation post—certainly one of the key functions of the real space station that is less than a decade away. The *Collier's* articles were later expanded into the classic book *Across the Space Frontier*, published in 1952 and edited by Cornelius Ryan. Its contributors included Wernher von Braun, Willy Ley, and Fred L. Whipple, and illustrations were painted by the wonderful space artist Chesley Bonestell. That same year, Arthur C. Clarke wrote a novel about space stations, *Islands in the Sky*, in which he described the elegant and popular space wheel.

Where did the space station wheel, more precisely known as the toroid shape, originate? In 1929 a book was published in German by an Austrian Imperial Army captain. The book, *Das Problem der Befahrung des Weltraums* (The Problem of Space Travel), was written by Hermann Noordung, a pen name for Herman Potocnik; it was the first book on space stations. The giant wheel he described, which he called a *Wohnrad* or "living wheel," was 100 feet (30.5 meters) in diameter, had a central hub and docking adapter, and rotated to create artificial gravity. This space station even had a huge parabolic solar collector to focus sunlight and create heat to run a steam generator for power. Wernher von Braun read this book when he was eighteen, and it no doubt influenced his life's dream of space travel to the moon and planets, and a space station in orbit above earth.

Although not in the same detail as Noordung, the two space pioneers Hermann Oberth and Konstantin Tsiolkovsky also wrote in the 1920s about orbiting space stations. In his *The Rocket into Interplanetary Space* (1923) Oberth described an orbiting manned satellite as a "space station," and wrote that it could be used as a navigation station, earth-observation post, rocket refueling station, and weather station, and could serve in other ways that the real space station will. One of its functions would be to spot icebergs and warn ships so that another *Titanic* disaster would not occur. Tsiolkovsky wrote in an article in 1911 about a large peopled space satellite that orbited the earth every two hours at a altitude of almost 2,000 miles (3,218 kilometers).

Before the twentieth century, there were no references to space stations in literature. There is one often-mentioned exception, and it is also unusual in that it was published in an American magazine, the *Atlantic Monthly*. A Boston clergyman by the name of Edward Everett Hale published a three-part serial article "The Brick Moon," in late 1869. The author, best known for his short story *"The Man Without a Country,"* told the story of a spherical space station, hollowed out inside, that was to be launched into space by a giant flywheel. Made entirely of bricks, which were whitewashed to increase visibility, and some 200 feet (61 meters) in diameter, the station was to be launched into a high polar orbit and used as a navigation aid.

The brick moon in Hale's story was accidentally launched early and ended up in orbit with thirty-seven construction workers and some of their families on board. Because they had no way down, plans were made to supply them with the essentials from earth.

Storyteller Hale (who was also the featured orator at Gettysburg the same day President Lincoln made his immortal address) had a novel way for his space station people to communicate with earth. It was not that they used Morse code, but rather how they produced the signals—by jumping up and down on the outer surface of their space station!

Early Space Stations

Almost two years before the United States would fly the *Skylab* missions, the Soviet Union launched their *Salyut 1* space station, on April

IMPORTANT RUSSIAN SPACE ACCOMPLISHMENTS

Date	Spacecraft	Event
10/4/57	Sputnik 1	First artificial satellite launched into earth orbit; beginning of the space age.
11/3/57	Sputnik 2	First living creature, a female dog named Laika, orbited the earth.
9/12/59	Luna 2	First robot probe to impact on the moon.
10/4/59	Luna 3	First photographs of far side of moon radioed back to earth.
4/12/61	Vostok 1	First human to orbit earth (Yuri Gagarin, 1934-1968).
8/6/61	Vostok 2	First full day in earth orbit by Gherman S. Titov (25.3 hours).
6/16/63	Vostok 6	First woman in space, Valentina Tereshkova, for forty-eight orbits.
3/18/65	Voskhod 2	First space walk by Alexei A. Leonov, lasting twenty minutes.
1/14/69	Soyuz 4	First linkup with another spacecraft (Soyuz 5) and transfer of crew.
11/10/70	Luna 17	First robot moon rover, Lunokhod 1; on moon.
4/19/71	Salyut 1	First space station launched into orbit, without crew.
7/15/75	Apollo-Soyuz Test Project	First American-Russian rendezvous and docking between two Soviet and three U.S. astronauts.
2/9/84	Salyut 7 and Soyuz T-10/11	Cosmonauts set human endurance record for spaceflight of 237 days from this launch date.
2/19/86	Mir	New modular space station launched, with six docking ports.
3/13/86	Soyuz T-15	First two cosmonauts dock with Mir space station.

19, 1971. Since that time, Russia has launched six more Salyut stations into earth orbit. As the United States developed and built its reusable space shuttle, the Soviet Union was establishing a permanent manned presence in orbit with the Salyut series 1 through 7, breaking their own space endurance records and upgrading their hardware. It was in Salyut 7 that three Russian cosmonauts set the new space endurance record of 237 days in space—from February 9 to October 2, 1984. The Salyut 7 space station, launched in 1982, was still operative in 1986, even after a complete power failure occurred in 1985 and all its systems, as well as food and water, froze solid. The crippled ghost station was saved by an extensive repair mission in June 1985, carried out by cosmonauts Vladimir Dzhanibekov and Viktor Savinykh. They charged the batteries, restoring electrical power to the station, and repaired the heat regulation system and instrumentation. Within ten days of the cosmonauts' arrival, the aimlessly drifting Salyut 7 space station had been restored to operation, waiting for a resupply spacecraft from earth.

Russia's Salyut space stations have evolved over time, and with the manned Soyuz spaceships, the Progress unmanned transports, and the unmanned Cosmos modules that can double the Salyut's habitable space, they comprise a long-duration space system for all Soviet activities in space, including research in materials processing, astronomy, and military space operations.

While Salyut 1 through Salyut 5 had only one docking port, Salyut 6 and Salyut 7 had two. Beginning with the Salyut 4 station, the solar panel configuration was revised from two panels on either end of the station to three steerable solar panels attached to the mid-section. Salyut 6 was the first of what was considered the "second-generation" Soviet space stations. Besides the two docking ports, fore and aft, it had a new propulsion system that could be refueled in orbit. A water regeneration device also became a standard feature of the life-support system, supplying the cosmonauts with a fresh supply of water.

The basic Salyut laboratory, without the

Soyuz, Progress, or Cosmos spacecraft attached, is 42.6 feet (13 meters) in length and has a maximum diameter of almost 14 feet (4.3 meters). It weighs about 19 tons and contains some 3,500 cubic feet (99 cubic meters) of usable space—enough for a crew of five cosmonauts. It is only about one-fourth the size of *Skylab*, in terms of weight and size. Two of the three sections (the docking-transfer section and the living-working section) are pressurized, while the instrument-propulsion section is not. When a Soyuz spacecraft is docked to one end and a Progress resupply ship is at the other end, the total length of the configuration is 95 feet (29 meters).

It was *Salyut 6* and *Salyut 7*, with their extensive design changes, that proved how productive the Salyut stations could be. The second docking port, used on *Salyut 6* for the first time, permitted thirty-three successful dockings of manned and unmanned vehicles. During its lifetime in orbit, which began with a launch on September 29, 1977, the station was occupied by five long-duration crews and visited by eleven other temporary crews (for a total of thirty cosmonauts), all of which added up to 676 days of manned operation. *Salyut 6* remained in orbit for four years and ten months, until July 1982, when it was replaced by *Salyut 7*.

Additional modifications were made on *Salyut 7*, launched on April 19, 1982. Some of these were:

- One of its docking ports was redesigned to accommodate larger spacecraft.
- The space station has more automatic modes than earlier Salyuts.
- New shielding protection for observation windows was added because previous portals were degraded by propellant firings and micrometeorites.
- The color scheme was changed to improve the interior.
- A refrigerator was installed.

During the first few years in orbit, the *Salyut 7* space station was visited by several Soyuz crews, who stayed on board for both short and long periods. The first crew, from *Soyuz T-5*, An-

atoly Berezovoy and Valentin Lebedev, boarded the station and remained for a record 210 days. They were visited by two other crews during their marathon in orbit: The crew of *Soyuz T-6* included the first French cosmonaut, Jean-Loup Chrétien, and the visiting *Soyuz T-7* crew included the second Russian woman in space, Svetlana Savitskaya, some twenty years after Valentina Tereshkova was the first to rocket skyward, in 1963. Also during 1982, four Progress supply ships docked with *Salyut 7*.

While there were problems with the Salyut space program in 1983, including one on-pad launch abort and the docking abort of *Soyuz T-8* because of a radar malfunction, the Soviet Union was building up its orbital infrastructure with several spacecraft that could dock with their *Salyut 7* space station and create new mission configurations.

The *Soyuz T-10* crew set the world's space endurance record in 1984 after spending 237 days in the *Salyut 7* space station. During this period, in July 1984, the crew of *Soyuz T-12* visited the space station, and one of its crew members, Svetlana Savitskaya, became the first woman ever to walk in space. After almost eight months in orbit the space-endurance winners (Leonid Kizim, Vladimir Solovyov, and Oleg Atkov) returned to earth on October 2, 1984. Soon thereafter, the systems aboard the abandoned *Salyut 7* began to degrade. It was during the winter of 1985 that the power failed and the water system froze up. Many experts believed that the end had come for *Salyut 7*, but the daring rescue and repair mission of cosmonauts Dzhanibekov and Savinykh in their *Soyuz T-13* spacecraft during June 1985, along with some resupply missions, brought the ghost ship back to life for future crews. In late 1985, a new Soviet module was attached to one end of the *Salyut 7* space station and the *Soyuz T-14* docked with the other end, bringing a fresh crew to orbit. The 929-class Cosmos module doubled the space station's size, and the full complex was 115 feet (35 meters) in length and weighed 103,000 pounds (46,720 kilograms). Because there were two military officers on board and because its orbit took it frequently over the United States, it was believed to be a dedicated military mission.

Future Soviet Space Stations

The Salyut space station program, begun in 1971 and evolving ever since, proves that the Soviet Union is seriously nearing its goal of having a permanently manned space station above earth. Even now the Russians have the capability of launching two or more Salyut stations and linking them together in orbit with their 929-class Cosmos modules to form a larger space station. And no one can question their commanding lead in crew hours in space, a lead they will retain for some time, even when the space shuttles fly again and make more frequent, short-duration flights. After the U.S. space shuttle had flown for several years, the Russians still had almost 70,000 hours in space while the U.S. had less than 30,000 hours.

Indeed, the Soviets have already spoken about their future space plans. In a 1984 Moscow ceremony celebrating the fifteenth anniversary of the docking of two Soyuz spacecraft in orbit, the head of Russian Cosmonaut training, General Vladimir Shatalov, spoke about plans to develop space stations that were much larger than Salyut, that would be "space complexes composed of separate units." He went on to say that the units would contain laboratories, research equipment, and workshops, and that some would be what the U.S. aerospace industry refers to as "free-flyers"—spacecraft that would orbit separately from the main space station most of the time but that could also dock with it periodically for repair, equipment upgradings, replacement of consumables, and perhaps unloading of products that have been produced on board such as purified drugs. These large space stations, Shatalov said, would be supplied by heavy transport ships.

Then, on February 13, 1986, the Soviets launched a new modular space station that had six docking positions for other spacecraft. Named *Mir* (the Russian word for peace), this space station was, many experts believed, the beginning of Russia's permanent presence in space. The first two-man crew, Leonid Kizim and Vladimir Solovyov, then docked with *Mir* a

few weeks later, on March 13, 1986, with their *Soyuz T-15* spacecraft.

Experts in the West have known for a long time that the Soviet Union had been developing a powerful new booster, perhaps in the same class as the Saturn 5 moon rocket. Such a new Russian booster could place some 300,000 pounds (136,080 kilograms) into orbit. The West also knows that Russia is developing and testing a reusable spacecraft like the space shuttle that could make routine flights to and from earth orbit. It is not a question of whether or not the Soviet Union has the capabilities to launch, build, and operate a permanently occupied space station; they do. It's only a question of when they will. Bets are they will beat the United States and their allies into space with a permanent space station facility. In fact, they may already have done so with the launching and manning of the modular space station *Mir*. After the first in-flight crew rotation took place aboard the *Salyut 7* space station in late 1985, Soviet space officials projected that they would have a permanently manned space station by 1990 but that it would not be the Salyut. The *Mir* station may be it, or it may be the next-generation station. If they do attain a permanent presence in orbit before we do, this should not hurt our spacefaring egos. Russia may get there first, but the United States and its international partners will build a better one—a high-technology, high-quality space station that will last and evolve for up to three decades, or to 2025.

What visions do the Russians have for their future space activities in the twenty-first century? A *Pravda* article published in 1983 provides a scenario. It describes an orbital complex of large facilities that would automatically monitor earth resources, detect weather and forest fire threats, track ships and aircraft, provide communications, and produce materials. The space structures, in orbits ranging from 125 miles (201 kilometers) to 2,500 miles (4,023 kilometers), would include research and housing facilities, refueling stations, workshops, power stations, large parabolic antennae, and construction sites. The article also mentions a base for ambitious space projects such as the construction of large mirrors in orbit that could reflect sunlight and provide night-time illumination to regions in the far north of the Soviet Union. Freight and passenger spacecraft would regularly service such an orbital complex. All of this sounds familiar, of course; perhaps the West and the East should read less of one another's aerospace literature.

If the Russians can put even a portion of this space vision into orbit by the turn of the century, their next adventure may be to pursue the dreams of their early space pioneers: cast off from an orbital mooring site and rocket their cosmonauts to Mars.

America's First Space Station: Skylab

After the triumph of the *Apollo 11* moon landing, enthusiasm for the expensive manned moon expeditions waned. The Vietnam War had been a major drain on the economy. Grandiose plans for America's post-Apollo space program, which included a space station, a reusable manned spacecraft, a space tug, and manned expeditions to other planets, were cut way back. Only the space shuttle program survived. Three Apollo moon missions—*18, 19,* and *20*—were canceled, and a series of ambitious near-earth space missions known as the Apollo Applications Program was cut in half. The cut-down program became Skylab, America's first space station, and the three manned missions to it.

The original concept called for two space stations built from Apollo hardware. Both in low earth orbit, the stations would be the space homes for seven different crews. As it turned out, only the single space station, Skylab, was launched, on May 14, 1973, and became the home for nine astronauts—three crews of three men each—between May 25, 1973, and February 8, 1974. The crews occupied America's first space station for a total of 171 days and 13 hours, and performed about three hundred scientific and technical experiments, which included extensive medical tests and detailed observa-

tions of the sun with several solar telescopes that took more than two thousand images of our planet's star. Even Comet Kohoutek was extensively studied and photographed in December 1973 and January 1974 by the third manned mission to Skylab. Instruments aboard Skylab gave scientists more data about a comet than ever before, even though the comet disappointed many observers on earth, mostly because of the way the mass media hyped the event, distorting expectations without providing accurate information.

Skylab was, and remains, the largest spacecraft the United States has ever put into orbit, about the size of an average three-bedroom house. On earth, it weighed almost 100 tons. The space station was built from modified Apollo equipment, and its basic structural component was the third stage of a Saturn 5 moon rocket, dry of fuel and converted into a two-story spacecraft, one a laboratory and the other the living quarters.

The astronaut crews followed the space station into orbit over a period of about nine months. They were launched in a modified Apollo command and service module atop a Saturn 1B rocket and rendezvoused with the Skylab in orbit by attaching their spacecraft's nose to a docking port at one end of the space station. This port led to the airlock module, which in turn led to the two-story living and working compartments.

The entire Skylab station, including the attached command module that brought the crews to and from earth, was almost 120 feet (36.6 meters) long. The orbital workshop area, which included the living and working quarters, was the station's largest section. It was just over 48 feet (14.6 meters) long, and 21.6 feet (6.6 meters) in diameter; its volume was 10,644 cubic feet (301 cubic meters), plenty of room for body acrobatics in weightlessness (for more details of life aboard Skylab, read Chapter 4, "Daily Life in Space").

The adjacent airlock module contained the extravehicular activity (EVA) port through which the astronauts would exit for their space walks and repair missions, including the all-im-

America's first space station, Skylab, was launched on May 14, 1973, and was about the size of a three-bedroom house—the largest spacecraft ever put into orbit by the United States. It was the home for three different crews of three men each in 1973 and 1974. In 1979 its orbit decayed, and it burned up over the Indian Ocean and western Australia. Courtesy NASA.

portant EVA repair mission carried out by the first Skylab crew in May 1973 that literally saved the space station from having to be abandoned. Because the micrometeoroid shield was ripped off during launch, tearing off with it one of the solar wings and jamming deployment of the other, astronauts Conrad, Weitz, and Kerwin had one of the most challenging repair missions of the Space Age.

Before their launch on May 25, 1973, scientists designed and built a parasol sunshade that went into orbit with them. Once docked, Charles Conrad and Joe Kerwin successfully de-

ployed the parasol, and the dangerously hot temperatures in the interior of the space station dropped to acceptable levels. The jammed solar panel was also freed during an EVA later in the mission, which gave Skylab 3,000 watts of electricity that was needed to continue the mission. With only one solar wing left and the parasol deployed over the orbital workshop area, one of the crew members commented that Skylab looked "less like a space station than like Uncle Wiggily's airship." Still, it is what works in space that counts, not the aesthetics of a space station's shape, and one of the most important

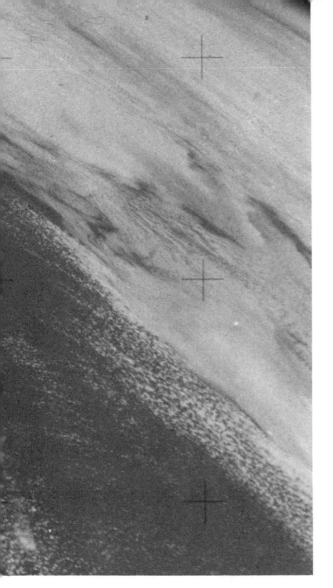

lative EVA time for one mission, with a total of 22 hours, 21 minutes for four EVAs outside the space station. The distance record for a manned space mission was also set: orbital distance traveled from November 16, 1973, to February 8, 1974, was 34.5 million miles (55.5 million kilometers).

From launch to the end of the third manned mission, the Skylab space station orbited the earth some 3,900 times at an average altitude of 270 miles (434.5 kilometers). Its orbital paths covered 75 percent of the earth's surface, and it took ninety-three minutes to complete one orbit. Before it burned up in the atmosphere, scattering space junk over the Indian Ocean and western Australia, on July 11, 1979 (long before the maiden space shuttle flight could get into orbit and save it), America's first space station completed 34,981 orbits around planet Earth in six years and almost two months. Most of this time it was unoccupied, however, because the three astronaut crews were on board for less than six months of its seventy-five months in orbit. Our next space station above earth in the 1990s will be occupied by crews, barring unforeseen circumstances, twenty-four hours a day for the station's full lifetime of twenty to thirty years.

The long-duration manned Skylab flights provided the foundation of knowledge that makes future, permanently manned space stations possible. More than 45 miles (72.5 kilometers) of magnetic tape provided billions of bits of data on astronaut health, earth resources, astronomy, and systems functioning. This data trove has been used in designing, building, and operating the space shuttle fleet; and the same 4,100 astronaut hours of experience in Skylab by all three crews will be mined during the design and development of the space station—a space venture offering the technological bridge between the twentieth and twenty-first centuries. Those who serve on the space station will have much to thank Skylab and its crews for, especially in creature comforts, including a tastier and more exciting menu and a shower that actually works.

lessons learned from the Skylab missions was that nine well-trained astronauts could perform many varied and complicated tasks in space over long periods of time without any serious physical effects.

Skylab 4, the third and final manned mission to the space station, set many records. It was the longest mission in space, lasting eighty-four days, a record that stood from 1974 to 1978, the year in which the Russians began setting space endurance records of their own, which they hold to this day. It was the first time major repairs were made on a spacecraft that saved an entire space program. And several EVA records were set: the longest single EVA in earth orbit, lasting 7 hours, 1 minute; and the longest cumu-

SKYLAB MANNED SPACEFLIGHT SUMMARY

Mission	Crew	Date	Mission Time Hrs., Min., Sec.
Skylab 2	Charles Conrad Jr., Joseph P. Kerwin, Paul J. Weitz	5/25–6/22/73	672:49:49
Skylab 3	Alan L. Bean, Jr.,Owen K. Garriott, Jack R. Lousma	7/28–9/25/73	1,427:09:04
Skylab 4	Gerald Carr, Edward Gibson, William Pogue	11/16/73–2/8/74	2,017:15:32

NOTE: *Skylab 1* was the unmanned launch of the *Skylab* space station in orbit on May 14, 1973. It was the world's largest payload put into orbit by a two-stage Saturn 5. The *Skylab* crews were launched by two-stage Saturn 1Bs.

The Space Station Program Is Launched

Although NASA had hoped early on to include a space station in the space shuttle program (it would, after all, seem logical to have a *place* for the space shuttle to go), the agency had to settle for development of the space shuttle only. Congress approved the space shuttle as NASA's down-to-earth space program, and the U.S. space station would have to wait for another twelve years.

It was on January 25, 1984, during his State of the Union address, that President Reagan initiated the space station program that would be at least a decade in the making. In his speech the President spoke of America's pioneering spirit and how the nation would continue to develop the next frontier—the high frontier of space.

"America has always been greatest when we dared to be great. We can reach for greatness again. We can follow our dreams to distant stars, living and working in space for peaceful, economic and scientific gain. Tonight I am directing NASA to develop a permanently manned space station, and do it within a decade."

These paragraphs of the President's address to Congress and the American people also gave some answers as to why the United States needed a space station, answers that went beyond the nation's past great achievements and the need to "reach for greatness again."

"A space station will permit quantum leaps in our research in science, communications, and in metals and lifesaving medicines which can be manufactured only in space. We want our friends to help us meet these challenges and share the benefits."

Almost six months later, speaking during ceremonies marking the fifteenth anniversary of the *Apollo 11* moon landing, the President spoke of other space-made marvels that would be part of America's future in space. "...We can manufacture superchips that improve our competitive position in the world computer market; we can build space observatories enabling scientists to see out to the edge of the universe; and we can produce special alloys and biological materials that benefit greatly from a zero-gravity environment."

While President Reagan's remarks on both occasions touch upon several important activities that will be served by the space station, the list of what the space station can and will do grows dramatically as the advanced design studies continue.

The Space Station Do List

What will the space station do once it is assembled in orbit? NASA's best answer would be: Whatever its customers and users want it to do. By defining what the space station will do, the users will also define the space station design. As the user needs and functions change over time, so too will the physical space station. Whatever the initial design, it must be an open one, able to change and evolve easily as the do list changes.

During the design phases of space station development, several studies have identified more than three hundred missions for the permanently manned orbiting complex. This long list would contain specific missions and opportunities, not such important general goals as: developing a permanent human presence in space; increasing national pride while at the same time increasing international cooperation among friendly nations; and advancing technology to new, state-of-the-art levels, especially in automation and robotics, as the nation enters the next century.

What follows is a brief list of general functions the space station might serve:

• Assembly area for large satellites, antennae, and spacecraft

• Repair and service station for spacecraft and satellites of all types on a much larger scale than is possible on the space shuttle (for example, the Hubble Space Telescope or a military reconnaissance satellite)

• Launch base or staging area for spacecraft taking satellites and other equipment to geosynchronous orbit

• Launch base for manned or unmanned missions to the moon, the planets (the vote is for Mars), or the asteroids

• Warehouse and storage area for rocket fuel, space parts, replacement parts, and so forth

• Permanently staffed laboratory for life sciences research

• Permanently staffed laboratory for materials processing research in a microgravity environment

• Advanced astronomical observatory that would supplement and eventually take over for the Hubble Space Telescope (this would include a program search for planets around other suns, detailed study of sun-earth interrelationships, and other programs at all wavelengths—infrared, optical, X ray, gamma ray)

• Advanced earth observatory that would manage all remote sensing programs (crop, fishing, and mineral resources; weather forecasting, including wave and wind data; and perhaps some

military reconnaissance, especially from a space platform in polar orbit

• Industrial center (as distinct from the research lab) for the processing of commercial materials in microgravity. This would include such products as drugs, alloys, crystals, and microelectronic components.

This modest sampling of what can be done on a future space station in no way reflects the hundreds of discoveries that will be made on the way to establishing a permanent station in space. But as a driver of high technology, as a way of lowering the costs of doing all kinds of business in space, and as a means of enhancing international cooperation in peaceful pursuits (as Arthur C. Clarke constantly reminds us in such films as *2001* and *2010*), the space station is the next important foothold for humankind's climb to the stars.

The Power Tower and the Dual Keel

When NASA awarded advanced definition space station contracts to aerospace firms in the mid-1980s, it gave them a basic reference design that had been painstakingly produced through extensive research and evaluation of many designs over a two-year period. It was referred to as the Power Tower. While it was in no way the final design, it was NASA's basic design of choice for the first year of design studies. Then a new reference design, known as the Dual Keel, was approved as the official design configuration in early 1986.

The Power Tower. By looking at a model, or an artist's rendering of this reference design, one can easily see how it came by its name. A 400-foot (122-meter) trusswork tower is the central element of this concept, and massive solar panels to provide power are grouped at one end on a large crossbeam about 300 feet (91.5 meters) wide. All other components—including the pressurized living and working modules (probably five in the beginning), the manipulator

The Power Tower space station design was the first reference architecture NASA gave to competing aerospace companies, but it lost out to a new, improved design in 1986. Courtesy NASA.

arms, radiator panels for heat dissipation, various instrument packages, radar dishes and antennae, docking modules, vehicle service hangers, and so forth—are attached to the central tower structure or the crossbeam.

Like the more recent, more symmetrical Dual Keel design, the Power Tower concept was open, accessible from all points in space, and provided room for growth and change over time. "It must be an evolutionary design," says Philip Culbertson, NASA's associate administrator for the space station. As user needs change, so will components of the station. NASA emphasized to their contractor teams that they had to design a "customer friendly" space station.

Another important advantage the Power Tower design had over earlier alternative designs such as the Big T or the Delta was that it would be more stable in orbit. And designers

believe that the Dual Keel space station will be even more stable in orbit because of additional trusswork and interlocking rectangular layout.

Some 250 miles (402 kilometers) above earth, the station will still encounter slight drag forces from the thin, outermost reaches of the atmosphere. Thrusters on the space station would be capable of moving it to a higher attitude or changing its attitude relative to earth, but the long, flat profile of the Power Tower was the first design to allow this to be done more efficiently and with less propellant. Because of the space station's long central beam (some three and a half times the length of the space shuttle), the earth's gravity would help stabilize its attitude by pulling along the entire length of the structure. This means that the Power Tower design would naturally align one end with the earth; it would be more stable and would wobble less. Such a configuration had the added practicality of two structural ends pointing in the right directions for two crucial areas of scientific research: earthward and spaceward. The astronomical instruments would be up, and the earth-sensing instruments would be down. The living modules, able to accommodate a crew of six to eight during the first years of the space station, would also always point toward planet Earth, presenting beautiful views through the portholes—the quick sunrises and sunsets, the rich blue expanses of entire oceans.

The Dual Keel. As design studies progressed in 1985, the Lockheed and McDonnell Douglas contract teams working under contract for the Johnson Space Center realized that the Power Tower design had serious drawbacks, many of which centered on the all-important goal of making the space station as customer friendly as possible. The Dual Keel space station, certainly much more pleasing to the eye because of its rectangular symmetry, addressed these design shortcomings.

If NASA's first reference design could be described in one word as a "tower," then the Dual Keel could be conceived of as a celestial "box kite" or, from above, a "catamaran." Two major structural beams, each about 360 feet (110 me-

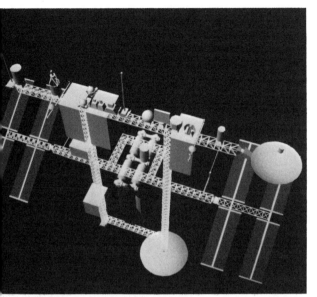

A computer-generated design and an artist's rendering of the Dual Keel space station, which became NASA's reference design in early 1986. The new design offers several advantages over the Power Tower and is considered more "customer friendly." Courtesy McDonnell Douglas Corporation; and Rockwell International.

ters) in length and parallel to one another, are crossed midway up an even longer crossbeam, which may be as much as 400 feet (122 meters) in length. Top and bottom beams close off the ends of the parallel keel beams and form the two large boxlike areas, which provide considerably more space than the tower design on which to mount the telescopes, experiments, satellite service bay, refueling bay, and docking areas for spacecraft that would boost satellites to geosynchronous orbit or conduct experiments away from the station complex. There would be much more open space in which to grow and change with the Dual Keel, and this would ensure that the station would always remain flexible to the changing needs and goals of its commercial and scientific customers.

With the Dual Keel design, the crossbeam, which also supports the eight solar panels, would be mounted close to the middle of the configuration, near the station's center of gravity. Top and bottom beams box off the parallel keel beams; they both would be 146 feet (44.5 meters) long and would have all the sun and deep space astronomical instruments mounted on them. They would house the spacecraft facil-

ities, with the exception of the shuttle docking module.

The docking, habitat, and laboratory modules, instead of being mounted at one end of the structure as they were with the Power Tower design, would be near the middle of the Dual Keel space station, suspended from the crossbeam. Their central position near the station's center of gravity would reduce the microgravity loads that the materials processing module would have experienced with the Power Tower design—one of the primary reasons for the drastic design change.

There were also some safety concerns about the Power Tower's modules, which were to be connected end-to-end in a racetrack shape and could not be individually sealed off in case of an emergency. The Dual Keel space station design calls for the modules to be connected on one or both ends with external airlocks. If a pressurized module were ever punctured by a meteorite or piece of space junk, it could be sealed off from the rest of the station modules.

Studies indicate that the station will be hit at least once in its lifetime by orbital debris large enough to cause serious damage. There are at least forty thousand small objects measuring 0.4 inches (1 centimeter) in low or middle earth orbits, and experts believe that the station must be protected against such high-velocity objects that could blast through the walls of a module and kill the crew members inside. The space shuttle *Challenger* had a small pit about the size of a sesame seed dug in its window by a tiny speck of paint only 1/125th of an inch in diameter (1/5th of a millimeter). Imagine what kind of damage an object of a few inches or centimeters could do.

Besides the protection given by structural design, one plan calls for deploying a space debris monitoring satellite that would warn station command when such destructive objects were in the vicinity of the station and posing a threat. The position of the pressurized modules on the Dual Keel design and the proposed airlocks confront this hazard directly before the hardware gets built.

The Modular Cluster

The center for space station activities during the first years will be a cluster of five large cylindrical modules (including a logistics module) with connecting passageways. The structural design of these so-called common modules will be standardized, but their interiors will be adapted and designed for their specific functions, such as living quarters or microgravity laboratory. These giant aluminum cylinders, about 15 feet (4.6 meters) in diameter by 44 feet (13.4 meters) in length will be suspended from the long middle crossbeam of the Dual Keel station, and one of them will provide a docking port for the space shuttle and perhaps the European Hermes spacecraft. Each of these pressurized modules will be about the size of a medium-size house trailer; their maximum size was one limitation already determined by the size of the space shuttle's cargo bay, in which they will be hauled already built and tested, one by one, into orbit. (See P. 116-117 for more details) There will be some 500 square feet (46.5 square meters) of floor space in each module, but in weightless-

ABOVE: *A view of the space station's modular cluster, to which the space shuttle is docked. The command center, living quarters, and laboratories are located inside these pressurized cylinders.* Courtesy McDonnell Douglas Corporation. FACING PAGE: *Full-scale mockup of the space station module with an astronaut mannequin in a manned maneuvering unit.* Courtesy Martin Marietta.

ness, the floor can be anywhere you wish it to be and all the interior surfaces can be efficiently used because they are equally accessible.

NASA's Marshall Space Flight Center was given the responsibility of coming up with the basic design for all the pressurized modules and choosing the aerospace teams to do the advanced design work. The space station will grow over time, and the number of modules may easily increase to twelve from the original five, as new laboratories are added, number of crew specialists increases, and international partners such as the European Space Agency and Japan plug additional modular laboratories and living quarters into the space station. The initial design calls for five modules adapted for the following uses:

• One for the living quarters and station operations, housing six to eight crew members (details of which will be described later in this chapter)

• Two for laboratories, perhaps one for life sciences (Europe) and one for materials processing research (United States)

• One for experiments in advanced technology (Japan)

• One for logistics, including a pantry and a storehouse for food and supplies (this module could also contain chemical toilets). It could be exchanged for a new unit by the space shuttle every ninety days, when a new crew arrives from earth.

Eventually there may be several unpressurized service modules attached to the top and bottom crossbeams, which would perform many of the same functions that the space shuttle's open cargo bay does on missions after its satellites or other payloads are launched—a space garage where repair and servicing of satellites and spacecraft take place.

But because change is inherent in the modular design, no one at NASA or the aerospace contractors will be surprised if advanced design studies change in some ways the apportionment of the space station's usable space. This is the beauty of an evolutionary design. Just as long as the station's structural integrity and stability in orbit are maintained, and the politics on earth (national or international) provide the funding, new facilities can be attached to the space station at any time the need arises. Real estate in space never has to worry about limitations to growth—at least not for another hundred years.

The Free-Flying Offspring

It has been the fleet of space shuttle orbiters that has received most of the public's attention as they carried out their dozens of complex and varied missions. Still, they are only one component of hundreds in the Space Transportation

System, which include (to name just a few) two launching sites, launch preparation facilities, the solid rocket engines that fall back to earth to be refurbished and used again, the tracking network, the Spacelab modules, the data relay satellites. The same will be true of the space station. In fact, without the Space Transportation System, which will haul its components into orbit, the space station could not exist.

The space station will be an orbital system carrying out dozens of functions, and many of its components will not be physically attached to it. These include the so-called free-flyers, spacecraft or platforms that will orbit near or far from the station, performing many specialized tasks (such as materials processing for private enterprise), but that in some way will be controlled by the space station or at least serviced and repaired there.

In fact, some of these free-flyers will be in orbit before space station construction begins by the mid-1990s. In the mid-1980s, NASA negotiated an agreement with Space Industries, Inc., of Houston, Texas, to launch for them two orbiting platforms on shuttle flights in 1989 and 1990. Space Industries will lease space on their platforms to industrial customers who want to manufacture products in space and take advantage of microgravity processing. Most of the launch costs will be carried by NASA at the start, and Space Industries will pay back the agency with 12 percent of its yearly revenues.

The space station will also be the home base for various types of rockets designed to accomplish certain tasks. One such spacecraft, which NASA refers to as the orbital transfer vehicle (OTV), will take satellites from the space station's low earth orbit to high geosynchronous orbit or bring them back for repair in the space station's open servicing bay. A new class of large satellite, parts of which will be brought to the space station by the space shuttle, will be assembled there and then hauled to high orbit by the OTV. Large space antennae, up to 328 feet (100 meters) in diameter, could also be assembled and deployed in this manner.

There will also be unmanned spacecraft that will work only in low orbit, servicing satellites and space platforms on site by remote teleoperation control from the space station. Called orbital maneuvering vehicles by NASA (OMVs for short), they will also act as low-orbit space tugs, bringing other space hardware to the station and deploying it again. They will not, however, look like a traditional rocket, but more like thick space Frisbees®. About 15 feet (4.6 meters) in diameter and 3 feet (.9 meters) wide, these free-flyers will be designed to fit at one end of the shuttle's cargo bay for launch into orbit. Their weight of 10,500 pounds (4.63 kilograms) will be more than half liquid fuel. They will be equipped with robotic repair kits that will be remotely controlled from the space station or from the ground. Even now satellites are being designed and built with modular components so that their operating lives in orbit can be extended by these remote-control mechanics that will be launched in the early 1990s. At least one of these maneuvering spacecraft will be included in the first shuttle flight carrying space station components because it will be an essential part of the initial construction phase—a multipurpose construction vehicle, part space crane, space truck, and space tool chest.

For some experiments and observations, space platforms that orbit away from the busy, heat-venting, and energy-loud station will be required. A platform in polar orbit carrying experiments and tracking antennae for the study of earth resources and ideal for military reconnaissance is planned as part of the orbital infrastructure. The polar orbit is uniquely suited to certain types of observation because a spacecraft can observe the entire earth twice each day. Such polar-orbit platforms would be out of reach of vehicles remotely controlled by the space station, and so they would be serviced by space shuttle crews or a formation-flying maneuvering vehicle deployed by a polar-flying shuttle mission. Eventually this unmanned platform could be replaced with a second, permanently manned space station.

At least one other free-flying platform will be deployed during the 1990s in an orbit parallel to that of the space station for certain contamination-free and noise-free experiments, observa-

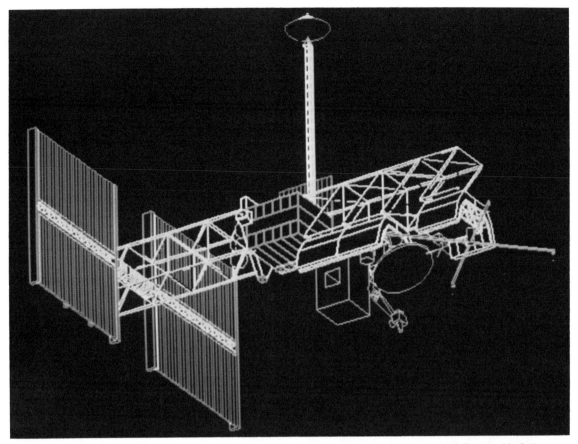

A computer-generated design and an artist's rendering of the Dual Keel space station, which became NASA's reference design in early 1986. The new design offers several advantages over the Power Tower and is considered more "customer friendly." Courtesy McDonnell Douglas Corporation; and Rockwell International.

tions, or industrial processing. But there will be more. Plans for the European Space Agency's participation as an active partner also include a free-flying platform in a parallel orbit within service range of the station. Europe's Columbus research laboratory, which will be attached to the U.S. space station during the first few years, may become one of the core modules of a separate European space station by the year 2000. If and when a full-service European space station becomes independent, additional free-flying spacecraft will be needed.

Eventually the space station will be like a busy port on earth, with its own traffic controllers managing a constant flow of spacecraft, of all designs and sizes, in and out of its inner harbor. There may even be a Chrysler, Ford or GM emblem seen on their exteriors, along with those of McDonnell Douglas, Rockwell International, Martin Marietta, TRW, Grumman, and Hughes. No doubt a few new Fortune 500 companies will be represented above the earth, companies that no one has heard of today.

Building the Space Station

The Dual Keel space station will be built on earth, its components delivered to orbit by the space shuttle, and then assembled at an altitude of 250 to 300 miles, (403 to 483 kilometers) above the earth by outstretched robot arms and astronauts in self-propelled maneuvering units. The size of the space shuttle's cargo bay, 60 feet

(18 meters) in length and 15 feet (4.6 meters) in diameter, is the one constraint that cannot be changed. Component design, and how the components will be assembled in orbit, is directly linked to the shuttle's cargo bay capacity, although unmanned expendable rockets may also be used as a result of the *Challenger* disaster. The pressurized modules that will contain the crew quarters, the laboratories, and the control center will probably be carried, one by one, to orbit by the shuttle fleet, and that is why their dimensions are about 44 feet (13.4 meters) in length and 15 feet (4.6 meters) in diameter. The same is true for Europe's Columbus module and the proposed Japanese module that probably will be delivered to orbit by the shuttle. Fully outfitted and finished for their station function before launch, the modules will be the easiest part of the challenging in-orbit assembly project. Plucked out of the cargo bay by long manipulator arms, they will be positioned near the mid-section of the central crossbeam and attached by astronauts who have had hundreds of hours of training in simulated in-space assembly.

The Space Station's Skeleton. It is the in-orbit assembly of the station's main trusswork beam system, more than 1,200 feet (366 meters) in all, that presents the most formidable design, engineering, and mission planning challenges. In the late 1970s NASA built and tested beam-making machines that formed continuous geodetic (triangle-shaped) trusswork from rolled aluminum stock. Because of their complexity, however, it has been decided not to use these robot machines for building the space station, although they may be deployed for future space-building projects.

NASA contractors are designing and testing several structural beams before a final choice is made. Ease of deployment and assembly, not material and strength factors, are the major concerns. One company, Space Structures International Corporation, is developing a hub-and-strut system that can be assembled by astronauts without tools, using only one hand. There are no nuts and bolts in this system; the pieces just slide together to form a triangular geodetic shape—the strongest shape in nature. Tests on

the ground have shown that two struts can be connected every thirty-eight seconds, and the first EVA space test on *Atlantis* in late 1985 proved that it is a viable building technique. As yet the question of whether or not the trusswork should be assembled piece by piece has not been answered. An alternative method would be to carry folded sections into orbit and unfold them like an umbrella at the construction site. The space workers, robots, and astronauts would do more unfolding and less assembling of the beam sections, which could contain more than one thousand struts.

The Convair Division of General Dynamics is developing such a preassembled truss beam system that would unfold in space once it was deployed from the space shuttle. The titanium beam, composed of hinged and folded struts, would be launched in a "flat pack," folded condition and then unfolded in a three-step sequence after deployment. Each beam section would be about 7.5 feet (2.3 meters) high and 5 feet (1.5 meters) wide after it was unfolded. The beam's weight could be as low as 1.44 pounds (0.65 kilograms) per foot or 4.7 pounds (2.1 kilograms) per meter.

The Convair beam design also has an extremely good packaging ratio—the ratio between the size of the folded structural section and its deployed, unfolded size. Its length increases nine times, and its volume increases forty-three times. Twenty-four of these beams, each one 500 feet (152.5 meters) long, could be hauled to orbit on a single space shuttle mission. If this system or an adaptation of it is approved by NASA, the future space station will quite literally be unfolded in orbit.

Unfurling the Solar Panels. Current design plans for the basic space station call for a total of eight solar panels, each one 30 by 80 feet (9 by 24 meters), to deliver 75 kilowatts of power. The maximum power the Skylab generated was only 23 kilowatts. As the station expands during the first decade in space, so too will its power needs; power generation is planned to reach 150 kilowatts by the year 2000. Four solar panels will be attached to each side of the central crossbeam that also holds the modules in the Dual Keel design. Two heat-dissipating radiators, the design of which involves an entirely new, lightweight technology, will also be attached to the crossbeam, but they will hang at right angles to the solar panels. Two more radiators will be attached near the module cluster.

This electrical power design, with its solar panels and their thousands of photovoltaic cells, will be combined with a solar collector thermal dynamic system that is also under development. Each solar panel would have an area of 2,400 square feet (223 square meters). All eight panels would add up to 19,200 square feet (1,784 square meters) of surface that would collect sunlight and convert it to energy. Because the photovoltaic technology is proven and has been used by spacecraft designers in the past, it will be the first power system to become operational on the space station and will provide 2.5 kilowatts of power. The thermal dynamic system will eventually provide an additional 50 kilowatts.

The deployment of the solar panels will probably be accomplished in a manner similar to the successful solar array test that was done on the space shuttle *Discovery* mission in September 1984. This large solar array, with its solar cells attached to eighty-four folding panels of thin, flexible sheets of Kapton, was successfully deployed several times by Judith Resnik, the second U.S. female astronaut to fly in space. At full deployment, after being unfolded from a long, 7-inch (18-centimeter) high containment box and run up with a movable mast, the 13-foot (4-meter) wide structure stretched 102 feet (31 meters) into space from the open shuttle bay. This made the solar array the largest space structure ever erected in orbit by the United States.

The solar array wing and its deployment equipment were designed and built by NASA's Marshall Space Flight Center and the Lockheed Corporation. The tests of the array and its deployment methods proved successful. In fact, the solar array panel was structurally more rigid than predicted—an important and favorable finding for the eight panels that will be unfurled on the space station and will collect sunlight for

fifty-nine minutes of each ninety-four minute orbit around the earth.

Delivery to Orbit. Will the space station be up in 1992, in time for the five hundredth anniversary of Christopher Columbus's westward voyage to the New World? The probable answer: Parts of it may be in orbit by that year. Certainly the "space-breaking" ceremonies will take place in 1992; during the International Space Year we will, no doubt, weigh anchor and begin the voyage as Columbus did five hundred years earlier. And Spain will be one of the many European nations that will participate in Europe's Columbus laboratory and related hardware.

As with so many major national commitments, politics and economics will play a more crucial role in the space station timetable than will research and development. But even with some *modest* funding delays, the basic space station should be built and operational by 1995. What follows is a brief scenario of how the space station *could be* delivered to orbit.

The first space station components could be launched by the space shuttle in the early 1990s. This mission would be the first of at least six shuttle flights to deliver the primary components, weighing more than 50 tons on earth, to orbit. Some 20 shuttle flights, however, will be necessary to build a completely operational space station.

Four solar panels, their connecting crossbeam, and two radiators could be delivered on the first flight. Using the space shuttle and its open cargo bay as the construction base, spacewalking astronauts and swinging robot arms will assemble and deploy these parts. The second shuttle flight could haul up all the trusswork sections for the dual keel beams and perhaps some of the supporting beams that would hold the pressurized modules at the center of the configuration. Two more radiators that will service the modular cluster might also be aboard this flight. On each of the next three shuttle missions, a fully outfitted habitation module, filling most of the cargo bay, would be flown to the orbiting construction site. The sixth mission could carry a fourth module and four more solar panels. These first six shuttle flights could be flown within a two- to three-month time frame, and the assembly of this basic station could be done within a few months. This would allow early man-tended, not yet permanent, operations. The full space station configuration, with the free-flyers, the other modules, including Europe's Columbus, and the spacecraft servicing facility could take as long as two years to be in place. But in the meantime, the ninety-day tours of duty will have begun on this catamaran in space. Besides all the important work and benefits the station will produce, its population will be learning to live full time in space. The space station will offer a new way of life. Within its first decade, there will most likely be the first marriage in space, the first zero-g consumation, and perhaps that first baby conceived, but not born, in weightlessness. Birth in space must wait until the 21st century.

The Space Station Life

Three months is a long time to live and work above the earth, but the accommodations aboard the space station will make the weightless life there familiar and comfortable for the first crews of six to eight men and women. While the living and work quarters inside the pressurized modules will not be luxurious during the first few years of the space station, they will certainly be more comfortable than the space shuttle. And compared to the cramped quarters of the early space days—inside the Mercury, Gemini, Apollo, and lunar module spacecraft—the living quarters will be rated first class.

Private Quarters. Designers envision a small individual bedroom for every crew member, a full galley, a recreation and workout area, and at least one bathroom. If married couples are on board, then two of the individual rooms can be transformed into a mini-suite, and the bed restraints can be zipped together just like sleeping bags.

The private bedrooms might more accurately be called cubicles because they will not be much larger than a phone booth—about 150 cubic feet (4.25 cubic meters). Still, in weightlessness any space becomes larger because all six surfaces can be used: there is no wasted space. Temperature, humidity, and air flow can be individually regulated by the occupant. Besides the bed (sleeping restraint), each sleeping area will also contain storage compartments, a porthole, a work desk with a keyboard and video terminal, and some wall space left over to hang personal pictures or favorite art work. Electronic letters can be written to loved ones; video tapes can be watched; music can be heard; or the beautiful, ever-shifting earthscape below can be watched and marveled at—the favorite pastime of the *Skylab* astronauts. It is this private life that will be important to morale on the space station. And the work will be their chosen work—difficult and rewarding. There may not be much leisure time after a hard day's labor, but what there is of it will be quality time off, with privacy, if desired, and a choice of activities.

Galley and Food. The galley will be about the size of a studio apartment kitchen, and its appliances will include an oven, a dishwasher, and a refrigerator. There will also be a food freezer which will provide more fresh foods, and a larger variety, for the crew members. An automated inventory control system will ensure that the astronauts do not run out of desserts. And the space station galley will even have a trash compactor and dehydrator unit.

Even with the larger selection, certain foods will still not get on the menu because of the mess they can make in weightlessness. Crackers that crumble are out, but bread is no problem. Any food that comes in small pieces, such as rice or baby peas, is difficult to handle unless it is held together by another food base. Free-floating bits of food are potentially hazardous to the station's equipment and to the breathing astronauts, and so space food designers will make sure they are kept to an absolute minimum. About 1,500 planned meals will have to be launched with each crew to sustain them for their ninety-day stay in space.

The galley table will contain tiny vacuum holes that will hold down the plates and drink containers, and low railings beneath the table will allow the eaters to remain floating in one position by hooking their feet under them.

One module design calls for the galley and wardroom to serve also as the meeting area, for group gatherings and discussions, and the exercise room, containing the health-maintenance equipment. Because cardiovascular tone is so quickly lost in zero gravity ("Your heart really goes on vacation when you are in space flight," says Dr. Douglas O'Handley, manager of the JPL Life Sciences Program), a fact that has been known since the Gemini days of manned spaceflight, exercise schedules will be taken seriously and maintained. Mission Control has devised some zero-g paddle games to help the crew members stay fit and healthy, and will continue to try out innovative exercises and games. Some of the classics from the old space days will also be brought back to life—like the chase-the-peanut-with-your-mouth-open game a *Skylab* crew enjoyed.

The Bathroom. The bathroom and shower will be located at one end of the living quarters. Earlier shower systems on *Skylab* and the shuttle proved almost worthless, and the space station shower will be an entirely new design—an enclosed stall with flat walls. Because of surface tension, up to an inch of water could cling to the walls of the stall in zero gravity. This will be avoided, however, with a mild air flow moving the clinging water down the walls to the floor, where it can be captured and recycled.

Advances in recycling technologies must be made because of the scale and permanency of the space station. Water must be reclaimed from showering, toilet use, and the dehumidification system. If such technologies are not developed and used, the extremely high costs of resupply payloads would probably prohibit the station from ever being built. For example, without a water recycling system, a full shuttle load of wa-

ter would have to be brought up every three months for a crew of six. Some 44,000 pounds (19,958 kilograms) of water would be consumed by a crew of eight every year!

It is likely that the chosen system for recycling water will produce water in two grades: one potable, for drinking and cooking; and one so-called gray grade for washing clothes and other needs. Water and atmosphere regeneration breakthroughs are so important to a permanently occupied space station that NASA plans to spend about $800 million on research and development of advanced recycling technology.

Each pressurized module will have the life-support systems built into its exoskeleton, and they will be designed as closed loops. Oxygen and hydrogen will be reclaimed through a distilling process, and the waste methane will be vented overboard.

The Work Modules. The laboratories—whether dedicated to materials processing or the life sciences—will be very similar in appearance to the various *Spacelab* configurations flown by the space shuttle. They will be packed full with specialized equipment—furnaces, electrophoresis processing units, an array of measuring devices—and only the highly trained mission scientists and doctors will feel at home in them.

The space station's command-and-control functions, possibly housed in a separate module, will be alive with lit multicolored instrumentation and the sounds of automation. Several monitor screens will show various portions of the space station's exterior (the solar panels, the antenna dishes, a docked shuttle) or the interiors of laboratories in other modules. Advanced automation and robotics will be designed into all systems on the space station. Personnel will not have to flip twelve switches in proper sequence to repoint the solar panels, for instance. Just by pointing to a lit symbol on a control panel, a crew member will be able to open or close a valve or a bay door, or even reorient the entire space station. Systems incorporating voice activation and other advanced forms of artificial intelligence will be used whenever possible be-

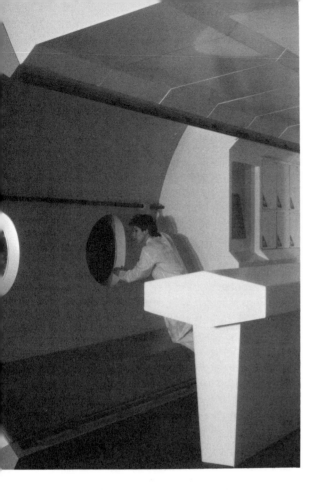

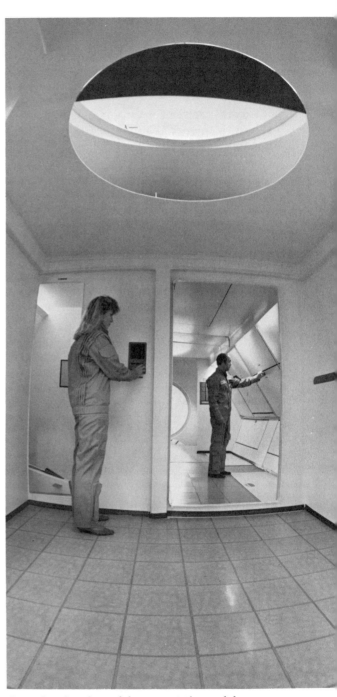

Seven interior views of the space station modules. These full-scale mock-ups are part of the design process, which is in its second phase. Work (the command center), eating, exercise, and sleep areas are shown. Courtesy Grumman Corporation; McDonnell Douglas Corporation; Lockheed Missiles & Space Company, Inc.

cause the space station must be, in one of its most important aspects, a spearhead project for technical revolutions, and it must continue to be at the cutting edge of technology during its growth and evolution over twenty to thirty years.

Expert systems of artificial intelligence will be able to plan space shuttle or other spacecraft missions and calculate their trajectories and orbital altitudes or completely oversee the system that controls the atmospheric pressure of the space station's interior and removes carbon dioxide from the atmosphere.

To support such advanced artificial intelligence, the computer power aboard the space station will be awesome by today's standard's—some 250 times more power than what the space shuttle carried during its first missions or about sixteen megabytes during its first few years. This would be about twice the power of a Cray-1 supercomputer, and even this could

double during the station's growth in the first decade.

An Observation deck? Space station design continues to progress, even after the basic reference design was changed from the Power Tower to the Dual Keel configuration late in the phase B design studies. More detailed module floor plans, none of them finally approved, are becoming available. One design question that everyone is curious about is whether or not there will be an observation deck, perhaps adjacent to the living quarters, where the space workers can gather and view their magnificent corner of the cosmos. Tired after their demanding concentrations, they could look out and see the blue and white daylit earth turning below, or, if it's during their nightside orbit, they could look out and see the bright white moon, or perhaps Venus, Mars, Jupiter, or ringed Saturn among the steady-burning, jewel-colored stars. It would be

A preliminary design for the space station galley, which has many features similar to the smaller space shuttle galley area. The choice of space foods will be greater than ever before, including many international dishes. Courtesy McDonnell Douglas Corporation.

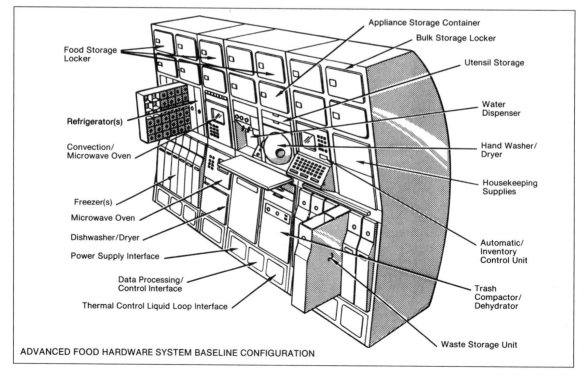

ADVANCED FOOD HARDWARE SYSTEM BASELINE CONFIGURATION

the spacefarer's equivalent of staring into the flames or glowing embers of a wood fire. They would become calmly mesmerized by the reflected sunlight from the planets or the distant starlight from those burning pieces of the universe that were born from tides of dust and gas. And hundreds of miles (kilometers) below, some of us on earth might look up and see the bright and fast-moving space machine fall across the sky. On earth or above, the awe will remain and drive us to new dreams, new inspirations.

Second-Generation Space Stations: Beyond 2020

From the beginning, the space station has been considered a technological bridge between two centuries, an open-ended space project that will evolve over a time space of several decades. While NASA has called it the next logical step in humankind's conquest and use of space, the world's first large space station will really represent a series of logical steps that will carry the technologies of the Untied States and its international partners (member nations of the European Space Agency, Canada, and Japan) into the future. As the space station celebrates its silver anniversary in the year 2020 and prepares for an early retirement (perhaps to be converted to the first museum in orbit), it will probably be sharing space with other space stations, more advanced and larger, that will be designed and built above and beyond the earth during the second or third decades of the next century.

In the spring of 1986, the National Commission on Space issued its report and recommendations to President Reagan. The commission, chaired by former NASA administrator Thomas Paine, had several Space-Age greats serving as members. These included Neil A. Armstrong (the first human to walk on the moon), Kathryn D. Sullivan (the first American woman to walk in space), and Charles E. Yeager (the first human to break the sound barrier and to fly at a speed of more than 1,600 miles, 2,574 kilometers per hour).

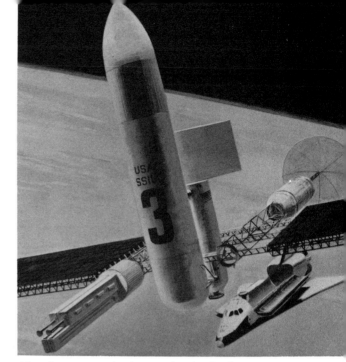

After the space station is operational, other large space structures will eventually be placed in orbit as part of the so-called space infrastructure. This orbiting fuel depot is one such project that was recommended by the National Commission on Space. Courtesy General Dynamics.

Among other things, the commission recommended that the United States make a major investment in the so-called space infrastructure in the decades after the space station is operational. These future space plans would include three or four second-generation space stations:

• One in low earth orbit to be used for a refueling depot and a maintenance and assembly station for large spacecraft. This would reduce the heavy workload at the first space station.

• A station in high, geosynchronous orbit above the earth, where it would help to support large antenna farms nearby, which will eventually replace the individual communications satellites of today.

• Another station that would orbit halfway between the earth and the moon and serve as a way station in the busy earth-moon space basin of the twenty-first century.

• And possibly one other space station that would be placed in a continuous Earth-Mars orbit and would serve as a means of transportation between Earth and Mars. During transit times, the station could be used for scientific studies.

These future space stations would probably be larger than the first. They would be like spaceports and would provide full service for the refueling, launching, repair, and maintenance of spacecraft. Perhaps the famous wheel-shaped space station of Hermann Potocnik, Wernher von Braun, Arthur C. Clarke, and others will be built someday, providing artificial gravity in its outer portions and zero gravity for manufacturing and research in its central hub. Such a space station would offer the best of all gravity worlds—a choice of gravities. Even the popular moon gravity—popular at least with the Apollo astronauts who landed and walked on the moon's surface—could be reproduced, and life spans for spacefarers could be increased because the body experiences less physical stress.

Commission chairman Thomas Paine expects to "see extensive population and utilization of low earth orbit early in the twenty-first century." But for this to happen, he adds, "The cost of moving goods into orbit has to be reduced to one-tenth of that of the shuttle, then reduced again to one-fifth to one-tenth of that level."

When payload costs are reduced by this degree, the spaceward migration will truly begin. Resources on both the moon and Mars will then be utilized to further reduce the costs of living and working in space and making the inner solar system of our yellow dwarf star the new enlarged home of humankind.

Solar Power Satellites

In the last half of the 1970s, when the energy crisis earlier in the decade was still a vivid memory for many people in the United States, the concept of large solar power satellites, some envisioned to be as big as Manhattan Island—13.3 miles (21.4 kilometers) in length and 3.32 miles (5.3 kilometers) wide and weighing some 100,000 tons, was formally studied as a possible long-term solution to the world's growing energy needs. The sun's energy would be collected by huge solar panels in orbit, converted to microwave energy, and transmitted to earth,

where it would be converted to electricity.

NASA and the Department of Energy conducted several joint studies on the feasibility of solar power satellites in the late 1970s and the general results indicated there were no insurmountable problems in the technology. The costs of the technology, the solar cells in particular, and the transport costs of the components into orbit, would have to be dramatically lowered before such an energy-from-space program could be seriously considered. James M. Beggs, NASA's administrator during the Reagan years, believes that solar power satellites are feasible but much too expensive as long as traditional power sources are dependable and relatively inexpensive. "There may come a time," he told *Omni* Magazine," when humankind is using so much power that a power station in space would become desirable, maybe not on economic grounds, but purely on the grounds of getting more energy without suffering any of the ill consequences of generating that energy." But if per pound or kilogram payload costs are lowered dramatically, solar power satellites start to look good even on economic grounds when they are based on an international market: The annual worldwide market for new energy—replacement of old plants and new energy needs—amounts to some $400 billion.

Space station technology and skills, especially the techniques of assembling large space structures above earth and new rocket hardware to lower payload costs, could be applied to solar power satellites if global energy needs ever make them essential to the future.

The Space Elevator

This concept, also referred to as the skyhook, originated with the Russian engineer Yuri Artsutanov, who wrote about it in a Soviet magazine article in 1960. It was Arthur C. Clarke who expanded on the idea and brought it to the attention of several generations of science fiction fans and space enthusiasts, many of whom are today's active space scientists and engineers.

Large communications platforms may also be built in orbit to support the growing orbiting commerce and manufacturing facilities as well as second-generation space stations that will be built in the twenty-first century. Courtesy General Dynamics.

Artsutanov and Clarke asked themselves if there was any way to build a bridge that would span the space between the earth's surface and a point in geosynchronous orbit some 22,300 miles (35,910 kilometers) above it. The speed of such an orbit equals the earth's rotation speed, and therefore the point in space or a satellite in that location remains in direct alignment with a corresponding point on the surface of the earth. If such a physical bridge could be built, the deep gravity well in which the human population dwells (the planet's surface) could be easily and inexpensively scaled with a space elevator, and we would no longer have to depend on sophisticated, costly rocketry to get us out of the well and into space. In theory, at least, once such a space elevator was operational, it would cost only about $10 in electricity to travel to orbit. Space would finally be accessible to everyone who wanted to go there, including the first generation of weightless entrepreneurs who want to be where the action is and don their space suits instead of rolling up their sleeves.

An extraordinarily strong cable, perhaps composed of crystalline graphite, would be anchored in geosynchronous orbit and would drop all the way to the surface of the earth at some point on or near the equator—perhaps in Clarke's adopted country, Sri Lanka, in which the author set his novel *Fountains of Paradise*. If a mountaintop can be found at the equator all the better—less cable and less of a climb to orbit. The space elevator's cable would have to be stronger than any material yet available, and because it would have to support its own weight, it would also have to be extremely thick at its mooring in geosynchronous orbit and thin at its position on or near the surface of the earth.

The weight of a descending cable car would pull another car up the anchor line to orbit, aided by electromagnetic forces. Estimates indicate that about 100 tons of payload or dozens of passengers could make the inexpensive trip to orbit in this way. If the destination was a low orbit, no higher than 1,000 miles (1,610 kilometers) above the earth's surface the trip could be made in less than half an hour.

If satellites were equipped with the same type of ultrastrong tethers, they could dip their lines into the atmosphere twice during each of their orbits, hook on to shuttle ships or spaceliners, and carry them into orbit for just a fraction of the cost of an expensive traditional launch.

When will such space elevators and skyhooks be built? Says Arthur Clarke: "About fifty years after everyone stops laughing."

Vacations in Space?

Some space experts believe there will be tourists in space before the year 2000. Even today there are wealthy people who are able and willing to pay almost any price to fly on the space shuttle. If the Cunard Line can sell tickets for their luxury cruise around the world at prices close to $100,000, there must be hundreds of people willing to pay that or more for the experience of a lifetime. The jet set would become the rocket set if NASA rules allowed them to purchase tickets and receive some pre-trip training.

In one sense, there already has been a tourist in space. Prince Sultan Al-Saud flew as a mission specialist on a space shuttle flight in 1985; under NASA rules, any nation that buys shuttle services has the right to fly its own mission specialist, and Saudi Arabia exercised this right to witness the deployment of the *Arabsat 1-B* communications satellite. While the prince did have some simple experiments to perform, most of these were accomplished by French astronaut Patrick Baudry, his flying companion. The prince was mainly an observer—a space tourist—and his country paid the high price of his ticket as part of its satellite launch package.

With only three space shuttles in the fleet until the early 1990s, however, and so much work to do in orbit, it is highly unlikely that NASA will sell shuttle tickets to the rich or the rich and famous in the next decade. Perhaps as the second generation of shuttle ships goes into service early in the next century, one of the aging shuttle orbiters will be outfitted with a passenger module and will carry paying customers to orbit. As many as

SPACE STATION CALENDAR

Date	Event
1984	United States initiates space station program.
1985	Canada, Europe, and Japan join United States in space effort.
1986	Basic Dual Keel design chosen.
1986	Russia launches *Mir* (Peace) Space Station.
1989–90	Space Industries, Inc., launches world's first processing facility in space.
1990	Soviets have a permanently manned space station.
1992	"Space-breaking" ceremonies for U.S. space station.
1992–94	First components of space station launched on space shuttle.
1994–95	Space station assembled in space.
1996–98	Space station fully operational.
2005-30	Second-generation space stations, including one that would orbit halfway between moon and earth.

350 people could be packed in such a modified shuttle spaceliner, which might be named the *Orbital Express*. Still, as long as the mode of transportation is the space shuttle, both the cost and the amount of space required to take a person to earth orbit will be prohibitive.

The ET Express

True space tourism, attracting large numbers of people, must await the development of a new generation of rockets or spaceplanes that will get costs down to the range of 10 to 30 U.S. dollars for every pound (0.45 kilogram) put into orbit. If a person and his luggage weigh 200 pounds (90.7 kilograms), then his ticket, under these ideal twenty-first-century conditions, might cost between 2,000 and 4,000 U.S. (1985) dollars. That opens up space for tens of thousands of people. It will be the beginning if a new era in space, one that will eventually make space ships as common as airplanes or automobiles. Spaceships will be used by everybody, and your children (perhaps even you, if you're under thirty) will take vacations in space.

Barron Hilton, the hotel magnate, has publicly stated that he will build a hotel in earth orbit once the transport costs of getting people there are reasonable—within the range just mentioned. When this happens the same people who frequently fly the Concorde will fly to orbit and vacation there. One study done by the Hudson Institute indicates that tourism above earth will someday become the biggest of all space industries, catering to people who are looking for *the* different vacation. Such orbiting resorts will depend on the advanced technologies and skills that were learned putting the world's first large space station in orbit during the last decade of the twentieth century.

If Mr. Hilton gets his way, there will be a Hilton on High. And since variety and choice are important to vacationers who are getting away from it all, the classic space station wheel or double wheel may finally get built, providing weightlessness in its central hub, partial gravity in its connecting spokes, and gravity simulating the earth or the moon in its outer areas. Novel recreations such as zero-gravity swimming pools and acrobatics, described in other chapters of this book, as well as old activities with new rules (such as handball in one-sixth gravity) will both delight and befuddle the tourists at the High Hilton. And honeymooners in their weightless suites will experience sex of a different kind. Perhaps the tempo of "The Blue Danube" waltz will be appropriate as they make love in slow, weightless motion high above the earth.

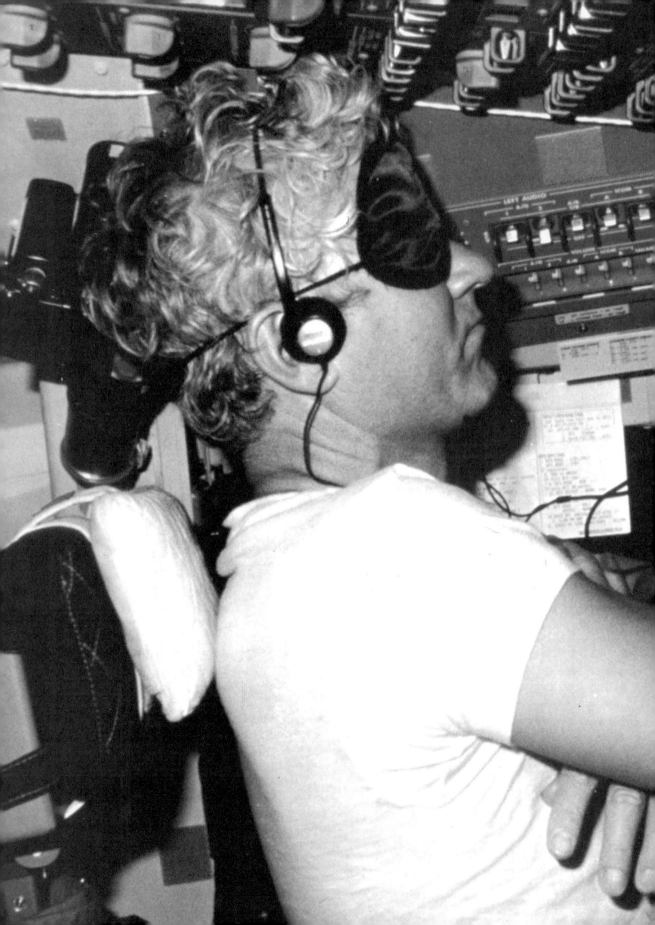

4

Daily Life In Space

The Wonders of Weightlessness

When television images of the astronauts are beamed down to earth from orbit, they usually show the fun side of living in zero gravity, where the human body weighs next to nothing. We see the astronauts floating and performing zero-g acrobatics; spoons and forks twirling above a food tray; perfect spheres of water or orange juice floating before an astronaut's eyes; a headstand on a fellow astronaut's finger; a walk across the ceiling. Floating, moving about, working, eating, sleeping, going to the bathroom— all these activities are done without gravity. Weightlessness is the single most unusual and ubiquitous reality of living in space. Just as gravity plays its frustrating tricks on us earthbound folk from time to time, so, too, does its absence play its tricks on novice astronauts. Living without gravity is both challenging and exhilarating.

"It's a marvelous feeling of power over space—over the space around one," said Joe Kerwin of Skylab 2. But this comment is from an astronaut who for the first time had plenty of space in which to live and work in the U.S. space program—Skylab had as much space as a three–bedroom house. Mercury, Gemini, and Apollo astronauts had much less room around them in which to experience such freedom. Their responses to weightlessness were nonetheless very positive.

America's first astronaut in orbit, John Glenn, even though he was squeezed into a small Mercury capsule, found weightlessness "extremely pleasant."

"You feel absolutely free," he said. "A person could probably become addicted to it without any trouble. I know I could." And his other "right-stuff" colleagues agreed, Alan Shepard referred to the experience as "pleasant and relaxing," and when Scott Carpenter arrived in orbit aboard his *Aurora 7* in May 1962, he exulted,

ABOVE: *One of the supreme joys of spaceflight (so far) is weightlessness. Here the crew of* Columbia's *first operational flight in November 1982 shows that the human body can assume any position.* Courtesy NASA. PRECEEDING PAGES: *Henry Hartsfield, commander of the last completed* Challenger *mission, in late 1985, takes a snooze in his position on the flight deck.* Courtesy NASA.

"I am weightless!" All of them loved the fact that they could feel no pressure against their bodies. Many things were easier. Not enough hands at a particular moment? Well then, just let the item you were holding float in front of you and free up a hand for something else.

Astronaut reactions to zero gravity have changed, however, as their spacecraft have grown in size. Early astronauts, because of their cramped quarters, had no room to learn the new techniques of moving through zero gravity. It may look easy when we see an astronaut "swim" from one location to another in a weightless cockpit, wardroom, or workshop of the space shuttle, but he or she has learned to do this. A space rookie on a shuttle mission, for instance, may take anywhere from several hours to a few days to learn the tricks of zero-g mobility.

middle space with no way to reach a bulkhead without the help of a spacemate. But within in a few days, they gain their space legs and demonstrate the zero-g ballet like their veteran colleagues.

If it were just this challenge of learning how to take these baby steps in space with grace and timing, that would be one thing. But as spaceflight missions became longer and the spacecraft roomier, the side effects of zero-g living became better known. When the human body experiences weightlessness it undergoes changes as profound as those a pregnant woman experiences and the side effects for many astronauts are less than pleasant. Read between the lines of what the ultimate space rookie, Senator Jake Garn, said after his mission on space shuttle *Discovery* in the spring of 1985.

TOP LEFT: *Living without gravity is both exhilarating and challenging. On a* Challenger *flight in 1983, astronauts Richard Truly and Guion Bluford catch a few zero-g winks. Who is right-side-up or down, as the case may be?* ABOVE: *The tangle of teleprinter paper in zero gravity. Astronauts Karol Bobko and Donald Williams are about to wrestle this paper snake into a storage area during their* Discovery *mission in April 1985.* Courtesy NASA.

"Especially in the first hour or two," says Sally Ride, "you feel very much out of control, very inefficient." But, she continues, "Once you adapt, it's fun. It's a completely new environment, and you've got a week to explore a different experience."

For many first-time astronauts, even with thorough training on earth, motion becomes a challenge at first, and many movements are miscalculated. Intended destinations are missed because the pushoffs are often too energetic. These neophytes are experiencing the baby steps of the Space Age, and they take their tumbles by bouncing off bulkheads, knocking loose stowed equipment, or getting stuck in some

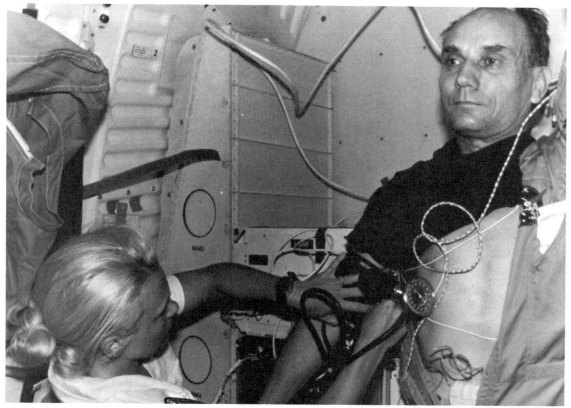

U.S. senator Jake Garn, like 50 percent of all astronauts who fly, experienced space sickness during his 1985 mission on Discovery. *Here Dr. Rhea Seddon applies a blood pressure cuff as part of the senator's participation in medical testing.* Courtesy NASA.

Space Sickness

When Senator Jake Garn returned to earth from his voyage on the space shuttle, he admitted that spaceflight was not for everyone. He told the press that would-be astronauts should have some aircraft flight experience as a prerequisite to living and working on the space shuttle. He also admitted that he had become physically sick during the spaceflight. Those of us who saw photos of him in orbit did not have to be told because Senator Garn *looked* sick.

Space sickness, called space adaptation syndrome by the experts, affects about half of all the astronauts who fly in space. It includes a complex of symptoms, and each astronaut reacts differently. One of the most unpleasant is nausea, which can occur each time an astronaut

turns or moves his or her head. Until the body is adapted to the zero-g environment, such basic movements often produce a strong sensation of vertigo that results in nausea. Some astronauts vomit; others just feel sick. Generally the symptoms disappear after about three days, but this is a good portion of a seven day flight, and that is why doctors and researchers are making a concerted effort to find answers and develop medication to alleviate the symptoms.

Senator Garn's primary responsibility while in orbit around the earth was to be a test subject for researchers who are trying to find a treatment for this malaise that can and has affected flight operations. The first test flight of the Lunar Excursion Module in earth orbit during *Apollo 9* had to be delayed because astronaut Rusty Schweickart was sick, plagued with nausea and vomiting.

The first politician to fly in space, Senator Garn was tested immediately after the *Discovery* reached orbit, when his body underwent several changes in response to zero-g. He also wore microphones around his lower abdomen which recorded the sounds of solids and gas moving through his bowels. Other tests included measurement of pupil size, heart and brain electrical waves, temperature, and blood pressure during reentry. Obviously, high status in Washington, D.C., is non–transferable to earth orbit. While Garn was up in *Discovery*, a joke memo was circulated at the Johnson Space Center: It offered to any astronaut the chance to become a U.S. senator, and it required only eight weeks of training.

A Cure for Space Sickness?

The profound changes that the human body undergoes during spaceflight have been recorded and analyzed for more that twenty-five years. With each spaceflight the list grows longer, and more sensors and sophisticated instruments will be probing the bodies of an increasing number of astronauts who will be flying in these last two decades of the twentieth century.

Entire space shuttle missions are in fact devoted to life-science studies. The body's metamorphosis in zero gravity includes hundreds of changes, even thousands if we go to the micro-level of human physiology. But the extreme complexity of the human organism means that many of these changes remain a mystery and scientists are not quite sure why the body responds as it does to the weightless environment. In many areas of space medicine, the experts are still seeking answers. Although they continue to narrow down the search, no cure has yet been found for space sickness and half of all astronauts still get sick during the first few days of their flight.

One of the major changes that takes place in the weightless human body is that blood shifts from the legs and lower part of the body to the upper torso and head. This blood shift is mistakenly interpreted by the body receptors as an increase in blood volume, and the body responds by increasing the urine flow and reducing the feeling of thirst. Astronauts therefore must take care not to become dehydrated.

Russian scientists contend that space sickness is caused by this fluid shift, but some U.S. space doctors believe it results from a form of motion sickness originating in the equilibrium sensors of the inner ear and a confusion in the eye-brain response to the fact that there is no definite up or down in space. Astronauts report that the condition worsens when they suddenly view the earth in a disoriented position.

Even though the body adapts after a few days, probably by repatterning the central memory network that has been conditioned over a lifetime on earth, the hope is to discover drugs that will suppress the symptoms so that future astronauts can spend more time working in and enjoying zero-g rather than getting sick in it.

Looking Not-So-Good in Space

Because of the redistribution of blood in the body, the physical appearance of astronauts changes in space, so much so for some that they don't even recognize themselves in the mirror! Here is a list of some of these less-than-flattering cosmic resculpturings.

- The face becomes puffy, the eyes bloodshot, and bags appear under the eyes.
- Veins in forehead and neck swell.
- Facial tissue shifts upward.
- Thighs and calves of legs shrink in size because of less blood.
- Body height increases because of spinal lengthening caused by fluid between discs.
- Internal organs shift upward, changing appearance of the waist.
- Breasts on female astronauts rise up.
- Penises float upward rather than hang.
- Long hair floats upward.
- Skin drys out and produces chapped hands and lips.

And these changes are just the obvious, external ones. Think of the psychological impact of not being able to recognize yourself in the mirror because of your puffed up, tissue-lifted face with bloodshot eyes. And think of what's going on inside those zero-g bodies? It's certainly more than what the hundreds of experts in space medicine now understand. But they are finding more answers each time a space shuttle or a Salyut goes into orbit, and when medical testing is done on or above the earth, such as NASA's testing of world-class gymnasts in 1985 to see if they were less susceptible to space adaptation syndrome because of their gymnastic training.

"It's great!" said Olympic medalist Kathy Johnson, after her thirty seconds of zero-g acrobatics aboard a KC-135 jet transport that was in a steep dive. During the short periods of weightlessness that such jet flights provided, the gymnasts felt right at home and looked forward to the day the Olympics would have a zero-gravity sporting event.

Inside the Weightless Body

Just as gravity affects the microphysiology of the human body, so too does the lack of it. The delicate balances are upset in zero gravity, and the body attempts to put them right again. Here are some of the internal body changes that occur in weightlessness:

• Loss of blood volume and red blood cells. After sixty days the decrease in red blood cells levels off.
• Increase in white blood cells.
• Decrease in heart size and blood output.
• Increase in the amount of fluid absorbed by internal organs.
• Loss of bone mass and calcium. Long Russian manned spaceflights indicate that this stops after several months.
• Loss of muscle mass, strength, and reflexes, especially in the large muscles of the legs. Little or no change in arms and chest. Exercise helps to slow this process.

• Head congestion, which can include blockage in the nasal, sinus, and ear passages. This in turn can cause moderate or severe headaches. The cause: more blood and fluid in the head. Exercise and eating can sometimes relieve that problem.

With all these bodily upsets and readjustments in zero-g, the accomplishments and record-breaking coups of the astronauts seem all the more amazing.

The Weightless Senses

The complex of body changes and the spaceship's environment also affect the way astronauts see, hear, taste, smell, and touch, which in turn influences every aspect of their daily lives in space.

• Vision of objects is enhanced, probably caused by lack of light scattering in the vacuum of space. Older astronauts have less ability to focus on nearby objects.

• Hearing on Skylab was limited to about 25 feet (7.6 meters) because the cabin atmosphere was about one-third as dense as on the earth's surface. This is less of a problem on the shuttle because the atmospheric pressure is higher.

• Changes in taste and smell occur, but they vary with each individual. Stuffed-up heads no doubt cause some of the bland taste problems and lack of smell that astronauts have complained about. Spicy condiments helped the taste problem. Russian research indicates that taste sensitivity is lost because of endocrine and metabolic changes.

• Changes in the sense of touch are very subtle, but it can be influenced by a dry atmosphere that produces chapped fingers. Coordination tasks, however, are not influenced by weightlessness. Skylab astronauts held mechanical pencils to write. It was the downward force needed to keep the writing hand on the table that required effort, not holding the pencil.

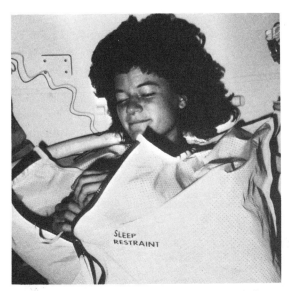

Weightless sleep is very different from earthbound slumber. If you don't anchor yourself down in some way, you'll drift off and bump into things. Sally Ride (top) sleeps during her historic mission in June 1983.

The Space Blues

Feeling the blues in space is a common experience for astronauts, especially during long flights. William Pogue, a Skylab astronaut, refers to this down-and-out feeling as "space crud." It has been described as a general fatigue and run-down feeling, and it occurs about three or four hours after eating. Pogue compares these zero-g blues to the kind of feeling you have when you're coming down with the flu.

The symptoms, it was soon learned, were relieved by eating, and the astronauts made every effort not to skip a meal, no matter how pressing their work schedules were. If they did not eat, those space crud feelings would come on and they were more apt to make mistakes. The one exception to this rule was during space walks, when the astronauts would go without eating or drinking for six or seven hours and still would not get the space blues. There has been no definitive explanation for this. One speculation is that the all-encompassing exhilaration of freely floating high above the earth kept the symptoms at bay.

Sleeping and Dreaming in Zero-G

On earth, a person is said to "nod off to sleep" when he or she falls asleep while reading in an all-too-comfortable easy chair. In space, an astronaut nods continuously during sleep. Why is this? This constant head nod during weightless sleep is caused by the pulse of blood flowing through the large arteries in the neck that feed the brain and head. While most astronauts notice the head nod, about 10 percent of them do more than that: They blame it for their symptoms of nausea.

In general, space folk require less sleep, probably because their bodies are not working as hard in weightlessness. In the Skylab crews, six hours of sleep was the average. Sleepers in zero gravity, however, spend different amounts of time at certain sleep levels (such as REM, rapid eye movement, periods) than they do on earth.

The heart also slows down significantly during space sleep. It is not unusual for an astronaut's heart to beat only thirty times a minute while sleeping in zero gravity.

Sleeping accommodations for the Skylab and space shuttle crews were a tremendous improvement over the cramped quarters of the Apollo command ships and lunar modules. The Apollo crews averaged much less sleep than the astronauts of these later manned space programs. Except for the lone astronaut in the command module during the lunar landing portion of each mission, there was no privacy. In Skylab, however, each astronaut had his own small private bedroom, where a sleeping restraint bag was hung vertically against the wall. Once zipped into the bag, the astronaut was secure and could not float away during sleep. A Skylab astronaut once attempted to sleep without the bag. He became just a feather moving about on the ventilation currents, however, and kept bumping into things.

The space shuttle has four sleep stations on mid-deck, and crew members take shifts sleeping. Each sleep station has a restraining sleeping bag that is attached to a padded board, which takes the place of a mattress when the astronaut is securely fastened to it. If he or she has trouble sleeping, and this is a common problem in weightlessness, an astronaut can choose to use earplugs and a sleeping mask.

Some space shuttle astronauts have been known to get their winks in any quiet corner of the spaceship. All they have to do is hook themselves on to a solid fixture, put on their sleeping masks and insert their earplugs if they wish, and go to slumberland.

There are several advantages to weightless sleep. Turning and shifting on an earthbound bed helps to spread out the pressure of gravity over various portions of the body and keep up good circulation. There's no reason for the body to accommodate gravity when there is none, and so there are fewer body shifts during sleep. Also, there is no snoring during space sleep because the soft palate floats and does not hang down in the air flow as it does when someone sleeps on his or her back in an earthbound bed.

And what of weightless dreams? Do the dreams slow down in proportion to the slower body motions necessary while living in space?

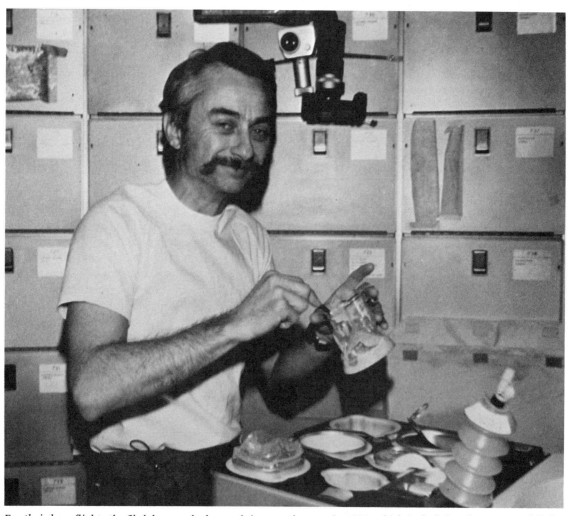

For their long flights the Skylab crews had a much improved menu, in 1973, which included such gustatory delights as prime rib and ice cream from the first in-space freezer. They had seventy-two food items from which to choose. Here astronaut Owen Garriott stirs up a meal. Courtesy NASA.

Sleep studies during the Skylab missions show that the REM period of sleep decreases slightly during spaceflight. But what was even more surprising was that the REM period increased significantly after the astronauts had returned to earth. Most dreams occur during these REM periods, so the postflight increase could indicate some kind of dream compensation time, although this is in no way conclusive. One Skylab astronaut reported that it was several weeks before he could recall a zero-g dream. But then, in a real sense, all astronauts are living their dreams.

Taste Buds in Orbit

Back in the early space days, no one knew for sure if astronauts could even eat in space. Would food be hard to swallow without the help of gravity? Would food collect in the throat and not get into the stomach? John Glenn's flight in *Friendship 7* put these concerns to rest. Eating in space proved easy for John Glenn, even if it wasn't all that tasty. The challenge for astronauts is to get the food into the mouth without pieces or portions of it flying off and floating around in the spacecraft's interior.

It was during the manned Gemini program that food began to improve. The toothpaste-shaped aluminum tubes of the Mercury days, which were filled with semi-liquid foods, were replaced with improved plastic containers filled with freeze-dried foods that were rehydrated by injecting water into the pack with a water gun. And the crumbling bite-size food chunks of the Mercury days were coated with gelatin to reduce the crumbs in the cockpit and the time it took the crew to snatch them out of zero-g so that their instruments wouldn't get fouled up. There was also an improved menu selection for the Gemini flights. On the longer flights, it wasn't necessary to repeat a meal for four days. The very first shrimp cocktail in space was eaten on a Gemini mission.

The food improvements of Gemini were carried on in the Apollo missions, and the moon-bound crews had a much larger selection of foods on their menu, as well as heated water for the first time in space, with which they could prepare hot drinks and hot foods. The hot water was 154 degrees F. (68 degrees C.), and this made it easier to rehydrate food quickly and thoroughly. Food tasted better as a result. The first two men on the moon, Armstrong and Aldrin, were blessed with hot coffee. They also ate bacon squares, frankfurters, canned peaches, and sugar cookies.

Another Apollo dietary innovation was the so-called spoon-bowl package. Once the food was rehydrated, a plastic zipper was opened, allowing the food to be removed with the spoon. The food's moisture caused it to cling to the spoon. The eater had to be sure not to move too quickly, however, or the food would float off the utensil and perhaps end up in somebody's eye. It was best to eat in slow motion.

Dining in the Skylab Room

The Skylab space station was the ultimate in dining pleasure during the first decade of manned spaceflight—which is not to say that the food didn't get its share of complaints, because it did. Still, no earthbound restaurant could compete with the view, which overlooked the *entire* Atlantic or any one of several continents. And because there were fewer weight and space restrictions than on the Apollo missions, galley conveniences such as a food freezer were flown in space for the first time. This provided gustatory delights such as prime rib and ice cream. Skylab astronauts could choose from seventy-two different food items.

The three Skylab crews also had a dining room table around which they could float and eat, and their menus offered a wider variety of food than on earlier missions. Astronaut William Pogue of the last Skylab crew enjoyed the hot scrambled eggs but found the chili disappointing because the oil had separated from the sauce and the meat and they could not be satisfactorily recombined. Members of the third and last Skylab crew benefited from the complaints of the earlier crews, who reported that most of the food was bland-tasting and boring, no doubt a result of their head congestion caused by the blood redistribution in weightlessness. One exception was the German potato salad, with its vinegar and onions, which was very popular with the first Skylab crew; all four cans on board were gobbled up.

Because food is so very important on long-duration spaceflights, just as it is on isolated military stations such as nuclear submarines, NASA made every attempt to initiate change in response to the complaints. A variety of spicy condiments was therefore brought aboard the third mission, and this helped perk up the taste buds. The crew especially enjoyed the hot sauce and liquid pepper.

To prepare a meal on Skylab, the galley-duty astronaut took out the food items on the day's menu, added water to the dehydrated foods, and then secured the containers in a special tray that had heating elements. Presto! The ultimate in convenience food.

The crews floated around the wardroom table for meals. There were parallel bars under the food trays designed to serve as space chairs, but the astronauts usually preferred to stand because attempting to sit in zero-g by bending at the waist placed a strain on the stomach muscles.

Eating without gravity had other problems besides bland taste, food tins and silverware floating off the trays, and bits of food wandering off from a bite. The bubbles in the pressurized water supply had no way of floating to the surface. Sometimes these gas bubbles could upset a meal. When water was injected into a food pack to rehydrate it, the plastic bags would burst and splatter food around the wardroom, making another mess to clean up. A mess in space, astronauts all agree, is worse than a mess on earth because it's more difficult to clean up.

The gas bubbles didn't just burst plastic food bags either; they also played havoc with the astronauts' innards. The bubbles caused an uncomfortable pressure to build up in the stomach, but food in the stomach does not settle in zero-g; instead it coats the stomach in a more uniform way. For this reason, a burp or belch in space can have a very unpleasant consequence: regur-

gitation. Astronauts therefore avoided burping or belching whenever possible and often accepted the discomfort of gas pains. The other way of getting rid of gas wasn't much better. As William Pogue said, "Farting about five hundred times a day is not a good way to go."

Making space food more tasty, as well as nutritious, is a continuing challenge for the zero-g food experts. It is important to the psychological well-being of astronauts, especially those future men and women who will make the longer space treks to Mars and beyond. And while the evolution of space food and its preparation took a giant leap during the Skylab missions, more culinary refinements and innovations have come out of space shuttle development and the first two dozen missions of the Space Transportation System.

The Shuttle's Automat

Space nutrition has come of age with the space shuttle missions. The daily menu is designed to replenish all the calories burned off during the hard work of space duty—three thousand calories, not including snacks. The food also contains ample amounts of potassium, calcium, nitrogen, and other minerals because space doctors know that the body tends to lose essential minerals in zero gravity. A diet of the right stuff will work against the known problems of weightlessness such as loss of muscle tone and bone mass, poor disposition, and the lack of ability to concentrate. The foods are also designed for easy digestion and have a minimum of roughage. No one wants heartburn in orbit.

There are more than one hundred food items on the shuttle menu, including twenty beverages and seven condiments (one of which is Skylab 3's favorite hot pepper sauce) to give the food some extra zip. Shuttle astronauts and mission specialists have a varied menu for six days before they have to repeat a meal. Beside three meals a day, there are plenty of snack foods to satisfy the munchies, some of which are natural foods. The snack foods include graham crackers, pecan cookies, almond crunch bars, nuts, hard candy such as Life Savers®, and gum. Besides coffee and tea, the beverage list includes tropical punch, lemonade, orange drink, orange-grapefruit and orange-pineapple drinks, strawberry drink, apple drink, and other choices. Even Coca-Cola and Pepsi, in specially designed zero-gravity cans, were tested on a space shuttle *Challenger* flight in the summer of 1985. The cola wars went even to the high frontier.

A sample supper menu on the space shuttle includes: cream of mushroom soup; smoked turkey; mixed Italian vegetables; vanilla pudding; tropical punch; and strawberries. The freeze-dried strawberries remain full size and have their color and texture intact. They can be rehydrated at the galley by adding water, or, if the astronaut prefers, a strawberry can be rehydrated with saliva in the mouth.

The two space meal items not found around an earthbound dinner table are special straws and surgical scissors. It would be impossible to drink from a glass in weightlessness; even if it were turned upside down, the liquid would remain in the glass. That's why there are dozens of straws that have a special closure clip to cut off the flow of liquid once it begins. If the straws didn't have this closure, the liquid would continue to flow after the mouth left the straw until the container was empty. This would make for shimmering spheres of juice, tea, or coffee floating around the cabin, which could foul up gear and faces.

Many of the food portions come in hard plastic containers that fit into the food trays and are held there by friction. The small surgical scissors, usually kept in a leg pocket, are used to cut open the plastic tops of the containers so that the usual eating utensils—spoon, fork, and knife—can get to the food. With proper care, food, much of which is served in a gravy or a thick sauce, stays on the spoon or fork because of surface tension. A spoon, for example, can be held upside down and still do its job of carrying food to the mouth. In the early days of the Space Age, many experts thought that any food in an open container would float away. This proved not to be the case, and they were happy to be wrong.

The shuttle's high-tech modular galley located on mid-deck has been compared to a big dispensing machine, but it's actually extremely compact, considering everything it contains. The galley dispenses hot and cold water, has a forced air convection oven to heat foods to temperatures as high as 180 degrees F. (82 degrees C.), and has a pantry in its top section. The food serving trays are stored in a compartment next to the oven, and the galley also contains some auxiliary equipment storage areas. Unlike Skylab, the space shuttle does not have a good freezer or refrigerator, but the space station of the 1990s will no doubt have room for these earthbound kitchen fixtures. On one side, the galley also features a type of space sink for washing the hands before meals. NASA calls it the personal hygiene station, and it includes a bubblelike washbasin with two holes for the hands, a soap dispenser, and a water dispenser. The sink is enclosed except for the openings for the hands, and it has an air flow to control the water so it can't escape.

While the shuttle galley may give the appearance of a contemporary automat, it still must be tended by the astronaut who has pulled kitchen duty for any particular day. Preparing a meal for a full crew of seven or eight takes no longer than thirty minutes, although meals are usually eaten in two shifts. Everything is prepackaged, so the chef simply gets out the pouches marked with the meal and day, adds water to those items that must be rehydrated, and pops them in the oven. As the food heats, water is added to the drink containers and the special straws are inserted. Then it's simply a matter of assembling the meal—the heated wet packs, the cans, the foil packs, and the drinks—on the food trays. It's mealtime above the earth.

Space food has come a long way since John Glenn squeezed some applesauce in his mouth during the three-orbit flight of *Friendship 7*. There will be new meal items going up on the shuttle in the next few years, and the space station will no doubt have an even greater menu selection in a galley that's more fully automated. In the 1990s, the menu will be truly international because of the participation of Europe and Japan. And if the French have their say, wine will finally be served in orbit.

Keeping the Weightless Body Trim

The two most significant body changes in zero gravity are the shift of body fluids to the head and the loss of weight forces on the bones and muscles. As the human body adapts to zero gravity, it dramatically becomes weaker in many ways, including loss of muscle and bone mass. Even the heart becomes smaller and its muscles weaker. Both the American and Russian manned spaceflight programs used exercise routines to counter this physical deconditioning, and it soon became apparent that astronauts and cosmonauts who exercised rigorously on a regular basis could adapt much better to earth gravity after their missions. Their postflight condition was directly linked to the amount of exercise they did in orbit.

Exercise programs came into their own during the Skylab missions of the early 1970s. The main reason for this is that the Skylab space station had room enough for astronauts to exercise on specially designed equipment. And exercise they did, especially the last two Skylab crews. Time spent exercising went from half an hour a day on Skylab 2 to one hour on Skylab 3 to one and a half hours on the final Skylab 4 mission of almost three months. All three Skylab crews had a stationary bicyclelike device called an ergometer on which they worked out, and the last crew also had the benefit of a unique treadmill device. It provided a means of more vigorous exercise.

The Russians also increased the exercise time for their cosmonauts. During the *Soyuz 9* mission of eighteen days in 1970, the cosmonauts exercised twice a day for one hour each time. On the longer mission of twenty-four days, their total daily exercise time went up to two and a half hours. After 1975, the Russian space exercise program was set; it involved three exercise periods each day for a total of two and a half hours of workout on a variety of equipment, including a treadmill device.

The primary exercise machine for the Skylab crews was the ergometer, a bicyclelike apparatus that can be pedaled with either the feet or

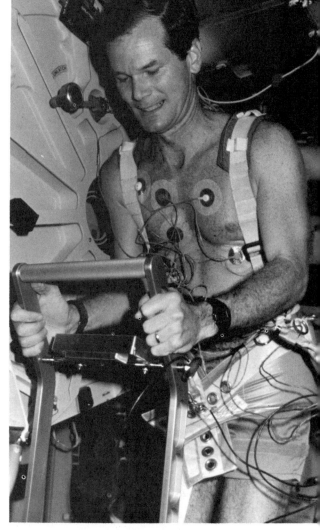

U.S. congressman Bill Nelson is all wired up for the first stress test in zero gravity during his Columbia *flight in January 1986. Could this set a precedent for all elected officials? Courtesy NASA.*

the hands. Each time an astronaut exercised on the ergometer, he was fully wired with electrodes and blood pressure cuffs, and breathed into a mouth piece and tube. The astronauts weren't just exercising; they were also subjects of an elaborate medical test.

When using the bicycle ergometer, the astronauts locked their shoes into its pedals, but they still tended to float off the seat, so they rigged up a pad mounted against the ceiling to counteract the force of pushing down on the pedals. On the first Skylab mission, the commander, Pete Conrad, once bicycled for ninety minutes, enough time for a full orbit around the earth. With such workouts, sweat built up on the astronauts' backs, and sweat in zero-g does not drip off the body. Instead it became a big puddle, as astronaut William Pogue described it, "as large as a dinner plate and about a quarter of an inch deep. It just sort of slithered around," he said. When the exercise session was over, an astronaut had to move slowly to get a towel to mop up the puddle on his back, otherwise this large sweat glob could fly off and make a real mess.

Treadmill exercisers also were flown on both the American and the Russian manned spaceflights. These were used to exercise the lower limbs more strenuously than the bicycle ergometer could. And the Russians used a so-called penguin suit, a special elasticized garment that opposes any movement to a degree and therefore partially compensates for the lack of gravity. Space doctors have found out that certain types of space exercises are good for different parts of the body. No single exercise can benefit the entire body and help reverse the general deconditioning that zero gravity causes. While several other exercise devices, such as the common chest expander and the stretcher known to the earthbound who work out, were also taken into orbit, they turned out not to be that effective because they influenced only a limited number of muscles.

The Skylab crews also had a strange contraption called the lower body negative pressure experiment, which was designed to pull blood from the upper portions of the body down into the legs. While not strictly an exercise machine, in that the astronaut is passive and lets the machine do the work, it nevertheless puts stress on the body and moves the blood into the lower limbs. Climbing into this big barrel and sealing one's lower body off with a rubber skirt so that the suction could pull the blood down was not a favorite activity for the astronauts. Joe Kerwin, the science astronaut for Skylab 2, climbed into this contraption that looked like one half an iron lung one day and immediately broke out into a cold sweat and got dizzy. That was all he needed; he turned the machine off and got out. He confessed afterward that he felt he might be sucked out of the barrel along with the air that was vented into space to create the vacuum around the lower body.

Because human legs appear to be completely useless in outer space, but still require a lot of calories to continue nourishment, one doctor seriously proposed that amputees should be chosen for future space missions. "The ultimate fuel cost of legs on long missions," he said at a life sciences symposium, "must be really staggering."

Shaping Up on the Shuttle and the Space Station

The space shuttle has its own specially designed treadmill on which the astronauts exercise, but there is neither the room to have more elaborate exercise equipment, nor the need for it until future space shuttle flights are longer in duration.

The shuttle treadmill is a rather simple device consisting of an aluminum base, straps, and harnesses. A piece of non-slip Teflon is attached to the base for the feet, and four bungee cords, made of rubber, come up from the base and attach to a belt and a shoulder harness, which hold the astronaut down and mimic the force of gravity. Once the tension on the bungee cords is adjusted, the astronaut then runs in place. If he or she wants to exercise harder, the tensions can be increased with a further adjustment. Depending on the duration of the shuttle mission, fifteen to thirty minutes of exercise each day is normal. Music and earth-watching are often enjoyed by the shuttle crew while they exercise.

The space station of the 1990s will have much more living space than the shuttle that will dock with it, and a full module could eventually be devoted to a workout room. For the scientists and astronauts who man the station for longer than a month, a rigorous exercise program will be essential to counterbalance the physical metamorphosis that the human body undergoes in zero gravity.

Sex in Space

When *Omni* magazine polled its readership about the future in 1984, with its second Delphic Poll, one of the questions asked was, when would the first baby be born in space? The majority of participants predicted that a child would be born in space by the year 2040. Only 17 percent thought that it could happen by the year 2000. Space expert G. Harry Stine thinks that the majority was too conservative in this case. He predicted that a child would most likely be born in space by 2000. "Any time we have a man and a woman in space in the right circumstances," he suggested, "it could happen."

But what are the right circumstances? Certainly having men and women together in space is an obvious essential, and space shuttle missions have been flying men and women together ever since Sally Ride went up in 1983. That takes care of one major prerequisite. Then there is the question of privacy and leisure time. When the space shuttle carries a full crew of seven or eight astronauts and missions specialists, where would the couple find any privacy? Today's space people are hard at work, don't forget. Of course if a Spacelab mission devoted to life sciences had a long module in its cargo bay, the room would be there, but in all likelihood the rest of the crew would have to know the mission was in part a honeymoon special. And this would imply that NASA would approve of this couple's sexual encounter in zero gravity and have some time for it written into the flight plan. This is very doubtful, and under these circumstances, it might be hard to find any consenting adults who would want their sexual activity put before the public.

It's much more likely that sexual encounters in space will take place in one of the space station modules in the 1990s. Besides the large amount of living space in the station compared with earlier manned spacecraft, there will also be more people living and working in orbit, as many as twelve at one time. But even with the right couple, the right place, and the right time, zero-g sex still has to contend with the dramatic physical changes that occur in the human body during weightless living. And what of the celestial mechanics of weightless lovemaking? How will two bodies remain intertwined in the strange environment without gravity? Newton's laws of motion still apply: Every action is opposed by an equal and opposite reaction. How will the lovers remain coupled?

One does well to remember how exercising in orbit is accomplished. On the shuttle treadmill, for instance, rubber bungee cords create some artificial gravity and hold the runner to the treadmill base. Future space lovers will probably wear some type of elastic belt to keep them from flying away from one another during intercourse. Just as several activities in weightlessness—eating and moving about, for example— are best done in slow motion, in a similar fashion sex in space may require slow-motion passion and not the more vigorous sort. Nice and easy does it may become the quintessential phrase for space sex.

Learning new sexual motions and techniques for zero gravity may be challenging at first, but it certainly won't thwart human desire and the physical side of love. It will, after all, be fun learning. Indeed, weightless lovemaking will probably renew many longtime, earthbound sexual relationships.

What may turn out to be the real sexual hang-up in zero gravity is the human body itself and all its adaptive changes, such as the blood shift to the upper portions of the body. Will there be enough blood left below the belt for good sex? And when it comes to procreation on high, what effect will zero gravity have on the sperm and eggs? Human blood cells change shape and number in orbit; sex cells will also be influenced in certain ways. And the menstrual cycle and ovulation in zero gravity will also undergo profound changes, as will the generation of sperm. Hormonal changes in men and women occur, but how the sex hormones are influenced is not known at this time. Much experimentation and research will be required.

But let us assume that love and desire conquer all in weightlessness as they often do on earth. The two floating bodies will have a freedom of movement unequaled on earth, and once the zero-gravity sexual techniques are mastered, there will be a whole new realm of sexual pleasures to enjoy. In the next few decades there will no doubt be a best-selling book on the joys of sex in space. For the first time since sex advice began appearing in print or on video, there will be some truly new sex positions offered and described to lovers.

Spacewear Fashions

The shirtsleeve atmosphere of today's spacecraft is a far cry from the bulky and restraining spacegear of the early Space Age, when the pressurized suits had up to fifteen layers of material. These early space suits, like the ones Alan Shepard and John Glenn wore on their missions, made motion difficult. Even the fingers were forced out straight by the pressure, and a simple chore such as grasping an object often proved difficult.

Space walks and emergencies are the only situations in which space suits must be donned today. The space shuttle's atmospheric pressure is equal to the earth's, and consequently, only a lightweight flight suit is worn, even during lift-off and reentry.

Men and women wear the same soft cotton, cobalt blue jacket and pants. The lined zipper jacket has many pockets, as do the matching pants. These pockets are prepacked on earth and contain such useful items as pressurized ballpoint pens and mechanical pencils, sunglasses, data books, surgical scissors for opening up the food packs, and a Swiss Army pocketknife. The routine habit of replacing these items in the right pocket saves the hassle of chasing them down or trying to find them floating around in the cabin.

A navy blue, short-sleeve shirt is worn under the cobalt blue jacket. It is a cotton knit with white buttons, and the NASA and flight patches are sewn on. Contemporary spacewear is functional and comfortable, not too tight and not too loose (loose clothing has been known to snag switches on control panels). All the clothing is fireproof. What we have today is a unisex fashion in orbit. The only different clothing article for men and women is the underwear. Women astronauts wear bras, obviously not for support in zero gravity, but to keep their breasts from moving around so much during normal flight duties.

The cobalt blue color is much more pleasing to the astronauts than the brown flight suits worn by the Skylab crews. The last Skylab astronauts complained a great deal about their spacewear. Edward G. Gibson, the science pilot, griped to ground control: "I just get tired of this darn brown! I feel like I've been drafted in the Army." He suggested that NASA supply different color T-shirts.

When the Skylab crews worked out or performed other strenuous activity, they often wore just their white t-shirts and briefs. If the temperature was cooler in an area of Skylab, then they put on their golden-brown jackets. But off-the-shelf underwear was often the uniform of the day during the Skylab missions. And why not? Comfort and utility are the essentials of spacewear design. The astronauts don't have to float at attention and pass a daily dress inspection.

Footwear, of course, doesn't receive much wear and tear in zero gravity. Sneakers are often worn by the crews on the Skylab missions, but so were just plain socks, which are also worn at certain times on shuttle flights. Special attachments have been developed for the shuttle footwear that elevate the heels and compensate for the natural toe-pointing position that the feet assume in zero gravity. Crew members attach these to their boots when they want to stand upright from the floor for certain activities. Two suction cups on each attachment anchor the feet to the floor, just as cleats were attached to the floor grids for Skylab astronauts to keep them in place for some tasks.

Diversity in space clothing will probably appear during the space station days of the 1990s. If the space station becomes a truly multinational effort as current international agreements suggest, the Europeans and Japanese will no doubt bring some of their own national fashions into orbit with them. These variations in dress will lead to a wider choice of spacewear for men and women, and above-earth fashion trends may someday influence those on earth. A Sally Ride designer jumpsuit anyone?

Seat Belts on the Toilet

Keeping clean in space is more difficult and time-consuming than it is on earth. None of the habitual motions that carry us through a day on

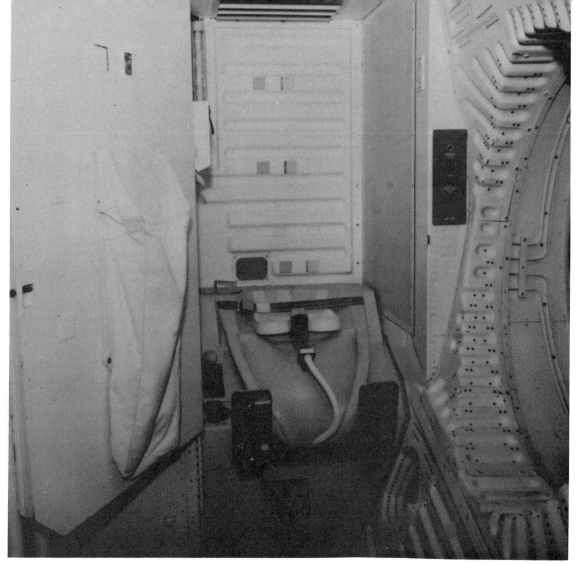

The space shuttle's unisex space potty (NASA calls it the waste collection system) has had its ups and downs. The challenge is to stay on the seat without floating off. A seat belt, foot restraints, and handholds help the astronauts remain seated. Courtesy NASA.

earth apply; zero gravity changes everything—the way hands and face are washed; the way teeth are brushed; and the way the bathroom is used. The most important rule seems to be: Take some time and do it right; mistakes can add up to frustrations that don't easily go away.

Just as the space shuttle's bathroom benefited from the experience of the Skylab missions, so the future space station's bathroom will benefit from the experience of the shuttle crews. The Skylab had the first shower contraption in space, which leaked water everywhere; valuable time was required to clean it up. Space shuttle crew members do not have a shower because of space limitations. They sponge them-

selves down instead; some things just never change. It is likely that the space station will offer the first adequate shower stall in space, but it won't dispense as much water as an on-earth shower. The reason for this is not just conservation of limited supplies, but the fact that less water is required because it clings to the body so well in weightlessness. And a current of air will flow around the body at the push of a button; this will force the floating water down the drain, so that an astronaut doesn't drown in it.

Skylab had the first space toilet to afford any privacy, because it had its own small compartment. NASA, without intending any humor whatsoever, called it the waste management compartment. The name has changed some-what for the space shuttle: Now the commode is called the waste collection system (WCS). Per-haps the word "system" helps to substantiate its $3 million development cost.

In Skylab's toilet, a funnel-shaped urine re-ceptacle collected the urine and a flow of air drew it into a collection bag, which was changed daily. For the solid waste, there was a com-mode-type seat mounted on the wall (it could have been mounted on the ceiling for that mat-ter and it wouldn't have made any difference in the zero-gravity environment). A seat belt held the astronauts to the toilet seat so that they wouldn't float off and create a mess. Such messes did escape from time to time, however, and caused some of the most unpleasant experi-ences of the Skylab missions.

This on-the-wall space potty also contained a bag, in this case a porous one, through which air flowed to settle the excrement. The bag was re-moved after each use and placed in a special chamber that dried and prepared it for its return to earth. That's right; it was all brought back to earth for analysis!

Space shuttle crew members are thankful that they can leave their bodily wastes in orbit, but they still have to strap themselves onto the seat and insert their boots into toeholds on foot re-straints. There are also handholds to grasp. The prospect of floating off the toilet seat hardly contributes to the relaxed privacy to which most people are accustomed. But there is a hatch win-dow through which the occupants can see the earth below. The designers will get no gripes about this one.

The shuttle toilet also uses the air flow of a fan to draw away the solid wastes and send them through the moving vanes of a unit called a slinger, which shreds them and then moves them into a compartment below, where they are dried and disinfected. What makes the shuttle toilet the first unisex bathroom fixture off the earth is a unique urine collection device that is mounted on the front of the toilet seat. Designed for both men and women, it also uses air flow to move the urine to a waste-water tank. The de-sign of this toilet feature even involved a group of women volunteers, whose urinary functions were photographed in detail.

The personal hygiene kit each shuttle astro-naut has in his or her locker contains all the ba-sics, such as toothpaste and toothbrush, comb and hairbrush, razor and shaving cream for the men, and nail clippers. On missions of one or two weeks which is what the shuttle will fly for several more years when new flights resume, the nail clippers will not get much use, because fingernails and toenails grow very slowly in space. Only on future missions of a month or more will the crew use the clippers.

Emotion and Behavior in Space

Now that the space station is being designed and outfitted for permanent occupancy in the 1990s, when as many as eight astronauts and scientists will live and work on the station for periods of up to three months, some scientists are becoming increasingly concerned about the psychic problems of extended space travel. It is generally agreed that the human factors of spaceflight have not received enough serious at-tention because of the tremendous engineering challenges of each new generation of spacecraft. As a result, the engineers and their approach to problems have dominated the space programs. As the Soviets continue to break their own world records for the longest spaceflights, with their cosmonauts remaining in orbit for seven months at a time, the space shuttle program, as amazing and versatile as it is, does not offer in-formation on the physical and mental welfare of astronauts on missions that last for several months. How can shuttle astronauts, on a mis-sion lasting a week or two, get bored? They can't; they've been too well trained and they're kept much too busy.

Skylab 4, the longest U.S. space mission, which lasted eight-four days, took place more than a decade ago—in 1974. Besides all its tech-nical accomplishments, the last and longest Skylab mission told the space experts some-thing very important: There is a limit to how much astronauts will cooperate with the flight controllers and scientists on the ground. The three reluctant and often irritable astronauts actually went on strike while in orbit, refusing for a short time to cooperate with what they believed to be unreasonable schedules and requests from the ground. Here were three humans—Gerald Carr, William Pogue, and Edward Gibson—who were in a metal test tube high above the earth for almost three months. Were their reactions indicative of future psychological problems that might occur in astronauts on long space journeys, be it in a space station above earth or on a spaceship heading to, or returning from, Mars?

About seven weeks into the Skylab 4 mission, after the three astronauts had fallen behind in an overly ambitious work schedule (this was, after all, the last Skylab mission) and had spent weeks unsuccessfully trying to catch up, commander Gerald Carr had it out with the people back on earth. The crew stopped working one day and did exactly what they wanted to do, and Carr had some straight talk with the ground.

"I'm also getting the feeling . . . the last few days that people there are beginning to hassle over who gets our time and how much of it . . . Where do we stand? What can we do if we're running behind and we need to get caught up? What can we do that's reasonable, and we'd like to be in on the conversation, and we'd like to have some straight words on just what the situation is right now. Commander out." Houston listened. The crew was assigned fewer experiments and was given more time to do them. But imagine what would have happened had NASA played it tough and uncompromising: a full-blown mutiny in orbit.

For the Americans, this Skylab experience was enough to wake people up to the fact that the psychology of space travel had to be seriously examined before long-duration missions were flown again. The tremendous physical changes that astronauts undergo, as well as the high stress of demanding duties and work deadlines, could jeopardize a future mission.

The Russian experience of *Salyut 7*, when three cosmonauts spent 237 days in orbit, mak-

ing it the longest manned spaceflight in history, emphasizes the need for intensified studies of spaceflight psychology. The diaries of the cosmonauts tell of the boredom and depression of long-duration spaceflights.

"Is it possible that some day I'll be back on earth among my loved ones, and everything will be all right?" wrote cosmonaut Valentin Lebedev after almost four months in orbit in 1982. In 1979, when two Russian cosmonauts eagerly awaited a rendezvous and docking with comrades during their 175-day mission above the earth, the docking was aborted because of a faulty engine firing. "We became incredibly depressed," wrote Valeri Ryumin in his diary on April 11, 1979. Before the docking abort, the two cosmonauts had longed for the company of their comrades so much that they meticulously followed every stage of preparation and launch. Then the terrible letdown of the mission abort. It was only the midpoint in their flight, and this has proved to be the toughest time for astronauts on long missions. The best way to fight off the gloom, Ryumin wrote in his diary, was to "work from morning till night so that we could not dwell on troublesome thoughts."

While the Soviet space program has a psychological support group, there is no such organization in the United States program. A NASA psychologist suggests that this is because the space agency managers have engineering minds and are not used to thinking about the astronauts in behavioral terms. The experience of the last Skylab mission bears this out.

Rigorous psychological screening must be initiated in choosing the space station astronauts. A technical genius who cannot work with people will remain earthbound. And if the space station becomes an international effort, which appears likely, then the cultural differences among the crew members could become troublesome. Intercultural sensitivity groups will probably be formed as part of the training program for those who are space station-bound.

The confinement of extended space travel can no doubt cause mental stress and could cause an astronaut to lose control. But if all crew members are trained in recognizing the symptoms—

uncooperativeness, withdrawal from others, antagonism and aggressiveness toward crewmates—such serious problems can be avoided.

Even if the hardware and the funding were ready for a manned Mars mission today, the astronauts are not ready. Such a mission would take at least three years, and space psychology has much to learn before we can send people on the demanding and lonely journey to the Red Planet.

Zero-G Leisure

The fun and games in weightlessness are limitless; it's the leisure time of the astronauts that's in short supply. Spaceflight is the most expensive way to travel, so work always comes first. Governments and their taxpayers always want results, and until the costs of travel drop dramatically, there will be no leisure class in space.

But once the work and the time-consuming house-keeping and personal hygiene chores are done, the astronauts spend their limited free time in several ways. Without question, earth-watching is the favorite pastime of the astronauts. The ever-changing earthscapes—the multicolored land masses and cloud formations flowing past—are a source of constant wonderment and pleasure. Skylab and space shuttle crew members did and do bring along leisure-time items for their off hours. They select their own favorite music tapes, books and other reading matter, diaries, and playing cards, but these activities don't enhance the amazing novelty of the weightless environment. New entertainment forms such as zero-g acrobatics and games are constantly being tried out, and some will become as standard in space as chess and baseball are on earth.

One fascinating game that the Skylab crews tried out was the catch-the-peanut game played whenever an astronaut discovered one of his or her dry-roasted peanuts had escaped its can and was floating around the space station. On discovering a wandering peanut, the astronaut would float over to a wall, open up his mouth, and shove off toward the peanut, trying to catch

it for a nibble. But his body trajectory and open mouth had to be just right to catch the peanut or it would bounce off his face and go twirling in another direction.

The Skylab crews also had a Velcro-covered space dart board and the Velcro-covered darts to match. But the game of zero-g darts didn't work too well because they wobbled through the thinner-than-on-earth atmosphere. The music tapes, books, and binoculars for earth-watching were used more than any of the other entertainment items that went into orbit with the men. Three small balls were also sent up in their so-called entertainment kits. But when played with, the balls just kept bouncing, bouncing, bouncing around all over the place and often got lost. Some type of zero-g pool or hoop and ball game will someday make its way into orbit, but the rule book has not yet been written. Whatever ball games are created for weightlessness, a method of stopping the ball's motion in either the game enclosure (table) or the cabin will have to be found; otherwise the ball or balls could go on bouncing for hours or days, and human patience also has its limits in space.

Body floating and acrobatics are certainly two of the most popular leisure activities in weightless space. Rarely did the Skylab astronauts go anywhere in a straight line, and they often did a flip or somersault along the way. This, of course, was not during the first few days in orbit, but after the side effects of adapting to zero-g had subsided. They could push off a wall sideways and glide across a room like a skater, "walk" across the ceiling, using both their hands and legs, or do a gainer or two—after a pushoff from the wall or floor or ceiling—on their way to the opposite surface. Some of these tumbling motions gave the zero-g-ers a dizzy feeling similar to what some of us feel on amusement park rides, but they experienced no ill effects such as nausea. While the space shuttle does not have the large amount of room that Skylab did, the astronauts always can find enough room for a graceful body twist or flip.

Larger space structures of the twenty-first century, even larger than the modular space station that will top off the first century of the

Space Age, will offer even more exciting opportunities for weightless creativity. Even human flight with wings or paddles attached to the arms could become popular. Centuries from now, space folk will have sports and sporting skills that we cannot even imagine today. The major earthbound sports of today will be historic curiosities to zero-g-ers.

Space Work

Time is *big* money in space, and the full-up, strenuous work schedules in orbit or during journeys to the moon reflect this fact. A tremendous amount of work must be performed by astronauts to get the highest possible return from scientific experiments and to successfully launch, and sometimes repair, commercial satellites. And there is the all-important work of keeping the spacecraft operating at its peak of safety and efficiency. Twelve-hour workdays for shuttle astronauts are the standard, and they often run longer. For the first Skylab crew, which had to make substantial repairs to the space station, including the deployment of a parasol sunshade over the spacecraft, the in-space workdays often ran to sixteen hours or more. These long hours, including working space walks outside Skylab, saved a multi-billion-dollar space venture that was in serious jeopardy because of spacecraft damage during launch.

The eighteenth mission of the space shuttle, launched in June of 1985, with a crew of seven including French and Arab astronauts as pay

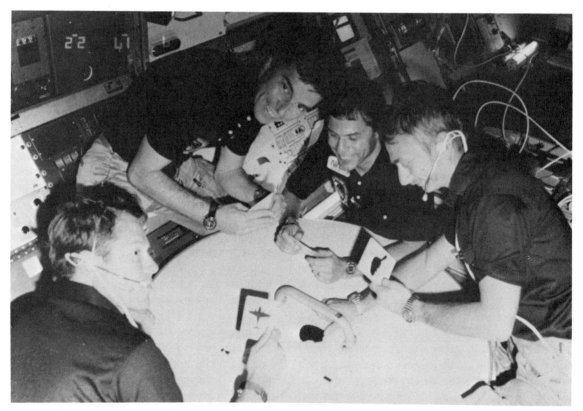

A space card game in progress? That would be everyone's guess at first glance. Actually these crew members of Columbia's Spacelab 1 mission are having some fun and games at our expense. If you really want to know what's going on, turn the picture upside down for the correct orientation. The "table" is really the Spacelab's airlock hatch, and the "cards" are visual targets for an awareness-of-position experiment. Courtesy NASA.

This job-well-done photo of the Challenger *crew was taken in April 1984 after the successful retrieval, repair, and deployment of the* Solar Maximum *satellite—the first in-orbit satellite repair. Left to right are: the late Dick Scobee, killed in the* Challenger *disaster; George Nelson; James van Hoften; Terry Hart; and Bob Crippen, commander of the mission. Courtesy NASA.*

load specialists, had a busy flight plan. It called for launching American, Mexican, and Arab domestic communications satellites. Three Getaway Specials were launched for West Germany, and three other mini-satellites, one of which contained nine student experiments in biological and physical science, were also launched. This mission alone, then, launched a total of nine satellites into earth orbit.

In addition to these large and small satellites, the *Discovery* shuttle also had a drop-it-off, pick-it-up-later satellite named *Spartan 1*, designed to carry a variety of experiments. On this particular mission, the *Spartan 1* was outfitted with astronomical instruments—X-ray sensors that searched for hot gas clouds in the galaxy and surveyed other celestial X-ray sources, including what may be a gigantic black hole at the center of the Milky Way. Unlike the other nine

satellites, this one was picked up toward the end of the mission and brought back to earth. The famous shuttle remote manipulator arm, itself controlled by a mission specialist, deployed it and picked it up. On each shuttle flight, one astronaut has the primary responsibility for operating the remote robot arm from the right side of the aft crew station at the back of the flight deck. This sophisticated giant arm, with two TV cameras mounted on it, is an extension of the human arm designed to do the big jobs required in the business of space. Time and time again, it has proved itself worthy.

As if the launching of all these satellites were not a heavy enough work load for this shuttle crew, they also had on board an experiment from the Strategic Defense Initiative Organization called the High-Precision Tracking Experiment. This tested the ability of a ground laser beam director, a component of the so-called "Star Wars" weapons systems, to accurately track an object in low earth orbit. And a space manufacturing experiment, using a zero-g solidification furnace, tested the feasibility of producing improved magnetic components for industrial, medical, and military use.

This example of a seven-day shuttle mission summarized only the major mission accomplishments and not the complex, step-by-step actions that were necessary to carry them out, nor the time-consuming operational checklist procedures constantly performed by the commander, the pilot, and other crew members. Add to these tasks the daily living chores—the cooking, cleaning up, exercising, bathroom time—and the astronauts *do* work, work, work in orbit. But that occasional look at the turning earth below and the spontaneous flip or tumble in weightlessness dramatically remind them that they are working high above the earth, traveling at a speed of more than 17,000 miles (27,353 kilometers) an hour, and that they were among the chosen few of the twentieth century, representatives in space of the 4.7 billion human beings on the surface of planet Earth.

Future Space Jobs

The year 1985 was a spacemark in the young history of the Space Age. A mid-year mission of the space shuttle put the one hundredth U.S. astronaut into space, although several of them had flown more than one mission. It has taken almost a quarter of a century for the United States to put one hundred of its citizens into space, but that number will soon double every few years as the human space adventure approaches the twenty-first century. Within just a few decades, those humans experienced in space travel will number in the thousands. By the year 3000, if our species survives, tens of millions of earthlings will be traveling and working in space.

The space shuttle has made—and its successor spacecraft will do even more to make—space travel possible for people of many different ages and from all walks of life. As the space

population increases, so will the space jobs they perform. New specialties will be required in the zero-g work market. As the large space structures are built with space robots, teleoperator skills in remote robot operation will be needed. Men and women trained in extravehicular activity in their independent maneuvering units will be increasingly in demand for robot repair and other space construction jobs. And there will be the scientific specialists, as there are today for the shuttle flights, in ever-increasing numbers. Besides the flying space doctors and the scientists from the Fortune 500 companies who are attempting to create future commercial products, there will be experts in microgravity and foreign passengers whose main purpose will be to dispense international goodwill. Some of the known earthbound job categories will make their way into orbit or on paths to the moon and beyond, but entirely new job specialties will also be created, including a space psychiatrist who will be trained in the particular psychological problems of spaceflight and adaptation to zero-g and readaptation to earth. Still, there will be the jobs that remain earthbound for the next decades. Why, after all, send cooks and dieticians up to orbit, when they can do their jobs very well and less expensively on the ground? They will eventually go, like the space farmers, when space structures dramatically increase in size and become villages, if not cities, in space.

Robots with advanced artificial intelligence will perform some of the routine and more dangerous jobs of space, but it will be people who build, operate, maintain, and reprogram them to carry out their jobs. Robot keepers (or robot groom, if you prefer) will certainly be one of the new job categories on the high frontier in the next few decades.

The distant future may well call for homesteading farmers for the planet Mars, heavy equipment operators for asteroid mining, or astronomers for an observatory on the far side of the moon. Whatever you job is today—cook, doctor, engineer, journalist, mechanic, secretary, teacher—there is an excellent chance that it will be a full-time space job within the next hundred years.

Time is big money in space, and workdays of twelve hours or more are standard. On this 1983 Challenger flight, astronaut Guion S. Bluford checks out the sample pump on the continuous-flow electrophoresis system experiment, which is separating biological materials. This work may lead to a future drug industry in orbit. Courtesy NASA.

5

Space Disasters & Close Calls

Facing Death in Space

Even Christopher Kraft—director of flight operations at the Johnson Space Center, whose familiar face and calm professional manner gave assurance to millions of TV viewers during the Apollo years—believed that astronauts would die in space. Kraft, who retired from NASA in 1982, admitted that at least twice in his career he thought the crews of troubled missions were doomed. He did not believe the *Gemini 8* crew would ever get back to earth, and he stated that the *Apollo 13* crew was saved by the "luck of God."

While the Russians lost cosmonauts on space missions during the early years of the Space Age, the United States lost its astronauts on the ground. Three Americans died in an Apollo spacecraft fire during a simulated test, their rocket still on the pad at Cape Kennedy. Four more astronauts died in air crashes. During the ten years it took the United States to reach the moon, no American astronauts died in space. Twenty-seven men flew to the moon, and twelve men walked on its surface without a single death in spaceflight.

After the tragic fire of *Apollo 1* on January 27, 1967, a cartoon appeared in the *Los Angeles Times*. Drawn by Paul Conrad, it showed the specter of death wearing a space suit. He held a Mercury spacecraft in one hand and a Gemini in the other. A smoldering Apollo spacecraft was in the background. The caption read: "I thought you knew, I've been aboard on every flight."

When President Kennedy said the United States was going to the moon, some respected experts gave the tremendous undertaking only a one in ten chance of success. Early during program planning, the question of odds came up, because reliability of any engineered system is directly linked to the amount of time allowed for design, production, and testing, and to the amount of money that can be spent. NASA had to decide early on what odds they would accept in their lunar venture. Should 50 percent of the missions be successful? Should 9 out of 10 astro-

CHARLES A. BASSETT II
PAVEL I. BELYAYEV
ROGER B. CHAFFEE
GEORGI DOBROVOLSKY
THEODORE C. FREEMAN
YURI A. GAGARIN
EDWARD G. GIVENS, JR.
VIRGIL I. GRISSOM
VLADIMIR KOMAROV
VIKTOR PATSAYEV
ELLIOT M. SEE, JR.
VLADISLAV VOLKOV
EDWARD H. WHITE II
CLIFTON C. WILLIAMS, JR.

ABOVE: *A commemorative plaque and fallen astronaut sculpture were left on the surface of the moon by* Apollo 15 *astronauts Dave Scott and Jim Irwin. The plaque listed in alphabetical order 14 U.S. astronauts and Soviet cosmonauts who had died while serving their space programs up to that time—July 1971. Courtesy NASA.*
PRECEEDING PAGES: *The shipwrecked crew of unlucky* Apollo 13 *was lucky to return to earth alive. There were times when no one, including the crew, thought they would survive the ordeal after the explosion. Their lifeboat was the moonship* Aquarius, *which they jettisoned just before reentry.*

nauts come back alive? Or should the reliability factor be such that 999 out 1,000 astronauts lived to tell about their flights in outer space?

This basic question of how safe Apollo hardware and software should be was debated for weeks, but no one was willing to make that decision. It was, after all, the base upon which the Apollo pyramid—the twentieth century's greatest legacy to the future of humankind—would be built.

Finally the engineering team sought the advice of Robert Gilruth, director of the Manned Spacecraft Center. Gilruth told them that a 50 percent success rate was too low and unacceptable. "We can make 9 out of 10," he said. "Maybe 99 out of 100, lose 1 man out of 100 on lunar missions."

Walt Williams, director of the Mercury program, responded: "That's ridiculous. Make it 1 in a million."

"How about three nines?" Gilruth asked. "How about a reliability of 9-9-9?" The decision was made. An astronaut would have a 1 in 1,000 chance of dying in space; and 999 chances out of 1,000 of returning to earth alive. Apollo would be reliable *and* expensive was the final decision reach.

But still there were deaths and many close calls during Apollo. The worst—the launchpad fire of *Apollo 1*—was the least expected place for tragedy to strike. In a certain way, it was analogous to having a car accident near your home instead of on a cross-country trip. You were safety conscious for the long journey, but your guard was down for the short hop to the corner grocery store.

As manned spaceflight increases during the last years of the twentieth century and the first decades of the next, what happens to the life-and-death odds for the ever-growing group of men and women who fly into space? The *Challenger* disaster proves that the odds increase dramatically as more missions are flown. Still, all the dangers begin on the ground; they did for the *Apollo 1* fire and for the *Challenger* explosion. Even Neil Armstrong had several close calls. Two of them were off the earth—one while he landed *Eagle* on the moon, the other while he orbited the earth. He also had a close call on earth.

Close Call in Orbit

Neil Alden Armstrong, the first human to walk on the moon's surface, experienced his most dangerous moments in space three years before his *Apollo 11* moon mission. It happened during the flight of *Gemini 8* in 1966. Armstrong and David R. Scott spent their first two orbits chasing an Agena target rocket that had been launched earlier that same day. Their goal was to rendezvous and then accomplish the world's first docking of two spacecraft in orbit. About six and a half hours into the flight, docking with the Agena was accomplished. Then twenty-seven minutes after docking, high above mainland China . . . big trouble. The joined Gemini and Agena spacecraft began to spin inexplicably. Scott noticed that the "ball" indicator on the control panel showed a 30-degree roll. But he knew they should be in level flight. Checking the horizon through the window was impossible because they were on the night side of the orbit. The spacecraft was also out of communication range of the tracking stations at this time. Armstrong attempted to stop the motion with the maneuvering thrusters, but this was only a temporary fix—the motion soon started again, this time faster. Physical damage to both spacecraft became a real threat.

Both crew members thought the problem was in the Agena target vehicle. They had tried everything, and nothing worked. Houston then ordered them to separate from the Agena rocket; this was the worst possible thing they could have done, but no one knew that until afterward. Scott hit the undocking button, and Armstrong gave the thruster a long burst to pull *Gemini 8* straight back. Once separated from the Agena, *Gemini 8* began to spin even faster—the problem was in their own spacecraft!

Armstrong told the *Coastal Sentry Quebec*, the tracking ship off the coast of China, that he considered the problem serious. "We're toppling end over end but we are disengaged from Agena . . . We can't turn anything off."

The ship communicated to Houston Flight Control: "He seems to be in a pretty violent tumble." Because the spacecraft was gyrating so wildly, the tracking ship couldn't get valid telemetry data or coherent voice contact.

This was the most dangerous time for Armstrong and Scott. *Gemini 8* was whirling around at least one revolution every second, sixty times every minute—almost twice as fast as a 33⅓ rpm record on the turntable. At first just dizzy, the two astronauts soon had trouble seeing the dials and controls overhead as their vision blurred. They were near their physiological limits, close to blacking out. If they lost consciousness, and the spacecraft continued to spin faster, it would mean disaster. Two dead astronauts in orbit.

Before that happened, Armstrong said, "All that we've got left is the reentry control system." They turned off the thruster system switches and activated the reentry control system. He broke a mission rule, but it was all he had left to save their lives. Soon the hand controllers responded! They slowly regained control of the *Gemini 8* by using the reentry control system direct. Later they would discover that the number 8 maneuver thruster had stuck open, "failed on," as space jargon would have it; this stuck thruster had almost killed them.

The firm rule of Houston Manned Spacecraft Center was that the spacecraft had to land immediately after the reentry control system was used, and when Armstrong and Scott came within range of the Hawaii tracking station, they were ordered to make an emergency landing in the western Pacific during their seventh orbit. At 10:23 P.M. Eastern Standard Time, two very lucky astronauts landed 500 miles (805 kilometers) west of Okinawa—they had cheated death in space.

The Dangerous Flying Bedstead

While the *Gemini 8* emergency in orbit was Neil Armstrong's most dangerous moment in space, his historic moon landing in *Eagle* was not without the tense moments (read Chapter 1, "A Space History: The Golden Years"). Even the earthbound training for the moon landing was dangerous for Armstrong and his fellow astronauts.

A free-flight trainer for the moon landing, built by Bell Aerosystems in the 1960s, was essential in preparing the astronauts for their final descent—the last 500 feet (152 meters)—to the lunar surface. Two versions were built of what many called the "flying bedstead." NASA called the first design the lunar landing research vehicle (LLRV), and the second, advanced version, the lunar landing training vehicle (LLTV).

The vehicle had two propulsion systems: A 4,200-pound-thrust fan-jet engine supported five-sixths of the craft's weight. The remaining one-sixth of the weight was supported by two pilot-controlled hydrogen peroxide engines and strut thrusters. These tandem systems were designed to simulate the one-sixth earth gravity of a lunar landing.

The flying bedstead was difficult to fly, however, and it became uncontrollable if it tilted too much. It made a lot of people nervous. Three pilots had to bail out of this cranky machine before it crashed in flames—one of them was Neil Armstrong.

Fourteen months before the historic moon landing, on May 6, 1968, Armstrong had to eject in his rocket-propelled seat from the flying bedstead before it crashed and exploded in flames. This was the astronaut's twenty-second flight in the trainer, and it had reached a peak altitude of 500 feet (152.5 meters) during the simulated lunar landing at Ellington Air Force Base. At "punch out" time, Armstrong had been airborne about five minutes and was flying at an altitude of about 200 feet (61 meters) when the vehicle malfunctioned. After he parachuted to safety, a physician examined him and pronounced him physically fit.

After this crash, there were two more with research pilots at the controls. Investigations were conducted into all of them, after which some NASA officials were inclined to scrap the spooky vehicle and depend on other training devices. But the astronauts protested, saying it was essential for realistic training. Months later, the landing crews of the first two Apollo moon missions, including Armstrong, said it accurately forecast a lunar landing.

In March 1969, the Flight Readiness Review Board decided to resume training flights with the advanced version of the trainer, which had been modified to overcome its unstable tendencies. On June 14, 15, and 16, 1969—just over a month before the mind-boggling moon landing—Armstrong rehearsed on the flying bedstead—eight times on the last day. "It's a strange, eerie sensation to fly a lunar landing trajectory—not difficult, but somewhat complex and unforgiving," said Neil Armstrong after *Apollo 11*. The same words could have been used to describe the unsightly flying bedstead that had also proved to be temperamental and unforgiving.

Apollo's *Darkest Hour*

Six years of manned U.S. spaceflight, from the suborbital Mercury flights of 1961 to the last Gemini flight in late 1966, and not one astronaut had died on a mission. Yes, there had been three deaths, all in T-38 jet crashes: Theodore Freeman died on October 31, 1964, in a crash at Ellington Air Force Base, in Texas; and Charles Arthur Bassett and Elliot McKay See were both killed in a T-38 jet crash near St. Louis, Missouri, on February 28, 1966. Manned spaceflight, however, had an excellent record: Nineteen Americans had flown in space, and seven of them, including Gus Grissom (the first man in

the world to do so), had flown twice. Together the Mercury and Gemini programs had logged about 18 million miles (29 million kilometers) in orbit. But the Apollo program was a significantly more complex technology. Many believed the odds for a mishap were increasing with the complexity of Apollo; they were right.

On March 21, 1966, less than a week after the aborted mission of *Gemini 8*, NASA announced the crew for the first manned Apollo mission: Gus Grissom, Edward White, and Roger Chaffee. In December of that year, NASA abandoned its hope to launch the first Apollo by the end of 1966 and rescheduled the flight for February 1967.

On Friday morning, January 27, 1967, everything appeared ready for the scheduled preflight launch simulation of *Apollo 204* (later to be designated *Apollo 1*). A launch crew of one thousand men was on hand to support the three astronauts on their preflight test—a "plugs-out" test to see if the launch vehicle could function on internal power alone.

As Grissom, White, and Chaffee left the Manned Space Operations Building that morning, they talked with their friend, Tony Broadway, the building superintendent, about a planned duck-hunting trip the following week.

By 1:00 P.M., after several delays caused by ground support equipment, the three astronauts crawled through the open hatch to take their flight positions—Grissom on the left, White in the center, and Chaffee on the right. For almost two hours, the crew worked with the hatch open, and then at 2:50 P.M., they were sealed in and the cabin pressure was increased to 16.2 pounds (7.3 kilograms) per square inch of 100 percent pure oxygen, which itself will not burn, but which at such a high pressure could rapidly feed any existing spark of fire.

The astronauts and launch crew continued the countdown and ran simulation tests for more than three hours. The countdown progressed slowly. A minor problem was discovered and corrected in the communications system. "Hey!" Gus Grissom said into the intercom, "how do you expect us to get to the moon if you people can't even hook us up with a ground station? Get with it out there."

Shortly after 6:00 P.M., at fifteen minutes before the simulated lift-off, the spaceship was switched over to internal power. And five minutes later, at T minus ten minutes, a hold was called to check the environmental control system's highly combustible glycol coolant and to check out some of the electrical equipment.

As part of the test launch, the crew was scheduled to practice an emergency egress from the command module through the double hatch. It was supposed to take about ninety seconds with the new six-bolt hatch, designed to replace an earlier quick-release, explosively charged hatch design used on Mercury and Gemini. When such a hatch prematurely blew off on Gus Grissom's suborbital Mercury flight in 1961, his *Liberty Bell 7* spacecraft took in water and sank to the bottom of the Atlantic. (For a more detailed account of Grissom's close call, read Chapter 1, "A Space History: The Golden Years.")

Florida's winter darkness had come to Cape Kennedy and launchpad 34. The astronauts had been in their couches some five and a half hours. It had been a long day; everyone wanted to wrap up the test soon and get home to hearth and family. Just before 6:31 P.M., telemetry from the spacecraft showed a momentary power surge, indicating a major short somewhere in the almost 20 miles (32 kilometers) of electrical wiring. Within seconds, the horror began.

"Fire! I smell fire!" The first warning probably came from Roger Chaffee, at exactly 6:31 P.M. Then, five seconds later, Gus Grissom yelled, "Fire! We've got fire in the cockpit." He must have known in that instant the horrible, hopeless reality of a fire in a pure oxygen atmosphere under pressure.

Some 16.8 seconds after the minute, the last frantic words came from Chafee: "We've got a bad fire—let's get out . . . we're burning up!" During this last transmission, at 19.5 seconds after the minute, the tremendous intense pressure, which had risen to 34 pounds per square inch, split the belly of the cabin open, and any remaining hope was gone. What had been a fire in the lower left side of the cabin became an inescapable inferno as a powerful flash blaze engulfed the spacecraft and bellowed out into the

surrounding white room and gantry areas. After one last cry of horrible pain, Apollo's darkest hour had come.

Some 1,200 miles (1,931 kilometers) away in Houston, flight director Christopher Kraft and his team helplessly watched the telemetry read out. Only Edward White had been wired for heart and pulse data on this launch simulation. The telemetry dial beat for fourteen seconds and then dropped to zero.

Three brave and dedicated astronauts were dead—the worst disaster of the U.S. space program. In the aftermath, many believed that the entire Apollo program was in ashes with them.

In the Ashes of Apollo 1

During the entire simulated test launch, a television camera was trained on *Apollo 1*'s command module window, and the image was monitored on the first floor of the launch complex by Gary W. Propst, an RCA employee. Suddenly the television picture showed a bright glow inside the spacecraft. Flames then flared around the window and increased steadily. Propst saw space suited arms behind the window fumbling for the hatch. "Blow the hatch," he cried, not realizing the two-hatch system had no pyrotechnics but was designed to be entirely mechanical.

Donald O. Babbit, a North American Aviation employee and the pad leader, heard the cry over the radio and yelled to James Gleaves, the mechanical technician, "Get them out of there!" But the sheet of flame flashed from the command module, and the concussion threw Babbit toward the door. He and three other men instinctively fled from the explosion, but within seconds they were back in the thick smoke trying to remove the hatches. All of them choked in the smoke and had to run in and out of the area for air during their attempts. Other workers joined the rescue effort. It took them five and and half minutes to gain access, but at first none could see the astronauts through the thick smoke.

Five minutes later, when the firemen arrived,

the air had cleared and the bodies were visible. Chaffee was in his couch, still strapped in. Grissom and White were intertwined, locked in their death struggle below the hatch, where they had spent their last energies in a vain attempt to open it. Doctors arrived fourteen minutes after the first cry of fire, but the bodies were difficult to remove because portions of the space suits had fused with the molten nylon inside *Apollo 1*. The true word finally went out. Earlier NASA had released an intentionally vague report stating that an accidental fire had killed "at least one person." Time was bought in this way so that the astronauts' families could be contacted.

The deaths were accidental, the coroner reported hours later, and were the result of asphyxiation caused by inhalation of toxic gases. All three astronauts died with their faceplates closed. But all three of the astronauts' space suits were badly burned, about 80 percent consumed by the fire, and the bodies of Grissom, White, and Chaffee had second- and third-degree burns.

The nation was shocked and saddened. "Three valiant young men have given their lives in the nation's service," said President Johnson. "We mourn this great loss, and our hearts go out to their families." Vice President Hubert H. Humphrey, then chairman of the National Aeronautics and Space Council, looked ahead: "The United States will push ever forward in space, and the memory of these men will be an inspiration to all future spacefarers." And this was true. Armstrong and Aldrin, the first two men on the moon, did not forget them: They left an *Apollo 1* shoulder patch at Tranquillity Base to commemorate Gus Grissom, Ed White, and Roger Chaffee.

Dr. Floyd Thompson, director of NASA Langley Research Center, was appointed to head a seven-man review board to investigate the accident and determine its cause. An interim report issued on February 22, 1967, stated that no definite cause could be determined, but the most likely origin of the fire was theorized to be an electrical malfunction. In April of 1967, a fourteen-volume report on the *Apollo 1* disaster was issued, and still the cause was not positively

identified, but it was strongly believed that a faulty conductor under Grissom's couch arched to another metal object and began the blaze. The review also admitted that the six previous years of safe spaceflights had dulled the alertness of many in the space program, and there were many specific examples of low-quality workmanship, substandard manufacturing procedures, and a neglect for safety measures that had no place in the manned spaceflight business.

Before the horrible fire, NASA had as part of its quality motivation program the Lunar Roll of Honor, a microfilm listing some three hundred thousand individuals who had performed exceptional service and made a significant contribution to the achievement of landing men on the moon. The microfilm was to be left on the moon by the first landing expedition.

After the tragic fire and deaths, however, the Lunar Roll of Honor was dropped. In such a gigantic undertaking of such mind-boggling complexity, it became almost impossible to fix individual blame or honor individual success. It was always called a "team effort," but it was more than that—it was an interwoven nexus of human effort from hundreds of thousands of people.

Because of the deaths of Grissom, White, and Chaffee, it was a year and a half before the first manned Apollo mission flew in a *completely redesigned* command module that included a quick-escape hatch. There is no question that their deaths led to new procedures and improvements and saved the lives of their friends and colleagues—the future Apollo astronauts who would orbit above and walk upon the moon.

Suffocation in Orbit

The first humans to die while actually *in space* were Russians—three cosmonauts during the flight of *Soyuz 11* in June 1971.

The Americans had successfully flown to the moon with *Apollo 14* early in the year, and they had another moon landing flight scheduled with *Apollo 15* in late July and early August. The doubts from *Apollo 13* were being put to rest.

The Russians clearly wanted to steal some thunder away from the U.S. Apollo flights that had dominated world attention for several years, and it appeared that they would succeed to a degree when they launched into low earth orbit the first *Salyut* space station, weighing 40,800 pounds (18,506 kilograms) on April 19, 1971—in the nick of time for May Day celebrations.

Soyuz 11 docked with the *Salyut,* and the world's first space station was boarded on June 7, 1971. The three-man Russian crew boarded it and remained aboard for more than three weeks, a mission of 360 orbits, lasting 569 hours and 40 minutes (June 6 to June 30, 1971).

The Soviets played up their success to the hilt, with live telecasts from orbit. The telecasts showed the cosmonaut comrades enjoying their spacebound home with zero-g acrobatics, jokes, and even a birthday party celebration for one of the crew. These space station days were a genuine first for Russia; it would be two more years before the U.S. *Skylab* missions flew. Russia took the opportunity to regain some lost prestige; it played down Apollo as an impressive stunt without much scientific value and bragged about the world's first space station, *Salyut,* and its far more valuable scientific mission. The Soviet population was given a full course in national pride and three cosmic heroes to love. The Russian space program, for the first time in years, was up, and the Russian people were proud. But what goes up . . . . The euphoria didn't last long.

The commander of the *Soyuz 11* mission was Georgi T. Dobrovolsky, a forty-three-year-old lieutenant colonel in the Soviet Air Force. He flew with Viktor I. Patsayev, thirty-eight, a test engineer, and Vladislav N. Volkov, thirty-five, the mission's flight engineer. It was Patsayev who celebrated his thirty-eighth birthday while in orbit.

On June 30, 1971, the *Soyuz 11* separated from the *Salyut* station to prepare for reentry into the earth's atmosphere. The cosmonauts did not don space suits for their reentry; in fact, the So-

viets were so confident about their spacecraft that the mission was flown without space suits.

After separation from the station, the cosmonauts prepared to separate from their instrument section, the equivalent of Apollo's service module, which had exploded on *Apollo 13*. Then it happened: Explosive charges, designed to fire sequentially to separate the spacecraft, malfunctioned and fired together. The additional shock popped open a seal inside one of two pressure equalization valves, and the cabin's oxygen-nitrogen atmosphere rushed out and was lost in space. In less than a minute, perhaps fifty seconds, the interior of *Soyuz 11* was a vacuum. In vain the crew struggled to close the valve manually, but they soon lost consciousness and died of suffocation.

Ground control attempted to communicate with the crew . . . nothing. When the spacecraft landed, two helicopters rushed to the scene. Doctors discovered the three dead cosmonauts, still strapped in their couches.

The Russian nation mourned. The bodies of the three heroes lay in state in Moscow, and for the first time in the young Space Age, an American astronaut was permitted to attend the funeral ceremonies. Thomas Stafford, later to become the American commander of the joint Apollo-Soyuz Test Project in 1975, was a pallbearer at the funeral. Russia's attempt to regain its momentum in space had ended tragically with three dead cosmonauts. Their names would not be among the names of their dead comrades on a plaque left in the moon soil by *Apollo 14*. They had died too late to be immortalized with that symbolic honor roll of fallen astronauts.

Dead on Impact

Almost three months after the tragic *Apollo 1* fire, on April 22, 1967, the Russians launched a new class of spacecraft from their Baikonur launch site. *Soyuz 1* was commanded by forty-year-old cosmonaut Vladimir M. Komarov, a colonel in the Soviet air force. Komarov was also the command pilot of *Voskhod 1*, and his maiden Soyuz mission made him the first Russian to fly twice in space; he was also the first cosmonaut to die on his return to earth.

Those privy to classified intelligence information may know exactly what happened during the flight of *Soyuz 1*, but no one else in the West is certain. It is generally agreed, however, that *Soyuz 1* was intended to link up with another spacecraft and crew, which was never launched because of systems malfunctioning in Komarov's spacecraft.

Probably *Soyuz 1* had major malfunctions in its stabilization and control systems—the new Russian spacecraft was their first capable of changing its orbital path. Several reports claim that one of the solar cell arrays also failed to deploy, resulting in an electrical power shortage, and that other systems also malfunctioned.

Whatever happened, *Soyuz 1* had serious problems by the fifteenth orbit, and Komarov attempted to reenter the earth's atmosphere at the end of the seventeenth orbit, but he was unable to align his ship for a successful retrofire. Finally, on the eighteenth orbit, the retro-rockets fired and *Soyuz 1* entered the atmosphere.

The official Russian explanation was that main parachute lines tangled and the parachute failed to deploy. The spacecraft, with no means of decelerating, crashed to earth, killing Komarov. But other rumors flew. For example: The cosmonaut had died in orbit when the life-support system failed. One account claimed that via monitored radio transmissions, Komarov was heard sobbing a farewell to his wife after he failed to bring his ship under control.

Russian experts in the West speculate that if the main guidance system failed, the cosmonaut could have put his ship into a controlled spin as an emergency backup guidance method. Such a procedure would have doubled the g forces. Did Komarov black out? Was he, therefore, incapable of stopping the spin? If so, the spinning *Soyuz 1* would indeed have twisted and tangled the shroud lines. There is no doubt that Komarov and spacecraft crashed into the ground at between 300 and 500 miles (483 and 805 kilometers) per hour. Some reports state the spacecraft burst into flames on impact, completely incinerating its interior and its cosmonaut. What-

ever the details, Komarov was dead and so were Grissom, White, and Chaffee. The Space Age was ten years old, and 1967 was a black-bordered year for the earth's early spacemen—be they from the West or the East.

Shipwrecked Apollo 13

The commander of *Apollo 13*, James A. Lovell, held the world's record for the most hours in space (572) even before the third manned flight to the moon was launched on April 11, 1970. His almost 24 days off the earth would seem short, however, compared with the 5 days, 22.9 hours of the flight of *Apollo 13*, in its command module *Odyssey* and its lunar module *Aquarius*, as men and machines were tested to their limits.

Lovell and his two crew members, John L. Swigert and Fred W. Haise, no doubt silently asked themselves many times during their journey around the moon, "Why us? Why *Apollo 13*?" The precision of spaceflight has no place for unlucky numbers. But *Apollo 13* was an unlucky mission, and the best that could be said for it was that the flight was a "successful failure." But in the beginning, no one knew if the three men would live or die.

More than two days out, on April 13, some fifty-six hours into the flight and 205,000 miles (333,000 kilometers) away from earth, something very unusual happened to the spacecraft.

"We didn't know what happened," said Christopher Kraft, retired director of the Johnson Space Center. "All we knew was that we were losing oxygen."

The crew had just completed a TV program to earth, giving us earthbound folk a tour of their moonship, the lunar module *Aquarius*, which was linked to command module *Odyssey*. It was shortly after 10:00 P.M. Eastern Standard Time.

Several hours after the mysterious event, the ground controllers finally knew how serious the situation was for the three astronauts. No one was able to believe that such a sophisticated spacecraft could be in such trouble—it took hours and an accumulation of data for the men at the Manned Spacecraft Center to believe that one of the worst scenarios imaginable had actually happened.

There had been a dangerous explosion in the command service module, but the airless outer space surrounding the spacecraft can play tricks on the human ears; a thunderous explosion on earth might sound like a thump in space. The sharp bang and vibration the crew heard and felt in the spacecraft were deceptively mild. Even the warning light did not indicate to the astronauts that they were shipwrecked, that their very lives were threatened.

But when they saw the pressure gauge in oxygen tank number 1 drop alarmingly, and then the steady, more gradual drop in oxygen tank number 2, the jabs of fear began, and they knew that all might be lost. Without oxygen, the command ship *Odyssey* could not produce electrical power and water because the fuel cells could not operate—they needed both oxygen and hydrogen to do their magic. Without the service module's oxygen and its generated power, the *Odyssey* would become a dead spacecraft with useless controls.

Thirteen minutes after the explosion, the crew noticed gas venting into space. It was oxygen. During this same period of time, all the backup teams on earth, including various contractor groups located all over the United States, moved quickly. Using the limited data, they calculated that the *Odyssey* did not have enough air, water, or power to keep the crew alive long enough for a return to earth!

About one and a quarter hours after the explosion, the number 1 oxygen tank was reading close to zero. Only fifteen minutes of power remained in the command module *Odyssey*. There were no more choices; they had to abandon the command ship and use the *Aquarius*, their lunar module, as their survival lifeboat. That was their only chance to avoid death in space. The flight of *Apollo 13* was now a survival mission; the moon landing was aborted, and the all-consuming drive of thousands of people on earth was to bring the astronauts back alive. Some five hours after the explosion, the prognosis for survival was only "fair."

Apollo 13's *Lifeboat* Aquarius

When one oxygen tank in *Odyssey*'s service module exploded, damaging another, some 300 pounds (136 kilograms) of liquid oxygen spewed forth into space. Out it came, a big blob at first, and then it formed a gaseous sphere that expanded rapidly, growing as the sunlight caught it and made it glow. Only ten minutes later, the glittering sphere was 30 miles (48 kilometers) in diameter. This expanding disk slowly disappeared. Only a powerful telescope in Canada observed traces of it an hour later.

The lunar module *Aquarius* was designed for two men for thirty-three to thirty-five hours; the emergency, however, required that the mooncraft accommodate three men for eighty-five to one hundred hours, as their crippled flight looped around the moon and returned to earth. The descent engine of the *Aquarius* also had to act as the propulsion system for the nose-to-nose configuration of *Odyssey* and *Aquarius*. None of this had been tried before. Everything was different. Simulators on earth were running around the clock. Detailed, lengthy checklists were created.

Once inside the cramped *Aquarius*, the three astronauts reduced its power consumption to one-fifth of normal. Commander Lovell also rationed the water to six ounces each per day—a fifth of normal intake—and they often drank fruit juices. There was no hot water after the explosion; their wet pack foods could not be reconstituted with cold water. This was worse than a bad camping trip and much more dangerous.

As they powered down *Aquarius*, everything began getting colder. As it got colder, the humidity rose because *Aquarius* could not take all the water out of the air of the joined spacecraft. Water condensed on the walls; windows frosted over; hot dogs froze. Astronaut Haise came down with a kidney infection and once shivered for four hours straight.

No waste was allowed to be vented overboard because its ejection would affect their life-

The damaged service module was also cast off before entering the atmosphere, and the crew saw the extensive damage for the first time. Courtesy NASA.

and-death trajectory and more fuel would have to be used to make the correction. "We kept the urine on board and had to figure out ways of keeping it," Lovell said. They used juice bags, water bags, and the unused moon rock bags.

"We had urine all over the place," Lovell recalled. "What to do with [it] taxed our ingenuity." The crew stacked it in places they had never considered before.

As Haise suffered with his kidney infection, Swigert suffered with cold, wet feet for two days after a water gun broke and leaked a quart of water into their cramped *Aquarius*. "It took six towels to sop it up," Swigert said. "Man, were my feet cold!"

The men used the darkened, all-but-dead command module *Odyssey* to sleep in, a man at a time, but it was horribly cold—near freezing at 38 degrees F. (3.3 degrees C.). Commander Lovell said it was "like a tomb." Everyone was very cold and no one slept well—less than an average of 3 hours a night for each man. The "sleeping bags" were made for keeping the bodies from drifting, not for warmth.

Without question, the *Apollo 13* crew suffered the most physical discomfort of any Apollo spaceflight, but even worse was the psychological burden of not knowing if they would return to earth alive. There were several serious scares on the flight after the explosion and departure from the dying *Odyssey*, and no one can deny the high stress involved in fighting for one's life after an unforeseen accident in space. There were several moments during the rescue mission of *Apollo 13*, Lovell confessed, when he and Haise and Swigert doubted they would make it back to earth alive. And these fears would come forth, perhaps even more often during the end of the flight, when all crew members were exhausted, dehydrated, and sometimes depressed. One crisis would be solved, and another would arise—it was a constant fight for survival.

For example, the buildup of carbon dioxide alone would have killed the crew had Houston not sent up detailed instructions on how to improvise and build a device to expel the noxious gas. The lithium hydroxide canisters in *Aquarius*, designed for two men for a limit of thirty-five hours, were not able to remove the gas sufficiently from the air, and the command module's canisters would not fit the round openings in the moonship. Lovell and Swigert built the contraption from command module canisters and other materials; it worked, and their own exhaling lungs did not poison them.

Mid-course corrections were also very tense. Because of the hundreds of pieces of debris from the explosion that followed the spacecraft and sparkled in the sun, it was impossible to find guide stars for navigation and proper, crucial alignment.

"Looks like I'm in the middle of the Milky Way," Haise said once. Ground simulations brought forth a new method that would align the spacecraft with the sun for proper firing of the *Aquarius* descent engine. Without accurate alignment, the ship was doomed.

The outcome of the first space rescue mission was uncertain; most of it was in uncharted territory, and much of the crew's extensive training was worthless. The difficult reentry procedures had never been attempted before; everything was novel. Several complex checklists were created on the ground, listing in order every switch position, every valve to be opened, every dial to check. The men were exhausted, dehydrated, cold. They made mistakes; the ground caught

and corrected them. Then, as they neared earth, there were new fears. Would *Odyssey*, the command module, and its reentry batteries power up after being down for several days and subjected to near-freezing temperatures? No one knew. And had *Odyssey*'s heat shield been damaged during the explosion? That was their last concern when they separated from their lifecraft *Aquarius*, the moonship that had kept them alive for three days, before reentering the atmosphere. They saw the extensive damage for the first time: The entire side of the service module had been blown out, and a tangle of twisted metal and wire was all that remained. If the heat shield had been damaged, the men would die inside the blaze of an Apollo meteor streaking toward earth.

The reentry was upon them. The *Odyssey* sped toward earth at 25,000 miles (40,225 kilometers) per hour and began to heat up to the point where communications blackout occurred. The blackout lasted almost two minutes longer than expected, and there were growing fears in Houston. As the crew neared the end of their survival ordeal in space, the icy cold water everywhere in *Odyssey* moved; inside the returning fireball, a cold rain fell on the crew.

Not Just a Bus Ride

The *Odyssey* splashed down in sight of the *Iwo Jima* at 12:07:41 Central Standard Time on April 17, 1970. The last moments of descent were televised because the spacecraft came down so close to its recovery ship. At the Manned Spacecraft Center, the hundreds of people who had struggled to save the lives of the astronauts saw *Odyssey* hit the water. They cheered and broke out the cigars.

In less than an hour, Lovell, Haise, and Swigert were on the deck of the carrier. They were weak, barely able to stand, and had lost more weight than any other spacecraft crew—Lovell alone lost 14 pounds (6.3 kilograms). All three were severely dehydrated. Astronaut Haise was sick with his kidney infection for three weeks after the flight. But they were alive, safely back on planet earth.

Robert R. Gilruth, then director of the Manned Spacecraft Center in Houston, remarked that the flight of *Apollo 13* reminded everyone "that flying to the moon is not just a bus ride." The sobering reminder that human spaceflight was a dangerous business came exactly in the middle of the Apollo flights to the moon, with four flights before and four flights after.

The cause of the explosion was determined during the course of a thorough investigation and was announced in June 1970. The oxygen tank that exploded had been checked before the flight; there was no obvious negligence involved. Just as in the tragic *Apollo 1* fire, a short circuit had been at fault. It had ignited electrical insulation in the pressurized oxygen tank number 2. The fire started, pressure rose dramatically, and the explosion occurred. But why the short circuit?

Defective thermostat switches were the root cause. Several years before the flight, oxygen tank number 2 had been dropped—only a few inches (several centimeters), however. Testing showed no damage at that time. Later, during further tests, it was discovered that it could not be emptied normally, so the ground crew turned on heaters and fans to empty the oxygen, which was not a problem in itself. Before the flight, on March 27 and 28, 1970, heaters and fans were turned on that applied 65 volts of current for periods of six and eight hours to pressurize the tanks. But the thermostat switches were designed for a maximum of 28 volts: their specifications had not been changed along with those for the heaters to speed up pressurization time. This excessive current welded the switches shut. As a result, the insulation burned off and the wires were bare. For three weeks before its launch, *Apollo 13* was a potential bomb. It could even have been blown up on the pad. An even worse scenario, however, would have been if the explosion had occurred when *Aquarius* was on the moon or after the lunar module was cast off before the return to earth. There would have been no lifeboat with its store of life-saving consumables. Three dead astronauts interred in space would have been inevitable without life-saving *Aquarius*.

But the lunar module *had* saved their lives. And Grumman Corporation, its creator, was proud: This spacecraft had been used beyond its intended use and its design limits, and had brought three astronauts back alive. As a joke, Grumman sent to Rockwell International, the command service module contractor, a bill for towing services. *Aquarius* had towed the CSM for about 300,000 miles (482,700 kilometers), and Grumman billed Rockwell $400,000 for this service. Rockwell wasn't laughing.

Ominous Gyrations and Vibrations

Unexpected, mission-threatening motions of the Apollo spacecraft occurred on two voyages to the moon: *Apollo 10*, the dress rehearsal for the first manned landing; and *Apollo 16*, a mission to the lunar highlands.

Snoopy's Gyrations. *Snoopy*, *Apollo 10*'s lunar module, descended to within 50,000 feet (15,240 meters) of the lunar surface. "We're down there where we can touch the top of some of the hills," astronaut Eugene Cernan told Houston. It was the first lunar module to fly to the moon, and all its systems were tested in the lunar environment for the first time. This included the firing of *Snoopy*'s descent and ascent rockets, and the first around-the-moon rendezvous with *Charlie Brown*, the command module.

After the lunar module crew, Cernan and Thomas Stafford, had descended closer to the moon than anyone before them, they prepared to fire *Snoopy*'s ascent engine, which would begin their return and rendezvous with *Charlie Brown*. This event, called staging, separated the lunar module's ascent stage from its landing stage. With an actual moon landing mission, it would take place on the surface of the moon and not in low moon orbit. The crew went through a detailed checklist, monitored by Houston, which covered every switch position in the cockpit.

"Son of a bitch," shouted Cernan, who began separation by firing the pyrotechnics. The mooncraft was out of control, gyrating wildly.

Snoopy was in a fast spin, pitching up and down. No one knew what was wrong, but there was no time to find out. If *Snoopy*'s wild motions weren't brought under control, it could mean disaster—crashing into the moon without an ascent stage.

Stafford then went to manual control. After a two-minute struggle, *Snoopy* was under control. "I don't know what the hell that was, babe . . . " Cernan told Houston. "We were wobbling all over the skies."

It was human error. The technicians who had prepared *Snoopy* before launch had left a control switch in the wrong position. Stafford had somehow overlooked it as he went through the detailed checklist. The switch in error had left the backup navigation equipment (known as the Abort Guidance System) in the automatic mode rather than the altitude holding mode. The wild gyrations were caused by the ascent stage radar hunting the skies for *Charlie Brown*, the command module, so it could lock on for rendezvous. The remaining top half of *Snoopy* was only doing what its computers and control panel switch told it to do. For the rest of the ascent, *Snoopy* flew smoothly and caught up with *Charlie Brown* for the first moon-orbit reunion.

***Casper*'s Vibrations.** *Casper*, the friendly command module of *Apollo 16*, got a minor case of the jitters after separating from the lunar module, *Orion*, which was ready to descend to the moon's Descartes highlands. These vibrations threatened to abort the entire mission, but earth base went into action and, after a six-hour landing delay, gave the crew a golden go for landing.

But the mission was in serious doubt when *Casper*'s pilot, Thomas Mattingly, reported to Houston that the backup steering system for the main rocket engine was causing the engine bell to oscillate. If the primary system failed and the backup system malfunctioned, the result would be three marooned astronauts facing certain death. The rocket engine and its steering controls were the only way that John Young, Thomas Mattingly, and Charles Duke could return to earth. For this reason, strict mission rules re-

quired both primary and backup systems to be normal before any go for lunar landing.

There was a crisis atmosphere in the Manned Spacecraft Center. The mission was on hold, and Houston gave it seven and a half hours, five moon orbits, to see if a solution could be found before the mission was aborted. Even if the odds against the primary system failing (and it was working perfectly) were 1,000 to 1, there would be no bending of the rules.

In Texas and California—at the Manned Spacecraft Center and at North American Rockwell, the contractor—hundreds of experts began analyzing the steering problem. After several hours, the cause was found to be a defective yaw gimbal actuator. This was making the engine oscillate back and forth, and the fear was that with a powered-up engine, these oscillations could damage or even break up the spacecraft. That was the new problem to solve. If the backup system *had* to be used, what damage would the oscillations cause?

Hundreds of man-hours later, the answers came forth. If the backup steering was used, the oscillations would not structurally damage *Casper* nor alter the spacecraft's attitude as the engine drove it out of moon orbit to return to earth. *Orion* could descend to the Cayley Plains of Descartes. *Apollo 16*, thanks to a lot of people of earth, had avoided an abort.

Out Cold

After a moon voyage of more than ten days, *Apollo 12*'s command module, *Yankee Clipper*, splashed down in the Pacific on November 24, 1969, landing only 4 miles (6.4 kilometers) from the recovery ship, U.S.S. *Hornet*. Soon the astronauts—Charles (Pete) Conrad, Richard Gordon, and Alan Bean—were aboard their recovery ship, facing the TV cameras. But what was on Al Bean's forehead? A Band-Aid®?

Beneath the Band-Aid® was the worst wound inflicted on any Apollo astronaut during the voyages to the moon, and it happened at splashdown, not on the lunar surface. Al Bean got whacked on the head at the moment of impact when a sixteen-millimeter camera bracketed to the spacecraft wall came loose and hit him on the head. His fellow astronaut, Pete Conrad, told the story: "It coldcocked him. He didn't realize it, but he was out to lunch for about five seconds. He was staring blankly at the instrument panel. I was convinced he was dead over there in the right seat."

The worst wound of the Apollo missions took six stitches to close up, but it was the Band-Aid® that concealed it from the world view.

The Space Race and Broken Hearts

America's race to the moon in the 1960s took its toll in astronaut lives, but many others suffered and paid with their lives and did not make the headlines. Why? Because no one knew until after the Apollo era had ended.

Who where these unknown victims of Apollo's push to the moon? They were highly specialized aerospace professionals at Cape Kennedy, all well educated. And they were young—their average age was only 31.1 years. The tremendous pressure to perform in the highly competitive environment at America's spaceport and fulfill President Kennedy's goal to land men on the moon, as well as to "beat the Russians" there, caused life-threatening stress.

An eight-year study revealed that excessive occupational stress caused heart attacks and sudden cardiac death in unusually high numbers of these space workers at Cape Kennedy. As a group, these men did not have the usual risk factors (for example, obesity and cigarette smoking) in their life-styles. What they shared was extraordinarily high stress.

And then, in the late 1960s, their strong performance was "rewarded" with budget cutbacks and layoffs—from about 65,000 employees in 1965 to 8,000 in 1976. This threat of unemployment for sophisticated professionals with what had become unmarketable skills (astronauts weren't flying to the moon!) only added to the stress, and these men who had given

so much were subjected to sustained fear, alarm, and feelings of inevitable loss. Many died.

A study of autopsy reports revealed that 85 percent of the victims had lesions in their heart tissue known as contraction bands—damage from a high quantity of stress-induced chemicals from the adrenal glands.

Those aerospace workers and their families were the unknown victims of the Space Age. During the Apollo years, there was more danger on earth than to be walking to the moon.

The Poisons of Skylab

Skylab 1, America's first experimental space station, was launched unmanned on May 14, 1973, just five months after the last Apollo mission flew to the moon. But there were mechanical failures during launch due to vibrations during lift-off and insertion into orbit. A critical meteoroid shield was ripped off, and one of Skylab's two solar panel wings was destroyed with it. Crucial insulation was also lost. The other solar panel wing was unable to deploy because a piece of the shield was wrapped about it. That left only the four rotary solar panels atop the station functioning.

Did all these problems have something to do with the fact that the launch had not been man-rated because the astronauts would follow it into orbit a few weeks later? Was this, after the technological pyramids of Apollo, a return to the early space days when mission mishaps were common?

The answer was no, and the reason was that the three Skylab crews could fix things in innovative ways and nurse this first U.S. space station back to health and on to fulfill its goals.

But it was not easy. The Skylab 2 crew—Pete Conrad, Paul Weitz, and Joe Kerwin—blasted into orbit on May 25, 1973, rendezvoused with the station on their fifth orbit, and finally docked their Apollo command module after many frustrating attempts. At least the crew had had the benefit of a crash training course on the ground for almost two weeks before their launch to prepare them for all the new and challenging activities they would have to perform to get Skylab operational again.

Mission Control knew that, without the insulation and the destroyed solar panel wing, the temperature had risen to 125 degrees F. (52 degrees C.) and higher in some areas of Skylab's large workshop area. The two big problems, diminished power and high temperatures, had to be solved. The high temperatures threatened to release carbon monoxide, carbon dioxide, and other poisonous gases into the interior of Skylab, and there was concern that these gases could poison the astronauts.

Because a toxic atmosphere was a real worry, ground controllers vented the air overboard several times before the crew was launched and renewed it with fresh oxygen-rich atmosphere from the tanks in Skylab's workshop. Equipment could also be irreversibly damaged if the heat was not brought down, and the Skylab 2 crew brought with them a specially designed umbrella, which they hoped to deploy.

After the astronauts docked, the men in Houston also began to pump up Skylab's oxygen/nitrogen atmosphere in anticipation of the crew's boarding the next day.

Shortly before noon on May 26, 1973, the crew opened up the forward hatch and inserted a special test tube in the docking adapter hatch. This probe measured for toxic poisons in Skylab's interior. Five minutes went by. The news was good; atmosphere in the airlock tunnel was safe. Astronaut Weitz had to wear a special mask, however, when he checked the large workshop area for the first time. The temperature there was still 125 degrees F. (52 degrees C.), and Houston continued to run tests for toxicity. Finally, the entire crew was given the go ahead. The poisons had been expelled from Skylab. The air was safe. The crew could move into their celestial house and get busy with repairs. Their first challenge: Rig and deploy the designed-under-deadline parasol. This would get the temperatures down and keep the potentially hazardous poisons of Skylab's polyurethane insulation locked in the materials and out of the men's lungs and bloodstreams.

The Hard-Hitting Hazards

Ever since TV's early space days of "Captain Video" and "Space Patrol" right up through Han Solo's adventures in *The Empire Strikes Back*, space heroes have been trying to dodge and manuever through dangerous debris, usually asteroids, not always successfully. In fact, "Duck!" episodes have been portrayed in space literature for over a century now. The hero must, after all, have a few problems to struggle with and overcome. Space fiction has usually depicted the threatening debris as large boulders or even mini-asteroid-sized objects, and certainly such big cosmic rocks are a potential danger for spacefarers who will journey to Mars and beyond. A state-of-the-art radar system, however, should protect them by early warning alarms.

Closer-to-earth space travel, in earth orbit or to and from the moon, is not likely to be threatened by such big chunks of cosmic debris. Nor will micrometeorites, so feared during the early Apollo program studies, pose a serious threat. The winged Pegasus satellites showed that not even immense wings in orbit get hit that often. No, what future space shuttle and space station astronauts have to worry about are all those thousands of small man-made objects and pieces of rocket booster debris up there with them.

The design of the U.S. space station that will orbit the earth in the 1990s has to take into account this potentially dangerous man-made debris. If even a small object a few inches or several centimeters in size were to penetrate a pressurized crew module, it could destroy the spacecraft and kill the crew members inside.

At the present time, the only debris capable of being *tracked* is about 4 inches (10 centimeters) in size or larger. But a U.S. Air Force GEODSS telescope system, designed to track objects in geosynchronous orbit, has discovered (not tracked) some forty thousand objects in low and medium earth orbits that are about 0.4 inch (1 centimeter) in size. This number is eight times the number of objects catalogued by the North

American Aerospace Defense Command/Space Command.

As these fragments collide with each other, their numbers will grow. Experts predict, based on data, that there will be a fair number of random collisions in the 1990s. Most of the debris has come from rocket booster and satellite breakups, over 90 percent of it from U.S. and Soviet missions.

The U.S. space station must be designed, say orbital debris experts, to withstand high-velocity strikes from objects 0.04 inch (1 millimeter) in diameter—at least.

For further study of the clutter on earth's orbitways, Johnson Space Center scientists want to fly a twin-telescope system in the shuttle payload bay to obtain photographic data on the de-

bris and refine their statistics for better design decisions. And some experts are recommending that a debris-monitoring-and-warning satellite be placed in a higher orbit than the station. Such a satellite would measure how much debris was accumulating and how much would likely sift down into the space station's orbit and become hazardous.

Even if such warning systems were in place, there is a good chance that the space station would be hit in its lifetime by an object 4 inches (10 centimeters) or larger in size. If the object is traveling at a high velocity relative to the station, it could blast through a module's walls and create an instant vacuum, suffocating those people within. There would be no chance to "duck," no chance to demonstrate the superior

A formal portrait of the ill-fated Challenger *crew, taken in November 1985. All seven died in the worst space disaster in world history on January 28, 1986, just seventy-four seconds after the space shuttle blasted off the surface of the earth. From left to right, front row, they are: Michael Smith, Francis (Dick) Scobee (commander), Ronald McNair; from left to right, back row: Ellison Onizuka, Christa McAuliffe, Gregory Jarvis, Judith Resnik.* Courtesy NASA.

piloting skills of Captain Video or Han Solo or any other space jockey hero from the golden days when space travel was only in the imagination. Above-earth clutter may soon make the crowded and dangerous asteroid belt of classic space yarns look like an empty four-lane highway.

The Challenger Disaster

They came from all over the United States. Some 2,500 guests, including the relatives and friends of the seven *Challenger* crew members, were invited to witness the launch of the twenty-fifth space shuttle mission in late January 1986. Their formal invitation read, in part:

•

The National Aeronautics
and Space Administration
cordially invites you
to attend a Launch
of the Space Shuttle
at the
John F. Kennedy
Space Center, Florida

R.S.V.P. As soon as possible

•

With enthusiasm and pride they came to be eyewitnesses to history, to watch their sons and daughters, mothers and fathers, the space heroes of a nation, ride a rocket through the sky and climb out of gravity's grasp to float in space and orbit around the planet at more than 17,000 miles (27,353 kilometers) per hour.

The Florida coast was clear and icy cold when the space shuttle *Challenger* blasted off at 11:38 A.M. Eastern Standard Time on January 28, 1986. The thousands of people who gathered to witness the launch saw the solid rocket boosters ignite and thrust the spaceship off the pad less than half a second after the countdown reached zero. It had all been seen many times before. Even to the trained eye, everything was normal as the *Challenger* lifted smartly off the ground, began its programmed roll maneuver at 9 seconds into the flight, and pointed toward its orbit

more than 100 miles (161 kilometers) above the earth.

The words from NASA's public affairs officer were all reassuring:

"Three engines running normally. Three good cells . . . Velocity 2,257 feet per second, altitude 4.3 nautical miles, downrange distance 3 nautical miles."

There was nothing like being there, standing on Florida's ground, watching this powerful spaceship ride its titanic flame. The thunderous shock waves physically assaulted the body. People *felt* the power vibrating inside them, something that TV images could never recreate.

At 59 seconds into the flight, the *Challenger* reached maximum dynamic pressure, which was calculated at 702 pounds per square foot for this particular flight.

"*Challenger*, go with throttle up," Mission Control radioed to commander Scobee.

"Roger." Scobee replied about a minute into the flight, "go with throttle up."

These were the last words transmitted from spaceship *Challenger*. Seconds later, at 73.226 seconds into the flight, when the spaceship reached a velocity of 1,977 miles (3,181 kilometers) an hour and an altitude of some 10 miles (16 kilometers), the tens of thousands of gallons of liquid oxygen and hydrogen in the huge external tank exploded in a gigantic fireball of death and destruction. Not even the computers, working at the speed of light, knew what happened. In a few microseconds of space and time, the ship exploded in an ascending pyre of debris and flame that for an instant rode out the energy created during the first and last 73 seconds of space shuttle *Challenger*'s final flight. Then hundreds of flaming spaceship fragments began falling from ten miles (16 kilometers) above the earth to the Atlantic Ocean below, leaving white streamers descending in the blue sky over the waves. They continued to fall for almost a half an hour, and rescue ships were held back. The crew compartment, still intact, also fell.

But in that tragic instant, in the time of a single heartbeat, when the *Challenger* exploded in midair space, the thousands of excited and cheering people on the Florida coast below did

Here the Challenger *crew members are shown in their actual flight positions, with one exception: Christa McAuliffe's backup, Barbara Morgan, is shown standing in the corner of the mid-deck during this preflight training session. On the flight deck are (top photo, left to right): Michael Smith, pilot; Ellison Onizuka, astronaut; Judith Resnik, astronaut; and Dick Scobee, commander. On the mid-deck are (bottom photo, left to right): Barbara Morgan; Christa McAuliffe, space flight participant for the teacher-in-space program; Gregory Jarvis, payload specialist for the Hughes Company; and Ronald McNair, astronaut. These are the actual positions in which the 7 astronauts died.* Courtesy NASA.

not understand that all was lost. It all happened so quickly, so far away, so high in the atmosphere. Human eyes and ears were unable to make any sense of this faraway reality. Everyone *knew* the *Challenger* was going to orbit. That's what this spaceship had done ten times before. There was nowhere else it could go. Even the NASA spokesperson kept reading his data display after the all-consuming fireball appeared on the images from the long-tracking cameras.

Soon the sounds of the explosion reached the Kennedy Space Center, and the spectator grandstands shook. The shock and disbelief came slowly, second by second, and then turned to fear and choking pain for the families and friends closest to Dick Scobee, Michael Smith, Judith Resnik, Ronald McNair, Ellison Onizuka, Gregory Jarvis, and Christa McAuliffe. Most of them instinctively knew that all was lost; some held out hope as they saw rockets flying away from the fireball (the solid rocket boosters) or sprouts and tendrils of debris flying free of the blast. But when NASA announced, "We have a report from the flight dynamics officer that the vehicle has exploded," many knew that hope wouldn't help.

When the *Challenger* exploded and fell from the sky, killing its seven astronauts, humankind once again was temporarily defeated in its struggle to escape the gravitational bonds of planet Earth. The disaster brought to mind the myth of Sisyphus, a metaphor for the human struggle that is rooted in existence. Sisyphus, a king of Corinth, was condemned in hell to an eternity of rolling a heavy stone up a steep hill, only to have it roll down again when he approached the top. We had slipped and fallen down the sides of the deep gravity well that encompasses our planet.

Gravity has always shown a Janus face to earthbound humanity. It has given us the foundation upon which to build our cathedrals and pyramids; and it has often thrown our bodies and spirits to the ground as we climb and reach into the future with our imaginations and dreams. As the chunks and pieces of *Challenger* fell out of the sky east of Florida, the spirits of millions on planet Earth also fell into despair and grief. An entire nation, and millions of other people around the world, mourned the deaths of seven brave and dedicated men and women. The worst single space disaster in world history reminded each and every one of us that human imperfection can still destroy in an instant our greatest works, our flying pyramids of technology, our future dreams. The degree of perfection required to control the immense energies and complex systems necessary for spaceflight is inhuman. But we know that, and our ingenuity finds ways to make them as perfect as they need be to fly to space and return. Then we are shocked into the reality: Unforgiving gravity still rules our planet Earth. We are still innocent children of the Space Age. Once again we have fallen and feel the terrible pain. As we get up on our feet again and start to climb once more, we promise ourselves always to remember them.

FRANCIS R. (DICK) SCOBEE
MICHAEL J. SMITH
JUDITH A. RESNIK
RONALD E. MCNAIR
ELLISON S. ONIZUKA
GREGORY B. JARVIS
S. CHRISTA MCAULIFFE

At the end of a CBS News Special Report, "Disaster in Space," on the evening of January 28, 1986, Dan Rather paraphrased the work of poet James Dickey:

Put them on the list of men and women who counted, these searchers and seekers, these astronauts and teachers who died today in what became the spaceship disaster; they died in the blue and silver furnaces of their space suits. Think about them, who they were and the way they were: dreamers, explorers, adventurers forcing themselves past the point of danger and deep fatigue to expand our understanding of what is up there and out there. They may never have known the nature of the trouble that killed them. For them, no more cries of "Wow, what a view!"; no more jokes with Mission Control, no

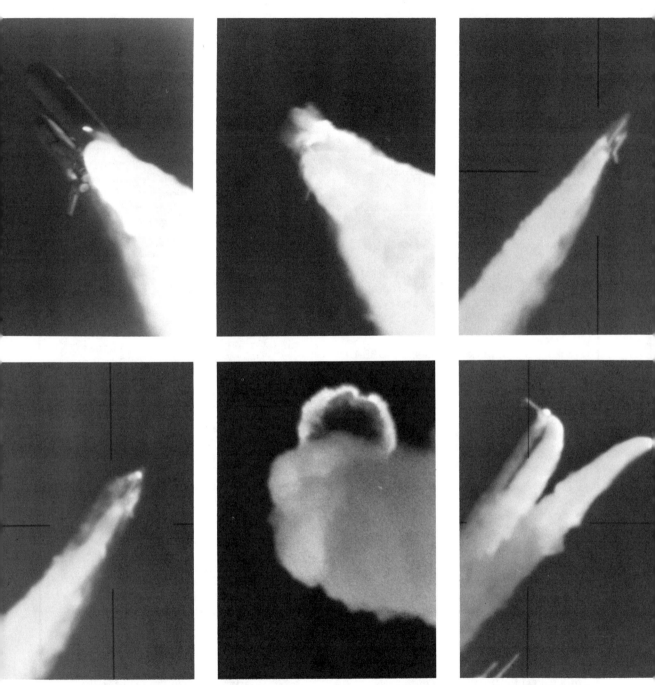

It was a fast death for the seven aboard Challenger; not even the speed-of-light computers gave warning before the fact. At less than a minute after launch (59.82 seconds), an unusual and dangerous plume appeared in the lower portion of the right-hand solid rocket booster. At 1 minute, 13.14 seconds, the explosion began. From another camera angle, the Challenger and crew were consumed during the seventy-third second of flight. All was lost when the giant external tank exploded. Both solid rocket boosters flew away from the explosion, including the one that leaked hot flame and caused the explosion (seen on the upper left here). The plume above the dark streak is coming from the rupture in the solid rocket booster. Courtesy NASA.

NASA

Shuttle Mission 51-L
Prelaunch Profile

Orbiter: OV-104 (Challenger)
Altitude: 153.5 n.m.
Inclination: 28.45 degrees
Mission Duration: 6 days

Crew:
Francis R. Scobee, Commander
Michael J. Smith, Pilot
Judith A. Resnik, Mission Specialist
Ellision S. Onizuka, Mission Specialist
Ronald E. McNair, Mission Specialist
*S. Christa McAuliffe, Payload Specialist
(NASA Space Flight Participant Program)
Gregory Jarvis, Payload Specialist
(Hughes Communications)

Payloads:
Tracking and Data Relay Satellite (TDRS-B)
Spartan-Halley

Experiments:
Acoustic Containerless Experiment System (ACES)
Monodisperse Latex Reactor (MLR)
Radiation Monitoring Equipment (RME)
Student Shuttle Involvement Program (SSIP)

Highlights:
*First Flight of a U.S. Citizen
Deployment of Second NASA TDRS Satellite
Deployment and Retrieval of Spartan/Halley

Principal Mission Activity
51-L will be a 6-day mission consisting of two major activities. The crew will deploy the TDRS-B satellite on Flight Day 1, with Flight Day 2 reserved for backup deployment. On Flight Day 3, the crew will deploy the Spartan-Halley carrier, with retrieval set for the fifth day of the mission. Entry checkout and cabin stowage is reserved for the following day with the landing at the Kennedy Space Center in Florida on Day 6.

Payload Data:

TDRS-B
The Tracking and Data Relay Satellite will be the second satellite deployed for NASA's Tracking and Data Relay Satellite System (TDRSS). TDRS-B will join TDRS-A in a geosynchronous orbit to provide high—capacity communication and data links with the Shuttle as well as other NASA spacecraft and launch vehicles. TDRS-B (WEST) will reside over the Pacific at 171° W. longitude. TDRS-A (EAST) is currently stationed over the Atlantic at 41° W. longitude. Both spacecraft will operate in the S-band and KU-band frequencies.

TDRS will be placed in the desired geosynchronous orbit by the IUS (Inertial Upper Stage). The three-axis stabilized upper stage will maneuver the TDRS-B to the desired altitude. White Sands Ground Terminal will then configure TDRS-B for on-orbit operations.

SPARTAN-HALLEY
This Spartan mission is intended to make observations of the ultraviolet spectrum of Comet Halley. The scientific objective is to measure the ultraviolet of Comet Halley as close to perihelion (that point in orbit which is closest to the Sun) as possible. This is the time when the rate of sublimation of volatiles should be greatest.

Two ultraviolet spectrometers will be mounted on a Spartan carrier. The NASA Spartan carrier will be deployed from the Shuttle and placed into its own independent orbit. The Spartan will position itself autonomously by first acquiring the Sun using the suntracker and then acquire the star Conopus with its startracker. On each orbit, Spartan will scan the tail of Halley just 5 degrees away from the coma. The Challenger will rendezvous with the Spartan 48 hours later and retrieve it with the Remote Manipulator System (RMS).

ACES

The Acoustic Containerless Experiment System (ACES) is a three axis acoustic containment furnace. When activated by the crew on-orbit, ACES performs a pre-programmed sequence of operations on a material sample for a period of two hours and automatically shuts off. The objective of ACES is to demonstrate a high temperature, three-axis acoustic levitation furnace and determine variations in the acoustic characteristics and capabilities. There is a major scientific interest in ex tending the present acoustic capability to high temperature in order to perform materials processing.

MLR

Assigned to its fifth flight, the Monodisperse Latex Reactor experiment package is a NASA-sponsored payload designed to produce perfectly spherical monodisperse latex particles of the 2-100 micron range in microgravity. The MLR hardware occupies the space of three mid-deck lockers and draws orbiter power through two power cables. It is composed of the Experiment Apparatus Container (EAC) and Support Electronics Package (SEP). The EAC is composed of four reactor chambers, each filled with a mixture consisting of a monomer, an initiator and an emulsifier. Particle growth is stimulated when heat is applied to each reactor chamber for up to 20 hours. Crew interaction with the MLR is minimal.

Following the flight, the National Bureau of Standards intends to acquire these latex particles to use as an improvement of the existing primary calibration standard in this size range.

RME

This is the seventh flight of the Radiation Monitoring Equipment. Its objective is to measure radiation levels in the orbiter's middeck. RME consists of a hand-held radiation monitor (gamma CN and electron dosimeter) and two pocket REM meters (neutron and proton dosimeter). Upon landing, the data will be provided to the U.S. Air Force Space Division for interpretation.

TEACHER IN SPACE

Christa McAuliffe will be involved in several activities which will be filmed and later used in educational products. Her activities include:

- Earth Magnetism—Photograph and observe the lines of magnetic force in three dimensions in a microgravity environment.

- Newton's Law—Demonstrate Newton's first, second and third laws in microgravity.

- Bubbles—Understand why products may or may not effervesce in a microgravity environment.

- Space Expressions—Generate from students creative works that reflect their interpretation of the space program/experience.

- Simple Machines/tools—Understand the use of simple machines/tools and the similarities and differences between their uses in space and on earth.

- Hydroponics in Microgravity—Show the effect of microgravity on plant growth, growth of plants without soil (hydroponics) and capillary action.

- Chromatographic Separation of pigments—Demonstrate chromatography in a microgravity environment and show capillary action (the mechanism by which plants transport water and nutrients).

Recommendations Made by the Presidential Commission

More than 6,000 people were involved in the commission's four-month investigation of the *Challenger* accident. Some 15,000 pages of transcript were taken during public and closed hearings, and another 120,000 pages of documents and hundreds of photographs were compiled.

The Report of the Presidential Commission on the Space Shuttle Challenger Accident was published on June 6, 1986 and delivered to President Reagan. Near the end of the 256-page report, four pages were devoted to recommendations. The nine basic recommendations set forth in the report are presented below in abridged form:

I

Design. The faulty Solid Rocket Motor joint and seal must be changed. This could be a new design eliminating the joint or a redesign of the current joint and seal . . . .

Independent Oversight. The Administrator of NASA should request the National Research Council to form an independent Solid Rocket Motor design oversight committee to implement the Commission's design recommendations and oversee the design effort . . . .

II

Shuttle Management structure. The Shuttle Program Structure should be reviewed. The project managers for the various elements of the Shuttle program felt more accountable to their center management then to the Shuttle program organization . . . .

Astronauts in Management. The Commission observes that there appears to be a departure from the philosophy of the 1960s and 1970s relating to the use of astronauts in management positions . . . .

• NASA should encourage the transition of qualified astronauts in management positions . . . .

Shuttle Safety Panel. NASA should establish an STS Safety Advisory Panel reporting to STS Program Manager . . . .

III

Criticality Review and Hazard Analysis. NASA and the primary Shuttle contractors should review all Criticality . . . items and hazard analyses . . . . An Audit Panel . . . should verify the adequacy of the effort and report directly to the Administrator of NASA.

more thumbs up for cheering crowds; no more phone calls from the President. They will not see their parents and their wives or husbands and their children meeting them—gone with the rest of the engines and the exploding sky—gone, but theirs were lives that mattered.

In Challenger's *Wake*

Why were they dead? Why did the $1.5 billion *Challenger* spaceship explode and scrub the United States space program for more than two years?

Just hours after the tragedy, Jesse W. Moore,

NASA's associate administrator, held a news conference. All data, visual records, even notes made by the launch team would be impounded. Moore also announced that he was forming an interim investigating board. He warned the press that speculation as to the cause of the disaster was not productive. The real answers would come only after extensive and detailed investigations of all the evidence, controlled tests on components, and hundreds of interviews.

But everyone knew that NASA should not be responsible for the primary investigation. A few days later, President Reagan announced the formation of a presidential commission on the space shuttle *Challenger* accident and named

SPACE DISASTERS & CLOSE CALLS

Recommendations Made by the Presidential Commission (cont'd)

IV

Safety Organization. NASA should establish an Office of Safety, Reliability and Quality Assurance to be headed by an Associate Administrator, reporting directly to the NASA Administrator . . .

V

Improved Communications. The Commission found that Marshall Space Flight Center project managers, because of a tendency at Marshall to management isolation, failed to provide full and timely information bearing on the safety of flight 51-L to other vital elements of Shuttle program management . . .

VI

Landing Safety. NASA must take action to improve landing safety.

- The tire, brake, and nosewheel steering systems must be improved . . . .

VII

Launch Abort and Crew Escape. The Commission recommends that NASA:

- Make all efforts to provide a crew escape system for use during controlled flight.
- Make every effort to increase the range of flight conditions under which an emergency runway landing can be successfully conducted in the event that two or three main engines fail early in ascent.

VIII

Flight Rate. The nation's reliance on the Shuttle as its principal space launch capability created a relentless pressure on NASA to increase the flight rate. Such reliance on a single launch capability should be avoided in the future. NASA must establish a flight rate that is consistent with its resources. A firm payload assignment policy should be established. The policy should include rigorous controls on cargo manifest changes to limit the pressures such changes exert on schedules and crew training.

IX

Maintenance Safeguards. Installation, test, and maintenance procedures must be especially rigorous for Space Shuttle items designated Criticality 1. NASA should establish a system of analyzing and reporting performance trends of such items . . . .

William Rogers as its chairman and Neil Armstrong as its vice chairman. Rogers took measures to ensure that NASA stayed honest and announced that some of the important tests would be conducted by independent firms.

Weeks of testimony began, both closed-door and public hearings. We heard from the engineers, from the senior NASA managers of the various centers around the nation, from some of the astronauts. We heard of the pressures to launch, the concerns of a group of fifteen Morton Thiokol engineers who were against a launching during such freezing Florida temperatures. Such a launch, they all agreed, was beyond their launch experience. We learned how their opposition to launch was not fully communicated, and how it was finally overridden by company managers who felt pressured from the NASA brass. We learned that there were serious safety concerns about the solid rocket booster seals for several years prior to *Challenger's* last flight, and that they were not communicated to NASA's top managers. We learned that the mating of the two rocket booster segments on either side of the suspect O-ring seal took longer than on any other flight. We soon all realized that technical flaws, combined with a lot of minor oversights, poor communication, busy launch schedule pressures, and bad decision-making contributed to the disaster.

After just a few weeks of testimony in Washington and several trips to the Kennedy Space

Center, Chairman Rogers contended that NASA's decision-making processes and lines of communication were flawed. Spokesmen for NASA countercharged that the Rogers commission was reacting before the evidence was in and that the media were putting pressures on all concerned. About one month after the tragedy, we all knew that the component failure on space shuttle *Challenger* was the right solid rocket booster, which had sprung a fiery leak. This hot plume broke through the seal or casing of the solid rocket booster and began the series of events leading to the explosion 1 minutes and 13 seconds after launch. Experts soon knew how the *Challenger* exploded and had a detailed microsecond by microsecond scenario of the events from ignition to explosion. The whys of this component failure, however, would be longer in coming.

The commission also urged that NASA submit a report to the president in June 1987 on the progress of implementing these recommendations.

Time Line to Disaster

After analysis of detailed tracking-camera photography and the telemetry streams that flowed for 73.631 seconds of *Challenger*'s last flight, a sequence of events was determined.

- 6.6 seconds *before* liftoff: The three main engines ignited in rapid sequence, which was normal.
- 0.0 seconds (11:38 E.S.T.): The two solid rocket boosters ignited, and *Challenger* lifted off the pad.
- 0.4 seconds *after* liftoff: Tracking cameras detected black smoke coming from the lower end of the right solid booster; smoke was seen for the first 12 seconds of the flight.
- 36 seconds into the flight: The main engines throttled down to 65 percent power.
- 58.7 seconds: Black smoke appeared again from the lower joint area of the solid rocket booster.

- 59 seconds: *Challenger* reached maximum dynamic pressure.
- 59.2 seconds: Intense flame burned through the O-ring seal of the right solid rocket booster.
- 60:16 seconds: The flame grew larger; the chamber pressure of the right booster fell.
- 66.17 seconds: The flame migrated; bright spots of fire were seen on top of the rocket booster.
- 72.14 seconds: Probable time that the lower attachment point between the solid rocket booster and *Challenger* failed because of fire or stress.
- 72.20 seconds: The lower end of the solid rocket booster broke away and swung out from the large external tank; the top of the booster moved in toward the external tank.
- 72.88 seconds: Probable time that the top of the booster ruptured the external tank.
- 73.175 seconds: A large cloud of oxygen or hydrogen gas fuel streamed out of the external tank's rupture area.
- 73.200 seconds: A flash fire appeared between the upper portion of the external tank and the forward portion of *Challenger*.
- 73.226 seconds: The powerful explosion occurred, caused by ignition of the external tank's liquid fuels.
- 73.631 seconds: Vehicle destruction; last data transmitted from *Challenger*; recognition of the disaster begins on the ground.

Had Challenger Reached Orbit

The launch invitations NASA sent to family members, friends, and colleagues of the crew contained an inserted orange card, printed in black ink on both sides, which summarized the activities on space shuttle *Challenger*'s mission 51-L. It even included a list of Christa McAuliffe's teaching activities from space that were to be broadcast to schools throughout the United States. There may be a memorial space shuttle flight in the future to honor the *Challenger* crew and their flight 51-L that never reached orbit. A summary of that mission plan is reprinted here.

We mourned them all. Their deaths gave real pain to millions. But because NASA had Christa speak more often in front of the media, her smile was in America's collective conscience. Courtesy NASA.

6

*Space
International*

Space Above All

Most of the 4.8 billion human beings alive today will live out their lives *on* planet Earth. Only a few hundred special people have orbited above the earth or traveled to the moon. Yes, the number will grow rapidly as travel to and from orbit becomes more frequent; we will soon speak of thousands of space travelers from many nations who have had the unique experience of living and working in space. Still, for the next few decades they will be the chosen few.

The earth exists *in* space and makes its yearly circuit around the sun. Space is above each and every one of us, no matter where we live. It transcends national borders, which is to say that space is inherently a global technology, outside the sovereignty of any single nation. A satellite orbits the earth, not a nation. And as the Space Age prepares to enter a new century, its international destiny becomes apparent. There will be both cooperation and competition among the growing number of spacefaring nations. The Big Four in space—the United States, Russia, the European Space Agency, and Japan—may soon be joined by China. Even now the Big Four are cooperating and competing. For the cooperative space ventures, the noncommercial scientific missions, a global space agency would seem to be inevitable. Joint space ventures now in existence between the European Space Agency and NASA are multinational efforts and point toward the future. Why not, after all, spread out the costs and share the benefits of scientific research?

Spacelab, for example, the highly complex, modular scientific laboratory that is flown on the space shuttle, is a cooperative effort between the European Space Agency (especially Germany) and the United States. And Canada is the proud creator of the space shuttle orbiter's manipulator arm that has proved itself as a major space tool on dozens of flights above the earth. Even Russia and the United States are cooperating in space research during a time when tensions remain high between the superpowers: Russia's *Vega 1* and *Vega 2*, the two spacecraft that took close looks at Halley's Comet, carried comet dust analyzers that were designed and produced in the United States.

Cooperation in space fosters cooperation on earth, and open scientific exchange will benefit all nations, rich and poor. The cooperative spirit has not always flourished in the Space Age. International politics does have a way of getting in the way of productive joint efforts, but this may be less a factor in the future as each successful mission emphasizes the international benefits of the global technology of space. But the military in orbit could make cooperation more difficult.

For the Benefit of All Mankind

When the United States formed the National Aeronautics and Space Administration in response to the Sputnik surprise, its mandate as stated in the 1958 National Aeronautics and Space Act emphasized international cooperation. The agency was to "engage in a program of international cooperation in work done pursuant to this Act, and in the peaceful application of the results thereof . . . . " NASA's activities would be "conducted so as to contribute materially to . . . cooperation by the United States with other nations." The act also declared that the United States was to exploit space for "peaceful purposes for the benefit of all mankind."

There was no doubt whatsoever as to the importance attached to this mandate by the newly formed space agency of the United States. Three decades later, NASA can point with pride to more than one thousand projects with over one hundred countries, and over two thousand joint ground-based space research projects. Also, more than fifty cooperative satellites have been placed in orbit. The future promises even more exciting international space efforts. An example in the immediate future: Canada, Japan, and the European Space Agency have signed cooperative agreements as participants in the U.S. space station of the 1990s.

The twenty-first century may see a world space program that will carry out an international manned mission to Mars by the year 2035. Arthur C. Clarke's *2010* had the Americans and the Russians going to Jupiter together. In truth, the planet most likely will be Mars, and the date is perhaps a few years too soon, but it will be international politics and the priorities of the superpowers that will ultimately decide.

The United States and Russia

The first twenty-five years of the Space Age were largely a story of competition between the United States and the Soviet Union, although there were some cooperative agreements. Even as the space race between the superpowers heated up in 1962, both countries signed an agreement that established a dedicated direct communications link for the exchange of meteorological data. The amazing fact is that these agreements are still in place.

The two countries also signed joint telecommunications agreements in the early 1960s. The passive communications satellite *Echo 2*, a huge aluminum-coated balloon inflated in orbit, was launched by the United States in January 1964. Under the agreements, Russia performed experiments with the satellite by using their antennae at the U.S.S.R. Zemenki facilities.

The crowning achievement in the U.S.-U.S.S.R. cooperation in space was the Apollo-Soyuz Test Project, the first international manned spaceflight. A study phase agreement was initiated in the fall of 1970, and the actual flight took place five years later, in July 1975. (A summary of the Apollo-Soyuz mission follows.) That the United States had already been to the moon twice was certainly a large factor in Russia's decision to cooperate with the above-earth rendezvous and docking between cosmonauts and astronauts.

The most important agreement between the two superpowers, which included the Apollo-Soyuz Test Project, was the 1972 Intergovern-

mental Agreement on Cooperation in the Exploration and Use of Outer Space for Peaceful Purposes. One of its main accomplishments was the development of compatible systems for the rendezvous and docking of manned spacecraft and space stations of the United States and Russia. Such hardware compatibility would also make possible a coordinated space rescue attempt to save imperiled astronauts and cosmonauts if time was on their side. Other major cooperative milestones of this agreement included rocket meteorology; satellite meteorology; exchange of remote sensing data of the natural environment; joint seminars and meetings to exchange lunar and planetary information, including data from the missions to Venus; exchange of lunar rocks for analysis; and jointly published papers comprising a review of research in space biology and medicine. The Russians also flew a total of thirty-five U.S. experiments on four of their unmanned biological satellites—*Cosmos 782* (1975); *Cosmos 936* (1977); *Cosmos 1129* (1978); and *Cosmos 1514* (1983).

This 1972 Intergovernmental Agreement was renewed in 1977, but because of the difficult political climate between the superpowers in the 1980s, it expired in 1982. Before it lapsed, however, a cooperative agreement between the Soviet Union and the United States, Canada, and France was signed and put into force, on August 13, 1980. The agreement called for the launching of three search and rescue satellites, two by Russia (COSPOS) and one by the United States, Canada, and France (SARSAT).

The first Russian COSPOS satellite (*Comos 1383*) was launched on June 30, 1982, and it was followed by the second (*Cosmos 1447*) on March 24, 1983. Then the single SARSAT spacecraft was launched four days later—on March 28, 1983. All three launches and orbital insertions were successful. By the end of 1984, this multinational search and rescue satellite system was credited with saving more than five hundred human lives in more than one hundred air and sea emergencies. Ask those people how happy they are the agreement went into force before the political climate changed and the 1972 Intergovernmental Agreement expired.

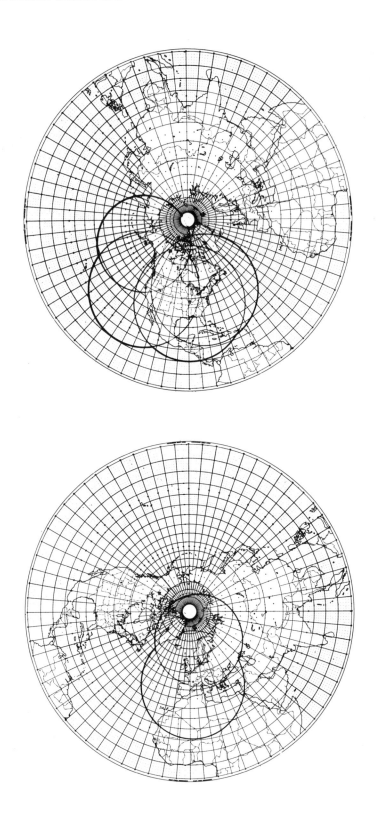

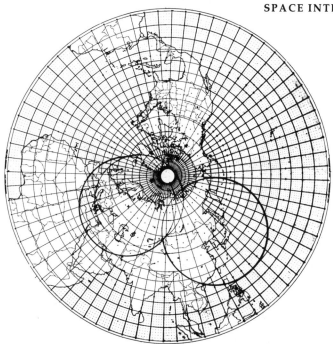

FACING PAGE AND ABOVE: *The United States, Canada, France, and the Soviet Union are cooperating in an international search and rescue project that uses satellites to locate people in distress on the land, at sea, or in the air. Called COSPAS/SARSAT, the system has already saved the lives of more than five hundred people, including two Canadians who had a river accident in a remote region of the Hudson Bay. Coverage regions around the world are shown in the three maps.* Courtesy Goddard Space Flight Center. PAGES 180-181: *Germany was the first European nation to fly a dedicated Spacelab mission in the fall of 1985. Space shuttle* Challenger, *on its last successful flight before the explosion, carried into orbit the first international crew of eight—the largest crew ever flown. Top row, left to right: Henry Hartsfield (commander), Bonnie Dunbar, James Buchli, and Reinhard Furrer (from Germany); bottom row, left to right: Ernst Messerschmid (Germany), Wubbo Ockels (from the Netherlands and representing the European Space Agency), Steven Nagel, and Guion Bluford.* Courtesy NASA.

The Apollo-Soyuz Test Project

A good question for a space trivia game would be: What was the mission of *Apollo 18*? The question might even prove difficult for space buffs, who would immediately clue in to Apollo-Soyuz but would probably look puzzled on hearing "*Apollo 18.*"

For the 1975 superpower linkup in space, the United States used the last Apollo spacecraft that would ever fly, one that was nearly identical to those that orbited the moon and carried the crews to America's first space station, Skylab. The original Russian plans in 1972 called for their using a new Soviet Salyut space station. But their schedule slipped, and they were not forthcoming in admitting it to NASA. Finally the truth did come out, and there was only one

option left: The Russians had to use a Soyuz craft, the same spacecraft they had been using since 1967.

Russia's *Soyuz 19* blasted off from the Baikonur Cosmodrome near Tyuratam on July 15, 1975, some seven hours before the Saturn 18 left the ground of Cape Kennedy. Because of the spacecraft involved, it was Apollo's responsibility to track and rendezvous with Russia's *Soyuz 19.*

Two cosmonauts made up the Russian crew: Alexei A. Leonov and Valery N. Kubasov. There were three astronauts in the American crew: Thomas P. Stafford, Vance D. Brand, and Donald K. Slayton. Spacecraft sizes were the only reason for the difference in crew numbers. Even though any remaining Apollo hardware would soon be in museums or mothballs, the U.S. spaceship was nevertheless considerably larger

ABOVE AND FACING PAGE: *The only time the United States and Russia met in space was for the superpower linkup in July 1975 called the Apollo-Soyuz Test Project. The two spacecraft, carrying three Americans and two Russians, docked in orbit, and handshakes were televised worldwide. It was the first wholly political space mission. After the docking of almost two days, the Russians and Americans took photos of one another's spacecraft.* Courtesy NASA.

than the Soyuz craft. Apollo weighed 32,549 pounds (14,764 kilograms), whereas the Soyuz weighed only 13,625 pounds (6180 kilograms).

After *Soyuz 19* and *Apollo 18* were successfully launched on July 15, the crews busied themselves with some of the rendezvous preparations, including orbital maneuvers. Both crews spent the second day in orbit doing experiments that were part of their flight plans. On the morning of rendezvous day, July 17, the U.S. Apollo crew made two orbital adjustments, which brought them within telescopic sighting of the Soyuz.

After some additional maneuvers by the American crew, the Apollo and Soyuz spacecraft rendezvoused in orbit and flew facing one another in a station-keeping configuration. The first international meeting of spacecraft in orbit occurred when Soyuz was in its thirty-fifth orbit and Apollo was in its twenty-ninth orbit around the earth.

"Three meters. One meter. Contact!" said Tom Stafford, who spoke in Russian for the occasion.

"Capture!" said Alexei Leonov in English.

The Russians and Americans had performed a mechanical mating of their spacecraft in orbit above the Atlantic Ocean some 640 miles (1030 kilometers) west of Portugal. Détente had made it all the way to orbit; above-earth politics had been born.

The Apollo and Soyuz spacecraft used different atmospheric pressures, so the mission called for an intermediate airlock chamber, which was built by the United States. As Stafford and Slayton worked through a detailed checklist to prepare the docking module for the historic meeting, nitrogen was added in the docking module atmosphere to equalize the pressure with that of Soyuz. Finally hatch 3 was opened, and Leonov's head appeared in the tunnel. Comrad Kubasov floated through next. Greetings were exchanged.

"Come in here and shake hands," Stafford called out. Handshakes were televised worldwide. A message of congratulations from Leonid Brezhnev was relayed to orbit. President Ford spoke to the crew members. Sometimes

two days. Then on Saturday, July 19, they undocked, pulled away from one another, and then maneuvered for a second docking. The next undocking was the final farewell separation, and it came before noon on July 19, 1975. Both spacecraft went off to their own orbits. The Russians returned to earth on July 21, while the American crew continued in orbit for three more days and landed safely on July 24, although not without some tense moments.

The failure to turn on the automatic earth landing system during descent produced a toxic nitrogen tetroxide gas that flowed into the cabin. Prompt use of crew oxygen masks may have saved the crew's lives and certainly prevented serious lung damage. When Apollo hit the water, it overturned into the stable position 2, with the conical top under the water. This prevented the blowers from venting the toxins into the atmosphere, and the crew would have breathed in the toxic fumes for at least five minutes if they had not used their oxygen masks while the flotation gear was attached to right the Apollo.

The potential tragedy was averted. The Americans and the Russians were down safely on earth. The world's first international crew had returned to their homelands, and it seemed likely that it would be the next century before these two nations met again on the high frontier above earth.

Future U.S.-Soviet Cooperation in Space?

The exploration of space, whether with unmanned or manned spacecraft, is extremely costly, and there are unquestioned advantages to joint space ventures. Space people dream of joint missions between the giant space powers of the United States and Russia. Apollo-sized programs that no single nation could undertake because of more urgent earthbound problems and shrinking space budgets could indeed be realized with major cooperative efforts. Joint exploration of the moon and establishment of a lunar base would be such a big space venture. And for many years a joint U.S.-Soviet manned flight to Mars has been discussed in space cir-

the ceremonies seemed as if actors reading their lines and it was a dress rehearsal instead of a final performance. But the image of cooperation was certainly there for everyone to see, and it *did* take real cooperation to make the Apollo-Soyuz mission succeed. It was July 17, 1975—almost five years since the joint project had begun.

During the period that Apollo and Soyuz were linked in orbit, the crews visited each other's spacecraft three times and took meals together, which included Russian sauerkraut soup and American beefsteak with barbecue sauce. A total of thirty-four experiments were accomplished, twenty-three by the Americans and six by the Russians. Of these, five were joint experiments. They were in the areas of astronomy, earth observations, and the life sciences, and included tests of crystal growth and electrophoresis in zero-g.

There were four crew transfers during the Apollo-Soyuz mission, including the first one after the famous handshakes, when Stafford and Slayton made their way into the Soyuz spacecraft and sat around a small table with the Russian crew for the TV cameras. The Russian and American spacecraft remained docked for

cles. Add the resources of the European Space Agency (ESA), and Mars is within reach.

But it all depends on international politics, and the first half of the 1980s saw mainly distrust and name-calling between the superpowers. What cooperation there was in space efforts was initiated informally through third parties such as France, which has its own cooperative agreements with both the United States and the Soviet Union. Examples include a U.S. instrument on board the two Soviet Vega spacecraft that went to Venus and Halley's Comet, and use of the U.S. Deep Space Tracking Network to acquire test data from the Vega spacecraft.

If and when the political climate improves, there are several joint projects that can be discussed between the two nations, ideas that have already been given serious thought.

The main one is a space shuttle/Salyut space station rendezvous, without docking, that could demonstrate some space rescue techniques with extravehicular activities. In one concept, a U.S. astronaut wearing a manned maneuvering unit would propel a space-suited Russian back to the space shuttle to demonstrate space rescue. Some PR and brotherly love.

Another possible mission would involve the U.S. and Soviet space shuttles in orbit together, but the Russian shuttle experienced development delays in the mid-1980s, and it appears likely that no such joint effort could take place before the 1990s. At the very least, it is probably that a Russian cosmonaut or two will visit the U.S. space station or ESA's Columbus facility before the year 2000. France, after all, owes the Soviets a ride in return for Jean-Loup Chrétien's flight aboard *Soyuz T-6* in June 1982.

The European Space Agency (ESA)

Formed in 1975, the European Space Agency grew out of the earlier European Space Research Organization. Its member countries include Austria, Belgium, Denmark, France, the Federal Republic of Germany, Ireland, Italy, the Netherlands, Norway, Spain, Sweden, Switzerland, and the United Kingdom. Canada has "observer" status.

While some of the member nations, especially France and Germany, have their own strong national space programs, ESA has proved that there is power in numbers and in pooling national resources for the good of Europe's future in space. The European Space Agency represents a major world space power and is extremely important to the future of scientific and commercial space efforts. The Agency's plans in the next several decades include heavy involvement in the U.S. space station through the design and building of a space station module called Columbus; the heavy-lift launch vehicle Ariane 5, which would eventually be man-rated and would fly the proposed French shuttle vehicle Hermes; and a free-flying retrievable space platform called Eureca that would be marketed internationally and is expected to fly its maiden flight in the late 1980s.

The ESA space program is vigorous and future-oriented, with long-term planning that goes beyond the turn of the century. As in the past, ESA will continue to cooperate with other spacefaring powers, but it will also compete with U.S., Soviet, and Japanese programs in getting its share of the $52 billion in annual commercial space revenues projected by the year 2000.

After ESA developed the Ariane rocket, Europe's competition for the space shuttle's launch business, it gave the project over to Arianespace, a private company that assumes full marketing, launch, and future development responsibility. This management and marketing group for Europe's Ariane launcher was established by thirty-six principal European aerospace and electronics firms, thirteen European banks, and the French space agency, the National Space Research Center, CNES. The French government is the largest stockholder in Arianespace. For this and other reasons, some private U.S. firms claim that Arianespace is subsidized and can therefore offer lower into-orbit costs to customers.

Dozens of ESA space projects have been designed, developed, and flown, some launched by their own Ariane rockets and others by the space shuttle and other U.S. launch vehicles. Financial contributions to the projects by the member nations vary. Germany, for example, funded the majority of the Spacelab, which was an ESA project, and France was the primary funding country for the Ariane rocket development before it went to Arianespace. Cooperative satellite projects such as the *International Sun-Earth Explorer 1* and 2 or the International Ultraviolet Explorer Satellite often contain experiments from several countries. The first Sun-Earth Explorer, for example, contained thirteen on-board experiments involving scientists from ESA, France, Germany, the United Kingdom, and the United States. The instruments studied the magnetosphere, interplanetary space, and their interaction.

The European Space Agency was very much involved in NASA's Hubble Space Telescope, scheduled for launch in 1988. It was responsible for developing the Faint Object Camera, which included a photon event counter. The European Faint Object Camera Team included scientists from Austria, Belgium, Chile, France, Italy, the Netherlands, the United Kingdom, and the United States—truly an international cooperative effort. The Space Telescope's Solar Array was also an ESA project, as was other scientific and engineering ground operations support. About 15 percent of the Space Telescope's observing time will be provided to ESA-sponsored astronomers.

The *Ulysses* spacecraft (formerly known as the International Solar Polar Mission) was built under the direction of the European Space Agency, while the Jet Propulsion Laboratory of the United States has management responsibility for the mission. *Ulysses*, after it takes a sharp turn at giant Jupiter, will journey out of the solar system's ecliptic plane to observe the never-be-

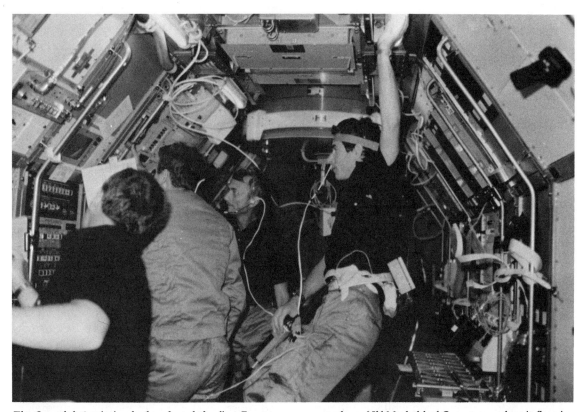

The Spacelab 1 mission had on board the first European crew member—Ulf Merbold of Germany—when it flew in Columbia's *cargo bay in late 1983. More than 70 experiments were conducted, and two hundred investigators from 16 countries participated in the flight. Seen here (left to right) are: Robert Parker, Byron Lichtenberg, Owen Garriott, and Ulf Merbold.* Courtesy NASA.

fore-seen polar regions of the sun and the interplanetary medium between itself and our star. (Read Chapter 8, "Robots from Earth," for a more detailed description of Project Ulysses.)

Another important future scientific mission on ESA's launch calendar is the *Hipparcos* satellite now in production and rescheduled for launch in the early 1990s. The mission's goal is to accurately measure the positions, velocities, and trigonometric parallaxes of about one hundred thousand selected stars. Such high-accuracy astrometry measurements will result in a tenfold improvement in the precision of existing observations and will provide a whole-sky catalogue suitable for detailed astrophysical studies. This will lead to a much greater knowledge of the internal structure and evolution of stars and will help scientists better understand such cosmic riddles as quasars and black holes.

But certainly one of ESA's most important contributions to our present-day and future Space Age is the development of Spacelab, a high-tech laboratory carried in the space shuttle's cargo bay. Its maiden flight aboard space shuttle orbiter *Columbia* was launched on November 28, 1983, and lasted ten and a third days. *Columbia* STS-9, the first shuttle flight to carry a six-person crew, landed safely on December 8. The flight also had the distinction of carrying the first non-American crew member, Ulf Merbold from Germany, the ESA nation most responsible for Spacelab's funding, design, development, and testing.

This first flight of Spacelab 1 was a very successful mission, involving over seventy experiments and two hundred investigators from sixteen countries. Experiments were conducted in the disciplines of stratospheric and upper atmospheric research, materials processing, plasma physics, biology, botany, medicine, astronomy, and solar physics, and in the technology areas of thermodynamics and lubrication. Some specific Spacelab 1 experiments on this inaugural flight were: the planting of sunflowers to find out about growth in zero gravity; an experiment in which an astronaut bounced while in a harness to determine the effect of zero-g on spinal reflexes; a silicone oil experiment to find out how it would behave in zero-g; and a common fun-gus-growing experiment to see if the absence of gravity affected its circadian rhythms. The series of Spacelab flights, four of which were flown through 1985 (including Germany's D-1 Spacelab mission), promises nothing less than a revolution in certain scientific disciplines and industrial processes.

The Evolution of Spacelab

In the 1970s, the United States approached the European Space Agency, headquartered in Paris, to see if its member nations would commit to the development and manufacture of a manned space laboratory that would fly in the large cargo bay of the U.S. space shuttle. ESA countries decided to take on the Spacelab challenge, and during its ten years of development, some forty companies in ten countries built components for this state-of-the-art laboratory that would provide ready access to space for a broad spectrum of experimenters in many scientific fields from many nations. Its modular design assured that Spacelab technology would have a long and active life in future space ventures, including its incorporation into the U.S. space station and the European Columbus module of the 1990s. The agreement between ESA and NASA also included a commitment by NASA to purchase a second Spacelab system, which, like the first, is designed to fly fifty missions.

West Germany became the ESA nation most involved in the Spacelab project, providing 53.3 percent of the ESA's committed budget, and it flew its very own Spacelab D-1 mission on the space shuttle *Challenger* in late 1985. Its experience in building Spacelab has given West Germany a competitive edge over the United States in some areas of technology, including certain industrial processes in zero-g. Italy was the major subcontracting nation for the module and invested 18 percent of the total ESA budget of $1 billion. More than two thousand Europeans were employed by more than forty contractors and subcontractors in Europe during the height of Spacelab manufacturing activities.

The basic structural unit of the Spacelab is a pressurized cylinder 8.85 feet (2.7 meters) long and 13.3 feet (4.1 meters) in diameter. But Spacelab can take on many configurations, depending on the needs of a mission. It's like a full-scale erector set for space, limited only by the cargo bay's dimensions (60 by 15 feet, 18.3 by 4.6 meters). Two modules join together to make the habitable "long module" workshop. With the end cones in place, these modules create a shirt-sleeve-environment working area for mission specialists and their experiments that is 23 feet (7 meters) long.

For automated activities in the outer space environment of the shuttle's cargo bay, there are space pallets (made in the United Kingdom) that can hold equipment such as solar, infrared, and X-ray telescopes, or various sensors or antennae. Space shuttle *Columbia*'s scheduled flight of Astro-1, which was canceled after the *Challenger* disaster, would have studied Halley's Comet with three ultraviolet telescopes mounted on a pallet in the open cargo bay. As many as five pallets can be flown on a mission, but then there is no room for a habitable module, in which case a smaller cylindrical igloo provides the needed electrical and temperature-control services. Most often, Spacelab missions have some combination of pressurized modules and exposed-in-space pallets. Spacelab 1, for example, flew with the long, two-module laboratory workshop and one pallet. It is this modular versatility that will give Spacelab technology a life that will extend well into the twenty-first century. The laboratory sections of the space station will be Spacelab modules with certain design changes, and it is likely that the same ESA member nations, Germany and Italy, will be primarily responsible for their development and manufacture.

Inside the laboratory module, Spacelab's environmental control system provides an atmosphere with the same air pressure as at sea level on earth: 14.7 pounds per square inch. The atmosphere is composed of about 20 percent oxygen and 80 percent nitrogen, and temperature can be controlled by the crew. Lithium hydroxide canisters remove carbon dioxide from the atmosphere.

Down the center of Spacelab's long module (with about the same space as a middle-size house trailer) is an 18-foot (5.5-meter) long center aisle, and the walls on either side are lined with laboratory racks that contain instruments and equipment that are custom-tailored for each mission. Tools, supplies, and equipment are stowed overhead, and there is a workbench for making minor repairs.

Spacelab has three computers: one to oversee the experiments; another to control Spacelab's environmental functions; and a third for use as an immediate backup if either of the other two fail. The computer mass memory can store 132 million bits of data, which, if converted into words, would equal 27 thick books. The Satellite Tracking and Data Relay Satellite is capable of transmitting data, voice, and video from the shuttle to earth at a rate equal to 10 good-size books every second.

On February 5, 1982, ESA formally delivered the first Spacelab flight unit to the United States in ceremonies at the Kennedy Space Center.

"It is indeed a proud day for us Europeans," said ESA director-general Erik Quistgaard to the assembled dignitaries and project workers. "Never before have we developed and built, as a joint European venture, such a major element of a NASA program. It is a great pleasure for us to hand over the first flight unit ...."

Vice President George Bush was on hand for the occasion. "We are returning to space together," he told his audience, "and that is no small achievement. Space Shuttle and Spacelab represent a bond, not just of transatlantic cooperation and friendship, but of a cooperation and friendship that will extend beyond the earth into space."

Spacelab's First Operational Mission

Spacelab 3 was launched into earth orbit inside the *Challenger*'s cargo bay on April 29, 1985. It was the first operational flight of the European-built manned laboratory, and it came only seventeen days after the *Discovery* returned to earth from its mission. In addition to the crew of seven astronauts, the *Challenger* also carried two squirrel monkeys and twenty-four white research rats to test the never-before-flown animal holding facilities and to test the animals' physical reactions to zero gravity.

The $220 million scientific expedition was one of the most productive missions of the shuttle program and produced discoveries of great value to the future commercialization of space and to space station operations in the 1990s. Among other things, the mission proved that scientists in orbit as mission specialists in Spacelab could work closely and productively with fellow scientists on the ground, solve problems as they came up, and also produce tremendous amounts of valuable scientific information. What Spacelab missions do during the next decade will provide an enormous data-base foundation for work done in the laboratories of the space station during the last few years of this century and the first few decades of the twenty-first.

Spacelab 3's experiments included an upper atmosphere study called ATMOS, which identified and measured more than forty different chemicals in the upper atmosphere to help scientists fight air pollution. ATMOS also found previously unknown molecules on the sun. Other experiments grew crystals from both fluid and vapor. The growth of red mercuric oxide produced a high-quality crystal much faster than expected, one that grew to twenty times the size of its original seed. Such crystals will be extremely useful as X-ray and gamma-ray detectors for the defense industry. Experiments in fluid dynamics in zero-g conditions were successful after some initial problems were solved with ground support. Such containerless processing of fluids by using sound waves to manipulate fluids in zero-g has the interest of some one hundred commercial companies. Significant fluid dynamics discoveries that will have commercial applications in the expanding free enterprise of space came about as a result of these experiments.

The experiments in life sciences, however, may soon produce benefits for future astronauts—despite the problems experienced with floating rat feces and animal food caused by an inadequate seal design on the animal cages. Constant physical monitoring of the astronauts, monkeys, and rats in zero-g, including urine

analysis for hormone changes, could result in methods to reduce the discomfort of the space sickness that about half of all flying astronauts experience. The flying monkeys also supported the averages when it came to space motion sickness. One monkey did not eat or drink normally for the first three days of flight because of space sickness. The other monkey had no symptoms of the so-called space adaptation syndrome. There may be some answers as to how to prevent or lessen the effect of space sickness in the mountain of data produced by Spacelab 3's life-science experiments.

The amount of scientific data produced by Spacelab 3 is truly mind-boggling. Three million frames of video images were taken on the mission, and this does not count the large amount of film brought back from orbit. The flight also collected some 250 billion bits of computer data, which could fill some 50,000 books of 200 pages each if converted into words. This is more than the annual book production of Europe or North America. This data-bit stream, if lined up in typewritten symbols, would stretch from the earth to the moon.

Nations in Orbit

Many of the participating member nations in the European Space Agency also have their own national space programs. In terms of active progress and budget, France and Germany are the most committed to the future promise of space.

The powers of the Far East, Japan and China, have rapidly expanding space programs, and India shops for the best launch deals in the international marketplace for her international satellites. Brazil is the most space-active South American nation.

Here are summaries of the space efforts of twenty-two countries, some of which will enter the ranks of the top ten spacefaring nations in the twenty-first century. Cooperative efforts with ESA and NASA area are also sometimes included.

Australia. Two domestic communications satellites—*Aussat 1, Aussat 2*—were launched in

1985 by the space shuttle. A third was put on hold in 1986 because of the grounding of France's Ariane rocket. Each Aussat satellite has fifteen channels, including TV broadcasting, and the three satellites will cover Australia's vast territory more efficiently and economically than any land-based communications network could.

Austria. A space-based broadcasting system, set up in cooperation with the Swiss, is planned to be operational before the end of the 1980s. Austria also contributed to ESA's Spacelab by building part of the mechanical ground-support equipment.

Belgium. This nation and its high-tech industries are involved in a variety of space projects. Belgian scientists sponsored six experiments in Spacelab 1's payload, and they have contributed to the Hubble Space Telescope as active members of the Faint Object Camera Team. Belgian industry has also participated in the development of *ERS-1*, Europe's first remote sensing satellite, ESA's Maritime Communications Satellite Program, (*Marecs*), France's *Spot* resource satellite, and the next generation of communications satellites, *Olympus*.

Brazil. This country's domestic satellite system for communications uses two Brazilsat satellites, both of which were launched in 1985 by France's Ariane rocket. Brazil also operates a *Landsat* ground station, which covers most of South America. The *Landsat* data has been used for mapping remote areas of Brazil's interior and for agriculture and other land-use applications. A weather satellite is also planned for launch before the end of the 1980s. Brazil's Institute for Space Research continues to develop and test its own rockets in an effort to gain an independent launching capability.

The United States has invited a Brazilian astronaut to fly on a future space shuttle mission.

Canada. In the early 1960s, Canada initiated its space program by having NASA launch its *Alouette I* and *II*, two satellites equipped for ionospheric research. The first in a series of Canada's Telesat satellites was launched by NASA in 1972, and the new generation of Telesats (once in orbit, designated *Anik*, the Eskimo word for brother) was put into orbit by the space shuttle during the 1980s. By deploying the Telesat satellites, Canada had the world's first satellite system to use geostationary satellites for domestic communications.

Each time a Canadian satellite is deployed in orbit by the space shuttle, it is done so with the indispensable robot arm (the Remote Manipulator System) that was designed and built by Canadian industry. It was first tested on the second flight of space shuttle *Columbia* in November 1981 and was declared operational by NASA a year later. Primarily designed to deploy and retrieve satellite payloads, the Canadian-built robot arm appears to be limited only by the imagination of the shuttle astronauts and their ground support teams.

In 1985 Canadian prime minister Brian Mulroney and U.S. President Ronald Reagan signed an agreement for Canada's participation in the space station that will be assembled in orbit in the 1990s. Canada's financial commitment is expected to total between $300 and $600 million, and definition studies are underway to determine what hardware Canada will design and build for the cooperative space station project. Canada's areas of interest are satellite servicing facilities, remote manipulator systems, and electrical power systems for the proposed free-flying platform that will orbit near the space station.

China. The People's Republic of China is both launching its own satellites and shopping the international marketplace for launching services to supplement its current launch capacity. At the same time, this nation is accelerating its international cooperative efforts in space and is offering its launch services to interested customers.

China is becoming more open about its space capabilities and is actively seeking space systems and technology from Europe and the United States. The country has recently signed cooperative agreements with France, West Germany, the United Kingdom, and the United States. Although China has dropped the idea of a manned space program for the present, it is solidifying its position as a world space power by aggressively expanding its launch capability. The Chinese have made rapid progress in liquid propulsion systems, and some experts consider it a remarkable development pace comparable to the U.S. efforts in the 1960s.

Four rockets make up China's family of launch vehicles, including the CZ-3, which was used to launch this nation's first geostationary communications satellite in April 1984. A second communications satellite is scheduled for launch in 1986. Since 1970, China has launched seventeen spacecraft, and the last few have utilized the advanced liquid oxygen/hydrogen booster upper stage, which can place 3,080 pounds (1,404 kilograms) into geosynchronous transfer orbit. This is the most powerful rocket being offered to international customers. Other recent Chinese space initiatives include a new launch site at Chengdu in southwestern China and two new tracking ships to support future space efforts.

In 1984 the Chinese Broadcasting Satellite Corporation paid NASA $200,000 earnest money for launch reservations for two of their domestic direct broadcast satellites, which will be built by Europe or the United States. They are scheduled for launch in 1988. For its space business, China will choose between the United States and Europe, who are increasingly in competition with one another.

China's new CZ-3 booster may offer the best deal in the international marketplace for small geosynchronous payloads and may eventually take business away from France and the United States. China is fast becoming a solid competitor in the world space market. In another decade, China and Japan may be neck and neck reaching for the benefits and profits of the high frontier. After the *Challenger* disaster, a U.S. company, Teresat Inc., contracted with China to have two U.S.-made communications satellites launched—*Westar VI* and *Palapa B*, the same satellites salvaged from orbit and brought back to earth by the space shuttle.

Denmark. The Danish Space Research Institute,

in cooperation with France's National Center for Scientific Research, is a principal investigator of the *High Energy Astronomical Observatory 3 (HEAO 3)*, which was launched on September 20, 1979. This sophisticated observatory was designed to scan the cosmos for cosmic-ray particles and gamma-ray photons, which scientists believe originate in supernova explosions and pulsars.

Danish scientists also contributed a crystal growth study to the Long Duration Exposure Facility, a satellite deployed by the space shuttle in 1984.

In cooperation with NASA, Denmark has an Automatic Picture Transmission (APT) ground station that has high-resolution picture transmission capability. Denmark is also one of seventy countries that has a scientific and technical exchange agreement with NASA.

France. In the mid-1970s, France and Germany joined in a cooperative effort and launched, using a U.S. Delta rocket, two communications satellites, *Symphonie A* and *B*. This enabled both these European countries to provide communications services to Europe, Africa, and South America. Just a decade later, France and Germany are the leading member nations in the European Space Agency, but they also have their own strong national space programs.

France's space agency, the National Space Research Center (CNES), was the prime mover in developing Europe's Ariane rocket, and it continues to develop a new generation of more powerful and versatile launch vehicles that will serve a worldwide commercial space market and compete directly with U.S. shuttle services. While Ariane 3 is operational and continues to build a substantial record of satellite launches, despite its in-flight failure and the loss of two satellites in 1985 and 1986, the more powerful Ariane 4 nears operational status. And the even more powerful Ariane 5 heavy-lift rocket, an expendable unmanned launcher, has been approved by the European Space Agency for development, which is expected to cost some $2 billion. Ariane 5, scheduled to be operational in the mid-1990s, will be capable of placing a maximum payload of 17,637 pounds (8,000 kilograms) into geostationary transfer orbit. A

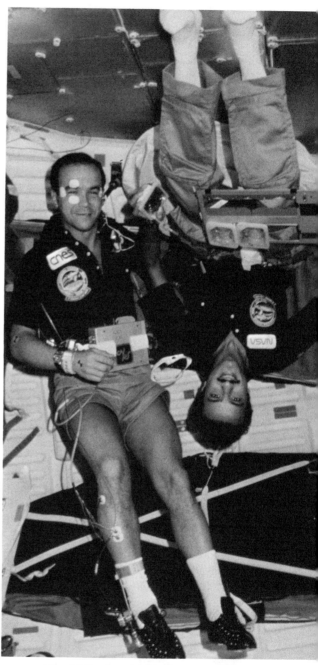

launch to low earth orbit could deliver approximately twice that weight, and France hopes to have its own reusable winged spacecraft called Hermes to put atop its powerful new Ariane 5 booster. Both the powerful Ariane 5 launcher and the Hermes shuttle spacecraft are the foundation programs that will take Europe into the twenty-first century. Europe's Columbus space station modules, which will probably orbit near the U.S. space station, require the heavy-lift Ariane, and the Hermes spacecraft will give European nations an independence in their space ventures that would not be possible if they relied on the U.S. space shuttle.

France's Hermes manned shuttle spacecraft will be able to transfer crews and equipment to the manned space station, service unmanned service platforms, and repair satellites in orbit. The French space agency, CNES, has determined the size of delta-winged Hermes to be about 50 to 60 feet (15 to 18 meters) in length and the payload bay to be some 9.8 feet (3 meters) in diameter. France hopes to begin full-scale development of Hermes in the late 1980s and looks forward to the first flight as early as

TOP LEFT: *Patrick Baudry, the first French astronaut to fly on the space shuttle, was a* Discovery *crew member in the summer of 1985. Here he is opening up a can of lobster, which anticipates the international menu that will be available on the space station.* ABOVE: *While Baudry does a medical experiment on himself, another international crew member, Sultan Salman Abdelazize Al-Saud of Saudi Arabia, eats his meal upside down.* Courtesy NASA.

1997. Unlike the larger U.S. space shuttle, which is attached to the side of its rocket boosters, Hermes would be launched atop the Ariane 5 heavy-lift rocket.

While the Hermes' initial design and development phases have been initiated and funded by the French space agency, France is seeking other European nations to participate in the manned vehicle program. It is expected that Hermes will eventually be supported by the European Space Agency, which has already approved development of the Ariane 5. To support its future space plans, France formed an astronaut corps in 1985.

France also continues its own active and varied satellite launch program beyond its ESA-sponsored projects. It launched its own *Spot 1* earth resources satellite in 1985, as well as the *Telecom 1B*, a data-to-telephone communications satellite.

With a continuing boost from the new generation of Ariane launch vehicles and the pooling of resources that the European Space Agency provides, France's future in space is bright.

Germany. As the leading participant nation in ESA's Spacelab program for the U.S. space shuttle, West Germany has proved that it has the management and industrial skills to remain a leader in the new generation of space projects.

In 1985 the West German cabinet approved its participation in the U.S. space station, and it will be ESA's leader nation in the development of Columbus, a three-module version of Spacelab that could eventually orbit close to, but separate from, the U.S. space station. The European-built Columbus will be a laboratory facility for experiments and manufacturing processes that will not be influenced by the movement and environment of the main station. As an independent orbiting habitat, Columbus would also have the capability of evolving into a separate ESA space station, an appealing prospect to member nations.

Germany was a prime mover in the shuttle pallet satellite (*SPAS-01*), which the space shuttle *Challenger* crew deployed on June 22, 1983. This was the world's first free-flying and reus-

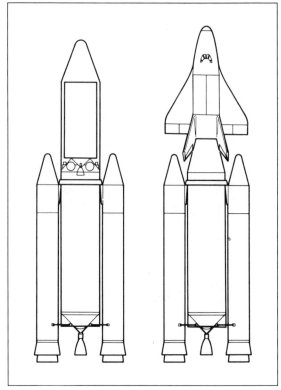

France is developing the powerful Ariane 5 rocket as well as the Hermes spaceplane that will take European astronauts into orbit by the turn of the century. Courtesy Centre National d'Etudes Spatiales.

able satellite, and it was Sally Ride, America's first woman astronaut, who deployed it and snagged it later in the mission to bring it back to earth. For this space first, Ride used Canada's Remote Manipulator System to launch and retrieve *SPAS-01*. It was this satellite that gave us the beautiful views of the space shuttle from above.

German industry is also the prime contractor for the future Eureca retrievable space platform, which has been approved by the European Space Agency. This first-generation space platform probably will be released into orbit from the U.S. shuttle cargo bay in the early 1990s for a six-month free flight. It will contain a payload of five microgravity facilities for materials sciences and life sciences, as well as communications and propulsion experiments. The payload weight limit will be approximately 2,200 pounds (99.8

kilograms), and it will have continuous power of 1,000 watts.

Once Eureca is released from the space shuttle, its internal engine will be fired and it will be moved into a higher orbit of 311 miles (499 kilometers) to begin its six-month operational mission, after which it will descend, rendezvous with a space shuttle for pickup, and be returned to earth for refurbishment.

Eureca represents an important development step between the Spacelab and the future European space station effort, Columbus.

West Germany has also contributed its space-engineering know-how to Project Galileo, which will probe some of the mysteries of our giant planet, Jupiter. In this cooperative effort between the United States and Germany, the all-important retropropulsion module that will inject the probe and the orbiter into their proper trajectories has been supplied by the Germans.

The *Rosat* X-ray telescope satellite, to be launched by the space shuttle in the late 1980s, was also developed and built by West German firms under a cooperative agreement with NASA. This state-of-the-art X-ray telescope is three to five times more sensitive than earlier X-ray satellites such as the *HEAO 1* and *2*. The reflecting surfaces of the large X-ray telescope are gold-coated. At least one hundred thousand new X-ray sources are expected to be found with *Rosat*, which will do a six-month survey of cosmic X-ray sources and then spend another year in closer examination of selected sources. Experts predict that *Rosat* will discover more black holes in its orbital lifetime than have been found since Cygnus X-1 was discovered in 1962.

West Germany will probably launch its new-generation direct-broadcast TV satellite, *TV-Sat* in 1987. The 4,409-pound (2,000-kilogram) TV satellite was developed under a cooperative program with France, which has its nearly identical *TDF-1* television broadcast satellite. This spacecraft design is being marketed by France and Germany in the Spacebus 300.

When the Space Age enters its second century, Germany will be an even stronger spacefaring nation because of mastering Spacelab technology and becoming a leader in zero-g indus-

trial processes. This expertise will be used to build Columbus, the separate European space station. The members of the Western Alliance will be neighbors in orbit.

India. A U.S. Applications Technology Satellite (*ATS-6*) was launched on May 30, 1974, by NASA on a Titan III-C rocket. This was one in a series of six ATS satellites designed to test new space instruments and demonstrate new satellite technologies. India used the *ATS-6* for the Satellite Instructional Television Experiment (SITE), which beamed educational programs to some 2,400 rural villages in India. The programs, produced by India, provided information on health, family planning, agriculture, and education.

Space shuttle *Challenger* launched *Insat 1B* into orbit in August 1983 for the government of India. Once the satellite was deployed from the shuttle's cargo bay, it was boosted into a geostationary orbit by a PAM-Delta booster. The satellite has telecommunications, community broadcasting, and meteorological capabilities. *Insat 1-C*, a sister satellite, was scheduled to be launched into orbit by the space shuttle and an assist booster in 1986, but the *Challenger* explosion made its launch date uncertain.

India shops the international market for its launch package and has used the European Space Agency's Ariane, the United States space shuttle, and Soviet Union rockets. In 1986, Russia launched the *IRS-1A*, a remote sensing satellite, for India. When the U.S. space shuttle launches the *Insat 1C*, an Indian mission specialist will be on board to oversee the payload and to perform experiments. Future decades will see India served by other satellites suited to this nation's unique needs. Competition in the international commercial space market may well help India get more in-orbit high tech for its rupees.

Indonesia. The first Indonesian communications satellite (*Palapa A-1*) was launched by the United States with a Delta rocket on July 8, 1976. Its sister satellite (*Palapa A-2*) followed it into orbit on March 10, 1977. After almost a decade, a second generation of comsats was developed to upgrade the service to the Indonesian islands, Thailand, Malaysia, and Singapore. These new satellites were designated *Palapa B-1* and *B-2*, and the first of them was launched by the space shuttle on February 6, 1984. Its deployment from the shuttle was successful, but the payload assist module (PAM) that was to boost it into geostationary orbit failed. The problem rested with its manufacturer, Hughes.

A multimillion dollar satellite, ready for its important task in geostationary orbit, and its secondary propulsion system did not work. It was an opportunity for the space shuttle to show its right stuff, and on *Discovery*'s mission in November 1984, both the *Palapa B-2* and *Westar VI* were recovered, locked into position in the shuttle cargo bay, and brought back to earth. When *Palapa B-2* returned to earth, it was owned by the insurance underwriters who, at high risk, had paid for the satellite retrieval. Indonesia has yet to get its *Palapa B-3* into orbit. The space shuttle was originally scheduled to launch it in the summer of 1986.

International Telecommunications Satellite Organization (Intelsat). Formed in 1964, Intelset is an international consortium including more than one hundred member nations and managed by the Communications Satellite Corporation (Comsat). Some thirty Intelset satellites have provided an international communications system from 1965 to the present day.

In the early 1980s Intelsat provided more than twenty thousand telephone circuits. This represented some 65 percent of the earth's total transoceanic telephone, telex, and data communications. The latest-generation satellite, *Intelsat 6*, has the capacity of forty thousand telephone circuits. The transmission capacity is doubled every few years, and it is projected that this will continue for decades.

Intelsat is *the* biggest success story in international cooperation in space. It has shown that more than one hundred nations from all over the earth can cooperate while also serving their own interests, as well as earn a profit doing it.

Ireland. The European Space Agency is the largest technological organization in Europe, and Ireland is its smallest and newest member. One of Ireland's main membership goals is to stimulate the development of high-tech industry in Ireland, and several new companies have been formed to accomplish such space work.

Serious consideration is being given to launching Ireland's own satellites, although no definite schedule has been established. A direct broadcast satellite would be the first off the launchpad. Ireland's scientists have also been involved in experiments on the Solar Maximum Mission and on the Long Duration Exposure Facility.

Israel. An agreement was concluded in 1984 between NASA and the Israeli Space Agency involving this nation's participation in the international Crustal Dynamics Project, which uses very precise laser ranging measurements to study the geodynamics of the solid earth, the deformation of the crust, and its rotational dynamics. Exchange of data between the more than twenty participating nations is important in understanding crustal hazards such as earthquakes and how the solid earth interacts with the oceans and the atmosphere.

Ground-based lasers use the *Lageos 1* satellite, launched in 1976, as an in-space reference point, and crustal movements are determined by repeated measurement of the transmission times of the laser beams reflected from the satellite. The moon and other astronomical radio sources are also used to obtain measurements in the Crustal Dynamics Project.

The Israeli Space Agency will measure tectonic plate motion and earth rotation using both a fixed and mobile laser ranging system provided by NASA.

Italy. An active member nation in the European Space Agency, Italy has made valuable and lasting contributions to many important space projects, including the Spacelab, and has ambitious plans for the next few decades of the Space Age.

Italy had 18 percent of the total European

Italy, in a joint undertaking with the United States, is building the Tethered Satellite System (TSS) for the space shuttle. The spaceship will trail the satellite up to a distance of 62 miles (100 kilometers) on the first mission. This reusable satellite will collect valuable data from the upper reaches of the atmosphere that has been inaccessible before. Courtesy Martin Marietta.

Space Agency investment in the Spacelab, and it was a major subcontracting nation for the pressurized module. The contract was worth $46 million. Italian workers at the Aeritalia aircraft plant rolled, shaped, and welded the cylinders for Spacelab, which were then trucked over the Alps to Bremen, Germany, on the North Sea for assembly. Their expertise will be put to use in building portions of Europe's Columbus, the space station laboratory.

On August 25, 1977, a U.S. Delta rocket launched the Italian *Sirio* satellite from Cape Canaveral. This was an experimental communications satellite that studied super high–frequency radio transmissions between Europe and North America during adverse weather conditions. In 1983 Italian scientists changed the orbit of the *Sirio* satellite. This extended its life which was scheduled to expire in 1979, indefinitely, and also allowed the satellite to fly over China. Italy and China have signed an agreement to increase cooperation in space projects. The change of *Sirio's* orbit involves the first cooperative effort between the two countries, whereby data from the satellite is shared. Italy built a mobile telemetry receiver and gave it to the Chinese, who positioned it in Beijing. Both nations now receive the satellite data. The age of Italian-Chinese joint space projects has begun.

Italy and the United States signed agreements in 1984 for two joint space efforts. The Italian National Space Plan (PSN) has agreed to develop and build a second Lageos (Laser Geodynamics Satellite), which will serve as a space reference point for the ground-based lasers used in the Crustal Dynamics Project. *Lageos* 2 probably will be launched from the U.S. space shuttle in the late 1980s or early 1990s. With two Lageos satellites in opposite orbits, the precision of current laser measurements will be inproved substantially.

The Tethered Satellite System (TSS) will also be built by Italy. The satellite will consist of science and service modules, which the Italians will build, and a mechanism that will deploy and retrieve the 1,102-pound (500-kilogram) satellite from the payload bay of the space shuttle, which the United States will build.

Trailing a tethered satellite from the space shuttle? In a way it's like fishing in orbit from the space shuttle, but instead of fishing for pike, the TSS is fishing for information about the upper atmosphere that is obtainable no other way. This so-called fringe-area atmosphere, approximately 60 to 90 miles (97 to 145 kilometers) above the earth's surface, is too high for balloons and aircraft, yet too low for satellites to remain in orbit for any useful time period because of atmospheric drag. The answer: a tethered satellite that will trail the space shuttle at distance up to 100 miles (161 kilometers). The TSS can be extended upward or downward from that shuttle's orbit. In the downward mode, the instruments of the tethered satellite will be rolled through the upper atmosphere and gather never-before-acquired information on the atmosphere and magnetosphere.

The first flight of the tethered satellite will probably take place in the early 1990s, and the shuttle crew will include an Italian as a payload specialist for the mission.

Europe's space station project, Columbus, which will be coordinated with the U.S. space station project, has the hearty support of Italy's space agency. It was, in fact, Italy and West Germany who proposed the Columbus project to the European Space Agency. The total ESA cost for the Columbus project is estimated at $2.12 billion during the years 1987 to 1995. If Italy had its way, Columbus would be flying by 1992—the five hundredth anniversary of Christopher Columbus's voyage to the new world.

Japan. The Japanese satellite, *OHSUMI*, was launched in early 1970, some seven months after *Apollo 11* landed two astronauts on the moon. While the Japanese space program started in earnest more than a decade later than the United States' and the Soviet Union's, it has nevertheless launched close to thirty applications and scientific satellites in its first fifteen years.

In the mid-1980s, Japan made a long-term commitment to space projects that will take the country into the twenty-first century. The plan includes development of a more powerful, two-stage rocket launcher called the H-2, the launching of more than seventy-five new applications and scientific satellites before the year 2000, and active participation in developing and building portions of the U.S. space station. Japan has also made an overall commitment to gaining the technology and spacefaring skills required for a more independent space program that will eventually make it competitive in the international commercial space market.

The Space Activities Commission of Japan has charted this nation's space activities to the turn of the century, and the required funding for the program will probably reach $10 billion by the year 1999. One of Japan's highest priorities is to develop the H-2 launch vehicle, which will be designed and built entirely with Japanese technology. Without the H-2 Japan would not be able to launch the larger, heavier spacecraft that it must put into orbit in the 1990s if the nation wants its share of wealth from the expanding high frontier.

The National Space Development Agency of Japan plans to have the H-2 launch vehicle operational by 1992. Its development cost is expected to exceed $770 million. The H-2 will be a two-stage rocket with two large solid–propellant strap-on boosters, and its advanced version will be able to deliver into geosynchronous orbit a payload more than three times heavier than that of the H-1 rocket: 4,409 pounds (2,000 kilograms) for the H-2 versus 1,212 pounds (550 kilograms) for the H-1. Once Japan's H-2 is operational, the nation will be able to meet its own launch requirements for the last decade of the twentieth century and will also be able to market its launch services to other nations without restrictions.

Future Japanese satellites include advanced applications satellites in broadcasting, weather, and communications, and scientific satellites, including *ASTRO-C*, scheduled for launch before 1988, which will provide detailed X-ray observations of celestial objects, including the mysterious central cores of active galaxies. Japan's tenth scientific satellite, launched in 1985, was the interplanetary probe *PLANET-A* (also named *Suisei*, meaning comet), which observed

Halley's Comet by ultraviolet imaging. Its relatively light weight of 275 pounds (125 kilograms) demonstrates why Japan needs the H-2 rocket.

A Japanese astronaut will fly on a space shuttle Spacelab flight in the future. He will conduct some thirty-four tests in materials processing in the Spacelab facility. This flight will anticipate Japanese participation in the United States space station in the 1990s.

In 1985 Japan and the United States signed an agreement of cooperation for the permanently manned space station, which may well turn out to be, with the participation of the European Space Agency nations and Canada, the greatest international cooperative space program in history. Japan is studying the development and construction of a main space station module with two independent and unmanned platforms. This main module would conduct a laboratory and supply store. The goal is to launch Japan's main module, two unmanned platforms, and a robotic teleoperator by 1995.

By the year 2002, Japan will probably be one of the major spacefaring nations on earth.

Mexico. The space shuttle launched two Mexican communications satellites in 1985. These Morelos satellites, each weighing approximately 1,500 pounds (680 kilograms) provide Space-Age communications services to Mexico for the first time. The first Mexican astronaut, Rodolfo Neri Vela, flew on space shuttle *Atlantis* in the fall of 1985.

The Netherlands. The Infrared Astronomical Satellite (IRAS), launched in January 25, 1983, was a joint effort between the Netherlands, the United Kingdom, and the United States. This amazing satellite, which made astronomical history with its many discoveries, was built by the Netherlands and contained a super-cooled infrared telescope that viewed about two thousand celestial objects a day in the infrared part of the spectrum that is invisible on earth because of the atmosphere. During the satellite's three hundred days of observations, it pinpointed the positions and intensities of more than

INTERNATIONAL SPACE PROJECTS

Date	Event
1988-1989	Olympus, Europe's new generation comsat, launched.
1990	Gamma Ray Observatory launched, a joint effort of Britain, the Netherlands, the United States, and West Germany.
1990	Russian space shuttle operational.
1990	Tethered Satellite System (built by Italy) launched by the space shuttle.
1991	Eureca, Europe's first free-flying, retrievable space platform, launched.
1992	Japan's H-2 launch rocket operational.
1995	Europe's Columbus module and Japan's space station module launched.
1995	Europe's Ariane 5 heavy-lift launch vehicle becomes operational.
1997	Europe's spaceplane, Hermes, built by France, is operational.
2004	Britain's HOTOL spaceship, a one-stage-to orbit, air-breathing scramjet, offers scheduled service to orbit.

two hundred thousand cosmic objects. One observation led to the discovery in 1984 of a circumstellar disk around the star Beta Pictoris—a possible solar system.

The Netherlands built the telescope with sixty-two detectors that are surrounded by helium, which keeps them at -456 degrees F. (2 degrees Kelvin). The IRAS telescope has a sensitivity that is equivalent to detecting a 20-watt light bulb on Pluto from the earth.

The Netherlands is involved in several European Space Agency scientific projects. For example, it contributed to the development of the Faint Object Camera that is part of the Hubble Space Telescope, and it developed an X-ray imaging spectrometer for the Solar Maximum Mission, in cooperation with the United Kingdom and the United States.

NASA plans to launch from the space shuttle sometime in the 1990s the Solar Optical Telescope, which will be capable of resolving details of the sun down to 43 miles (69 kilometers) across. The Astronomical Institute in Utrecht, the Netherlands, is contributing to the technical development of the project.

Norway. Norwegian scientists have participated in a Spacelab 1 experiment and in developing the instrumentation for two early European satellites before ESA was formed. Norway also participates in the international satellite-aided search and rescue program begun in 1979, COSPAS/SARSAT.

The Andoya rocket range in northern Norway, which began launching sounding rockets in 1962 in cooperation with NASA, is now part of the Esrange Special Project, ESA's sounding rocket program, which has launched more than four hundred research rockets.

Spain. The National Institute for Aerospace Technology (INTA), Spain's space agency, launched the *Intasat* satellite on a U.S. Delta rocket in 1974. This satellite transmitted radio signals to ground stations and provided important data on the earth's ionosphere.

Spain is host country to several tracking stations that are part of the global network essential for manned and unmanned space missions. The Canary Island facility, no longer operational, was often referred to by the early Mercury and Gemini astronauts—now a bit of trivia in the growing album of space nostalgia.

During the first twenty-five years of the Space Age, Spain has also participated in remote sensing projects, testing of early communications satellites such as Telstar and Syncom, the International Satellite Geodesy Experiment, the laser-ranging Crustal Dynamics Project, the earth resources programs of *Landsat 1* and 2, and the lunar sample studies from the Apollo missions.

Sweden. In addition to its involvement in ESA space projects, Sweden also has its own national space program that is carried out by the Swedish Space Corporation, a state-owned limited corporation under the Ministry of Industry.

Sweden's *Viking*, a magnetospheric research satellite designed to gather information on the earth's magnetic field, was launched with France's earth resources satellite, SPOT 1, on an Ariane rocket in 1986. It was Sweden's first satellite. *Tele-X*, an experimental telecommunications satellite for direct TV broadcasting and new specialized data and video services, is scheduled for launch in the late 1980s and will serve the East Nordic region.

The Swedish Space Corporation has also initiated studies to evaluate the possible development of *Mailstar*, a low-cost electronic mail relay satellite. *Mailstar* would operate from polar orbit and would require very little, if any, ground tracking support. If the studies show that the project is economically feasible and the traffic demand is high enough, a second Mailstar satellite could follow the first and shorten message relay time.

Also under study for possible development is a commercial launch base for small polar orbiting satellites. If there is enough commercial demand to support it, the launch base would be built at the Esrange facility in northern Sweden, which is above the Arctic circle. The facility already provides tracking, telemetry, and command services for the Viking satellite, France's SPOT earth resources satellite, and other spacecraft.

For a small nation, Sweden has a big space program, and its geographical location may well give it an advantage in certain commercial space services.

Switzerland. In cooperation with more than a dozen other nations, Switzerland participates in the Crustal Dynamics Project by acquiring and exchanging data derived from laser ranging instruments. The University of Bern contributed

to the development of the solar wind experiment that flew on the *International Sun-Earth Explorer-3*, which was launched by NASA in 1978. This spacecraft was renamed the International Cometary Explorer (ICE) after NASA flight controllers began changing the spacecraft's trajectory in 1982. This old spacecraft that was called upon to perform new tricks rendezvoused with comet Giacobini-Zinner in September 1985 and became the first spacecraft to encounter a comet.

The Swiss also contributed several experiments, along with those from the United States, to Spacelab's Space Life Sciences 1 flight of the space shuttle. This was the first Spacelab mission dedicated entirely to life sciences investigations, which included the proliferation of white blood cells in zero gravity and cardiovascular changes during spaceflight.

Swiss scientists at the University of Bern also had the honor of contributing a solar wind experiment of five Apollo missions, including *Apollo 11*. (For more details, refer to Chapter 1, "A Space History: The Golden Years.") The foil sheet of this experiment can be seen next to the *Eagle* moonship in many of the first photographs from the surface of the moon.

United Kingdom. Britain is one of the top four most active member nations in the European Space Agency and is involved to some degree in almost all of ESA's space projects. It also has a long history of cooperation with the United States in space projects that dates back to the *Ariel-1* satellite launched in the spring of 1962, which measured cosmic ray energies, solar X-rays, and solar ultraviolet rays. The cooperative Ariel satellite series continued until 1974.

Britain also built the hardware and ground support for the *International Ultraviolet Explorer Satellite*, launched in 1979. This spacecraft's ultraviolet telescope studied stars, vast clouds of gas, planets, and comets. The United States and the European Space Agency were also involved in the project.

The new generation of European communications satellites, considered an essential part of Europe's future space program, grew out of ESA

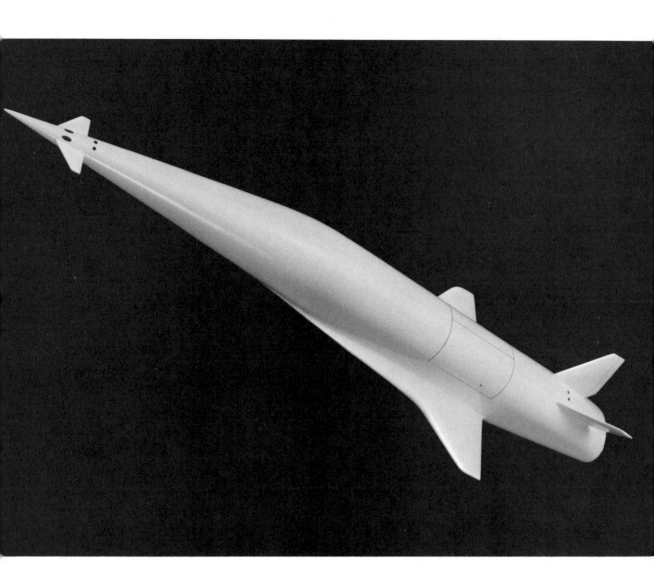

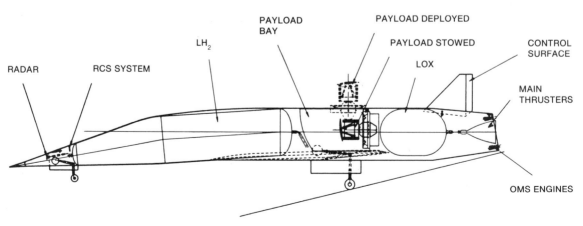

RADAR   RCS SYSTEM   LH₂   PAYLOAD BAY   PAYLOAD DEPLOYED   PAYLOAD STOWED   LOX   CONTROL SURFACE   MAIN THRUSTERS   OMS ENGINES

studies. This new class of large, multipurpose satellites, named Olympus, is now under development, and British Aerospace is the prime contractor for the program, which has a supporting international team of some forty major industrial subcontractors. The first launch of Olympus on an Ariane 3 rocket will take place in the late 1980s.

British scientists are involved—along with those from West Germany, the Netherlands, and the United States—in developing the Gamma Ray Observatory, which will be deployed from the space shuttle in 1990. The satellite will allow close observation of the nuclear processes occurring near neutron stars and black holes.

Europe's *Giotto* spacecraft, which encountered Halley's Comet in March 1986, had as its prime contractor the British Aerospace Dynamics Group, which managed a team of specialist contractors from ten European countries. A total of eighty-seven participating institutions and over two hundred scientists were involved in the mission. This same international team was responsible for *Geos*, the world's first scientific geostationary satellite, and *Ulysses*, the first spacecraft that will orbit the poles of the sun.

Giotto carried ten experiments and weighed 2,112 pounds (958 kilograms). The cylindrical spacecraft spins at a rate of fifteen revolutions per minute and has a tripod mounted on the top, which carries the magnetometer sensor and the antenna feed. *Giotto*'s power supply comes from a solar cell array and four batteries. A dual-sheet bumper shield on the comet side of the spacecraft protected it from high-velocity dust impacts during the flyby. *Giotto*'s encounter with Halley's Comet in 1986 was the European Space Agency's first interplanetary mission.

Great Britain also contributed the hard X-ray imaging experiment for Spacelab 2, which collected X-ray data from great clusters of galaxies billions of light-years away. It can also share the credit with the Netherlands and the United States for the outstanding success of the *Infrared Astronomical Satellite*, which opened up a new window that looked onto the cosmic sky.

For the beginning of the twenty-first century, British Aerospace has proposed to the European Space Agency a new launch vehicle, which could be operational by the year 2004. This single-state reusable launch system, called HOTOL, would be a horizontal takeoff and landing vehicle, which would operate like an airplane on runways that were built for the Concorde. HOTOL would use liquid-fuel rocket engines as well as air-breathing engines for extra power, and it would be capable of putting a 15,400-pound (6,985-kilogram) payload, such as an advanced communications satellite, into low earth orbit.

HOTOL, unlike the Ariane 5 heavy-lift vehicle scheduled for operation in the 1990s, does not have European Space Agency approval at this time, but preliminary studies continue.

The United Kingdom knows the importance of the high space frontier to its own future, and it will maintain a balanced and diversified space program through ESA projects and its own national program. If the commercialization of space fulfills its promise, there may even be a Harrods in orbit within the next century.

The United Kingdom is developing an advanced spaceplane called HOTOL, which will take off and land horizontally, using the very same runways that the Concorde currently uses. Its revolutionary propulsion system will be a combination of air-breathing and rocket engines. Courtesy British Aerospace and the British Embassy.

7

Rockets
& Spaceships

By Saturn

The great Saturn 5 moon rocket was the most powerful rocket ever launched off the face of planet Earth. To this day, the Russians have not been able to match its awesome power. Its first stage alone, which has a cluster of five F-1 engines, delivered a maximum thrust of 7,750,000 pounds (3,515,400 kilograms), some one hundred times more powerful than the Redstone rocket that boosted Alan Shepard and his Mercury space capsule into America's first manned suborbital flight. The thrust power of this first stage equaled 180 million horsepower. Add the thrust power of Saturn 5's upper two stages, and this giant 363-foot (111-meter) high monument to human ingenuity, intelligence, and organization becomes 115 times more powerful than the first manned rocket, the Redstone.

When the first Saturn 5 was launched on a test flight on November 9, 1967, sending a dummy Apollo command ship into earth orbit, there was no mistaking its power as the air roared and the ground trembled. That noise and its shock waves actually represented 1 percent of the Saturn 5's energy, which was converted into the thundering noise that witnesses felt throughout their entire bodies.

The world had never seen such power balance a great ship on its column of fire and thrust it skyward. In just 160 seconds, Saturn 5's first stage burned up 4.6 million pounds (2.09 million kilograms) of propellant. Another 1 million pounds (453,600 kilograms) of fuel were consumed by the second stage in a six-and-a-half minute burn. And just as amazing as its incredible power was the high reliability—more than 99 percent—of its millions of working parts. This, the greatest rocket ever made, sent twenty-seven American astronauts around the moon and twelve of them to its surface.

The Saturn 5 was launched for the last time on May 14, 1973, when it carried the huge unmanned Skylab space station above the earth to await its first three-man crew. In all, fifteen Saturn 5s were built. There were two unmanned test flights; eleven manned Apollo missions; and one Skylab space station launch as the finale. But that only adds up to fourteen, so where is the other giant rocket? It is in mothballs. If the space shuttle can't lift something extra big and heavy into orbit, then there are only two choices: Wait for the next generation of heavy-lift rocket vehicles, perhaps a decade away; or bring the Saturn 5 out of mothballs, assemble it, and hear its unforgettable roar once more.

Blasting Off the Shuttle

If the Saturn 5 gave the impression of being not quite sure it wanted to leave its launchpad during its initial power buildup, as huge clamps held it to earth before releasing it skyward, the space shuttle leaves no doubt in anyone's mind. The two solid rocket boosters, strapped on to the external tank, deliver their power quickly and smoothly, and the first reusable spaceship is off the ground in about three seconds on the plus side of countdown zero. The solid rockets deliver a combined thrust of 5.3 million pounds (2.4 million kilograms), and the shuttle's three main engines, mounted on the stern of the orbiter and fed propellant by the external tank, deliver a total thrust of 1.1 million pounds (499,000 kilograms). So when the space shuttle blasts off, its five engines (the two solids and three main orbiter liquid-propellant engines) deliver a total of 6,425,000 pounds (2,914,380 kilograms) of thrust—just a mere million or so less than the great Saturn 5. And while the Saturns were usually going all the way to the moon, the shuttles (with all that power) are just traveling a few hundred miles (kilometers) above the earth.

The launch experience is less rigorous for the shuttle crew members than on the earlier manned rockets. While pre-shuttle astronauts were pushed hard against their couches during launch, often experiencing up to 8.1 g's (a force eight times their body weight), g forces do not exceed 3 during the shuttle's ascent to orbit, an experience similar to taking a corner fast in a

PRECEEDING PAGES: *The Saturn 5 and space shuttle* Columbia *blast off from earth. Atop the Saturn is the* Apollo 11 *spacecraft, and inside are the men who will be the first to walk on the moon. This maiden launch of the space shuttle took place on April 12, 1981.* ABOVE: *The great Saturn 5, no longer flying, can still claim to be the greatest rocket the world has ever seen—in power* and *reliability.* Courtesy NASA.

sports car. These maximum forces are experienced only twice during a trip to orbit. The first time occurs about two minutes into the flight, when the solid boosters burn out, separate, drop off, and fall by parachute into the Atlantic or Pacific. This 3-g period is short. The second and final 3-g load during ascent comes about seven-and-a-half minutes into the flight, when the space shuttle is some 105 miles (169 kilometers) above earth and just before the big external tank separates and falls back to earth, disintegrating during the descent. This triple gravity load lasts for about a minute. Then the two orbital maneuvering rockets kick in for the final orbit insertion. All that power is spent in less than ten minutes, the time it takes for the shuttle to achieve orbit after leaving the Florida coast and the gravity of earth behind.

Atop the Rocket's Red Glare

It was the Redstone rocket that literally got the United States space program off the ground. A modified version, Juno 1, launched America's first satellite, *Explorer 1*, after the Vanguard failures, and then went on to boost Alan B. Shepard and Virgil I. Grissom into their suborbital flights in 1961. The rocket, named after the U.S. Army Redstone Arsenal in Huntsville, Alabama (later to become the Marshall Space Flight Center), was developed by the Wernher von Braun team, and its technology developed directly from the German V-2 rocket of World War II.

After the war, North American Aviation developed an improved rocket engine from the V-2 engines that came back to the United States. The German V-2 engine was the foundation on which both the Americans and Russians would build their ICBMs and their people-rated space rockets. It was the U.S. Air Force that wanted the new engine for an unmanned bomber called the Navaho, which never did fly. But the work was valuable and paid off for North American when the U.S. Army gave the company a contract to modify the Navaho engine. The end result was the H-1 engine that powered the Redstone rocket. Several H-1 engines, in a cluster,

The early rockets—Redstone, Atlas, and Titan—that began the manned space program for the United States. The Redstone took Alan Shepard on his historic suborbital flight; the Atlas put John Glenn into his three orbits around the earth; and the Titan put Gemini 8, *with Neil Armstrong and Dave Scott aboard, into a mission that came close to disaster. Courtesy NASA.*

Manned spacecraft of the United States, beginning with Mercury (upper left), then Gemini, Apollo, Skylab (lower left), Apollo-Soyuz, and the space shuttle. Nine Mercury capsules could easily fit inside the shuttle's cargo bay. Courtesy NASA.

eventually became the power plant for the Saturn 1's first stage.

The Redstone that fired Shepard and Grissom to an altitude of 118 miles (190 kilometers) above the earth on their suborbital missions was 83 feet (25 meters) high with the capsule and escape tower attached. This compared with the German V-2 rocket of 46 feet (14 meters). The Redstone's thrust power of 78,000 pounds (35,380 kilograms) compared with the V-2's thrust of 56,000 pounds (25,402 kilograms). Redstone represented the first growth spurt for rocketry power which would continue until the giant Saturn 5s began to fly.

Other Peopled Rockets

The Atlas. The Atlas rocket that sent John Glenn and other Mercury astronauts into orbit around the earth was originally built for the United States Air Force as an ICBM. This was also true for the more powerful Titan booster that lifted the Gemini astronauts aloft. It was good news for the world that these rockets blasted off with people aboard instead of bombs.

The Atlas and the Atlas/Centaur are among the most reliable and versatile rockets of the United States. Various modifications of the Atlas make it an entire rocket family, and about five hundred have been launched since 1957, delivering men and satellites into orbit; Surveyor spacecraft to the moon; Mariner spacecraft to Mercury, Venus, and Mars; Viking to Mars (with a Centaur atop a Titan 3-E), Pioneer to Jupiter; Voyager to Jupiter and Saturn (also with a Centaur atop a Titan); and on and on into the future, as this rocket family launches future space probes to the planets and secret satellites for the Department of Defense. These rockets will be used even more frequently in the wake of the space shuttle disaster, as will other unmanned expendable rockets under development.

The early Atlas that put John Glenn into his orbit around the earth was almost four times more powerful than the Redstone that took Shepard and Grissom on their up-and-down suborbital curves that plunked them into the Atlantic. Its thrust was 308,000 pounds (139,708 kilograms) versus the Redstone's 78,000 pounds (35,380 kilograms). Compared with the German V-2 (fired successfully for the first time on October 3, 1942), the Atlas was five and a half times more powerful when it was shot into orbit during December 1958 to make a mission called Project Score a success and to let the world know that the United States, like Russia, had a powerful rocket in its ICBM arsenal. In Project

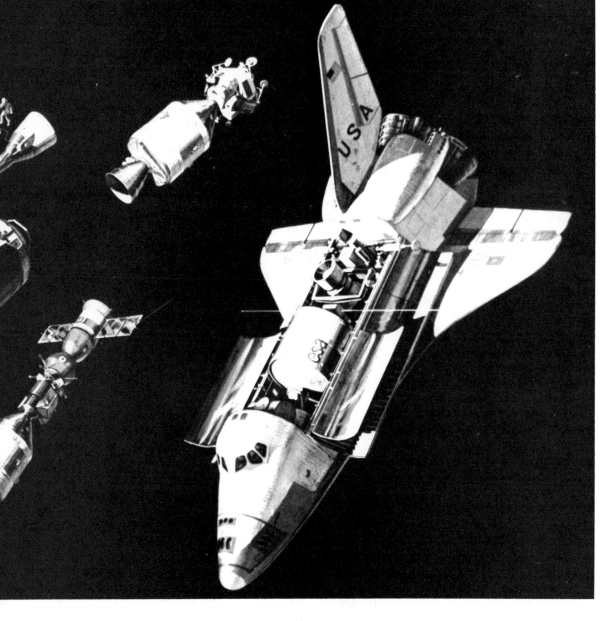

Score, the Air Force sent the entire rocket, less its fuel, into orbit. Included was a tape recording from President Eisenhower that transmitted a Christmas message to the people of earth for almost two weeks. This Atlas flight was launched with rocket fuel *and* the power of politics.

The Titan. Like the Atlas, the Titan rocket is a family of rockets, with many configurations, which was generated from the original Titan 1, an ICBM designed by the United States Air Force, which became operational in 1962. It was the Titan 2 that flew the two-man Gemini missions in the 1960s. Titan 2 was 90 feet (27.4 meters) tall, and its two stages delivered 530,000 pounds (240,408 kilograms) of thrust—almost seven times the power of the first American manned rocket, the Redstone.

The most advanced Titan rocket, the Titan 3E, has a Centaur as its upper stage, as well as two solid rocket boosters that are strapped on. This Titan can fire 38,000 pounds (17,236 kilograms) into low earth orbit or 8,000 pounds (3,629 kilograms) of payload off to the planets, as it successfully did with the marvelous Viking missions to Mars and the Voyagers to Jupiter, Saturn, Uranus, and beyond. Titan's fire played a major role in bringing back the new worlds of Mars, Jupiter, Saturn, and Uranus for earth folk to see and marvel over.

Nova: The Giant that Never Flew

After President Kennedy's famous to-the-moon speech of May 1961, NASA scrambled to define the methods and hardware needed to reach the moon. What rocket or combination of rockets would take the astronauts there and return them to earth? There were several proposals before the decision was made to build the Saturn 5, and one of these was called Nova. Like the Saturn 5, Nova would have used a cluster of F-1 engines for its first stage, but the cluster would include eight of them instead of Saturn 5's five. They would have given Nova's first stage of thrust of some 12 million pounds (5.4 million kilograms), some 4.5 million more than Saturn 5, and Nova's second stage would have delivered a thrust of 4.8 million pounds (2.2 million kilograms), almost five times the power of Saturn 5's second stage.

The Nova giant had the power for a direct ascent to the moon, but when NASA decided on either an earth-orbit or a moon-orbit rendezvous approach, which would require a less powerful rocket, the proposals and work on the Nova went into the history files. Perhaps it was a good thing, too; given the vibrating thunder that Saturn 5 produced, the gargantuan Nova might have been too loud for human ears, even with the NASA-supplied earplugs that reporters and invited guests near to the launch received and wore.

The Russian Workhorse Rocket

For the first quarter-century of the Space Age, the Soviet Union has had one dominant rocket to carry out its space program. Known in the West as the A-class launch vehicle, its Russian creators call it *semyorka*, meaning Rocket 7 (R-7). Good old Russian rocket 7 has every reason to be remembered affectionately by Soviet rocketeers. It began the Space Age by launching *Sputnik 1* in 1957 and went on to launch the first human being into orbit around the earth—Yuri Gagarin in 1961. It has blasted skyward more

than one thousand times in the last thirty years, and its various modifications, with new upper stages, have carried a total of 6,000 payload tons into orbit, including the manned Vostok and Soyuz spacecraft.

This workhorse Russian rocket was designed as the first Soviet ICBM by Sergei Korolev, who persuaded the Kremlin to finance the project in 1955. Two years later, in August 1957, it made its maiden flight, just weeks before it captured world attention by blasting *Sputnik 1* into orbit.

After it carried Gagarin and his Vostok spaceship into orbit, it also became known as the Vostok rocket. Its twenty rocket engines, configured into five clusters of four engines each, deliver about 1 million pounds (453,600 kilograms) of thrust. With the spacecraft on top, its two and a half stages stood 126 feet (38.4 meters) high. Russia's old mark seven rocket, the *semyorka*, has earned a secure place in the history of the Space Age.

Russia's Proton Rocket

Russia's moon rocket, probably intended to send cosmonauts around the moon if not actually land them on it, is known as the Proton rocket and carries a D-class designation in the West.

The Proton, with a thrust of about 3.5 million pounds (1.6 million kilograms), less than half that of the Saturn 5 moon rocket, sent four unmanned Soviet Zond spacecraft around the moon and back to earth during the period from 1968 to 1970. The first flight of the Proton took place in July 1965, when it placed into orbit a 26,896-pound (12,200 kilogram) physics laboratory, the Soviet's heaviest payload to that date. Another such payload was launched in November 1965.

When the Proton began launching unmanned Cosmos spacecraft to the moon in 1967, there were several failures. The rocket became somewhat more trustworthy when the Zond missions began in 1968. And then in 1969, the Proton sent to the moon the unmanned *Luna 15*, which touched down one day after *Apollo 11*'s historic moon landing.

But while the Saturn 5s carried men to and from the lunar surface, the Proton never carried cosmonauts, not even into earth orbit. In April 1971, it launched the first Salyut space station, which weighed some 20 tons, but it destroyed another Salyut station in the summer of 1972, when it failed.

Russia's big rocket could make some unmanned circumnavigations of the moon and place Salyut space stations in orbit, but it was never considered reliable enough to carry people into space, so it was a distant second to Saturn 5. Indeed, the Proton may have been one of the major reasons, along with the spectacular successes of the United States, that the Soviets never sent cosmonauts to the moon.

If there are delays in the debut of the Soviet superbooster that has been expected for some time by experts in the West, today's more reliable Proton, still used to place heavy payloads into earth orbit, may yet fly people.

Future Superbooster

Rumors about a Russian superbooster have been circulating in the West for several years. In 1981 the Pentagon confirmed that such a large booster was being built by the Soviet Union and said that it would be capable of placing into orbit some two hundred tons of payload—six to seven times the payload capacity of a single space shuttle flight and twice that of the once great Saturn 5.

While the West still awaits its debut, the record shows that a very large booster, perhaps comparable to Saturn 5, was tested three times (in 1969, 1971, and 1972), and all of the tests ended in failure—one on the ground and two in flight. If and when the so-called G-class superbooster becomes operational, it will be capable of placing a 100-ton-plus space station in orbit, sending cosmonauts directly to the moon, or placing large electromagnetic space weapons in orbit.

The power of such a superbooster would be awesome, the most powerful machine ever blasted off the earth. Its launch thrust could amount to some 12.3 million pounds (5.58 million kilograms), more than one and a half times the power of Saturn 5. And if one of them should blow up on the pad, it would not be a Russian-kept secret. Dozens of high-flying satellite eyes will see its great fireball and report it to the entire world.

The first Ariane rocket launch from the Kourou, French Guiana, launch site of the European Space Agency. The Ariane continued to fly its payloads into orbit after the space shuttle Challenger *disaster.* Courtesy NASA and ESA.

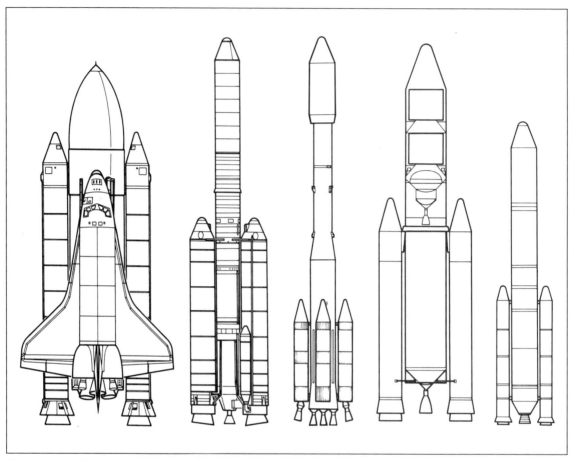

A comparison of present and future rockets. From left to right: the U.S. space shuttle; the U.S. Titan 34-D Centaur; Europe's Ariane 44 L; Europe's Ariane 5, which will fly in the 1990s; and Japan's H-2 rocket, now under development and scheduled to fly in the early 1990s. Courtesy Centre National d'Etudes Spatiales.

Coming Soon to Your Local Launchpad

Before or shortly after the year 2000, several newly developed rockets will point skyward on the launchpads of spaceports around the world, as the space frontier becomes truly international in scope.

France and the European Space Agency plan to have their powerful Ariane 5 rocket ready to fly by the mid-1990s. This heavy-lift expendable rocket, which will be capable of putting as much as 17,000 pounds (7,711 kilograms) into geosynchronous orbit, is expected to place between twenty and fifty commercial payloads

above the earth each year. Lift-off and initial ascent of the three-stage Ariane 5 will be powered by two large solid rocket boosters, each one fueled with 170 tons of ignition powder. Together these boosters will produce a maximum thrust of 2 million pounds (907,200 kilograms). Eventually the Ariane 5 may be used to launch the Hermes spaceplane, a mini-space shuttle, with four to six astronauts aboard.

Japan is developing a more powerful rocket launcher called the H-2 that is scheduled to be launched in 1992. This two-stage rocket will be capable of launching some 4,400 pounds (2,000 kilograms) of payload into geosynchronous orbit—three times the capacity of Japan's current launcher. The development of the H-2 will not depend on high-tech components from other nations, but will use only Japanese technology. This rocket will make Japan an independent spacefaring nation able to compete in the international space market.

The British have on the drawing boards a unique spaceplane concept called HOTOL that they hope to fly in the first years of the twenty-first century. The HOTOL, a single-stage-to-orbit spaceship, based on a secret engine design, would take off horizontally like the Concorde and land at existing airports. A launch trolley would be utilized for the horizontal lift-off to save weight. Like the space shuttle, HOTOL would be a reusable spacecraft, but its initial version would not carry astronauts; it would be a low-cost payload launcher for satellites.

The British spaceplane is designed to carry payloads of 15,000 to 24,000 pounds (6,800 to 10,900 kilograms) to orbit. It would be about 200 feet (61 meters) in length and have a 25-foot (7.6 meter) long cargo bay. If the price is right, the British HOTOL could give the French Hermes spaceplane a tough race to orbit—and to the financing institutions that will underwrite the development costs.

China and Russia may also produce new launch vehicles to compete in the commercial space market, but specific goals and rockets have not been announced. Even by upgrading available rocketry, these two countries could be in the race for all those dollars and other monetary units that will be paid for transport service to orbit during the next few decades.

In response to the *Challenger* loss and President Reagan's call to develop a private launching industry, the United States will have several new unmanned expendable rockets by the 1990s. Among the many aerospace firms developing or redesigning unmanned expandable rockets are Hughes Aircraft and Boeing Aerospace. They plan to build a three-stage medium-sized rocket that would be capable of launching several satellites or spacecraft into different orbits during one mission. Named the Jarvis, after the Hughes engineer Gregory Jarvis who died in the *Challenger* disaster, this rocket could be flying by 1990.

Shuttle Fever?

Reusable spaceships are in vogue for the future, even after the *Challenger* loss, and there is one primary reason for this: economics. By getting the payload cost per pound down to a more attractive level, there will be more space business for everyone.

Besides France's Hermes spaceplane and Britain's HOTOL unmanned reusable spacecraft, both of which are in the beginning design phases, a shuttle-type space vehicle is being developed and tested in Russia. The Soviets have tested a small-winged spaceplane that some Western experts contend could be the prototype for a larger 10- to 20-ton version, which would carry a crew. Drop tests of the Russian spaceplane, similar to those carried out during the U.S. space shuttle tests, have also been reported.

According to the Department of Defense, the Soviet Union is also building a heavy-lift space shuttle that would be capable of inserting into orbit twice as large a payload as the U.S. shuttle—perhaps as much as 130,000 pounds (59,000 kilograms). Imaging reconnaissance spacecraft have shown that the Russians are testing two large space shuttle vehicles, but it is believed that their development has run into problems. The best estimates as to when such a heavy-lift Russian shuttle would fly put it somewhere in the early 1990s. The Soviet Union, of course, will be certain to tell the world when it makes its first successful flight. (For more information on the space program outside the U.S., see Chapter 6, "International Space.")

After the Space Shuttle

By 1995 space shuttles *Columbia, Discovery,* and *Atlantis* will be more than halfway through their flying lives. Only the replacement shuttle for *Challenger,* approved by President Reagan in August 1986 and not on line until the early 1990s, will actively fly into the twenty-first century.

What will replace the space shuttle? In 1984 President Reagan signed the National Space

Strategy, which directed NASA and the Department of Defense jointly to study future launch vehicle technology. Future U.S. launch vehicles that will become operational in the late 1990s and 2000s will most probably include a second-generation space shuttle system, which will incorporate advanced technology and will have a larger payload capacity, in the range of 150,000 pounds (68,000 kilograms). Such a follow-on shuttle system would also have the ability to remain in orbit for longer periods of time.

As well as this advanced and larger Space Transportation System, the United States also plans to build a new expendable heavy booster that would at least equal the power of the Saturn 5 moon rocket and could be even more powerful. The design of this giant rocket will take into account the future launch needs of the Strategic Defense Initiative, and it would probably be capable of placing into orbit payload weights of 300,000 pounds (136,000 kilograms). It would be used to boost large and heavy payloads to the space station and could launch future manned missions to the moon to build a permanent base there. This U.S. superbooster could also send a manned expedition on its way to the planet Mars, an Apollo-size space venture that has been receiving serious attention and evaluation for several years. Once a wild idea of science fiction, a Mars mission could actually begin a human migration to the Red Planet that would take place over the next few centuries—a hard-won sanctuary for humankind that would offer us some extraterrestrial insurance.

The United States may also develop a small, manned, winged spaceplane for short space missions. It would have a quick-reaction launch time and could reach any target in the world within ninety minutes of takeoff. While such a spaceplane would most likely be used for high-priority national security missions requiring flights of one or two orbits, it could be redesigned and used for orbital hops between space stations and other orbit-to-orbit transportation needs. This would be the same technology as the "Orient Express" which President Reagan referred to in a 1986 speech.

Shuttle design changes during the last years of the twentieth century, including those generated by the many *Challenger* investigations and their recommendations, will be internal and not noticed by most people. The major enhancement of today's space shuttle will probably be an advanced rocket engine.

Just as earthbound transportation vehicles evolved during the twentieth century, so too will the space transportation spacecraft evolve during the next hundred years. After the original space shuttles will come new advanced shuttles, the designs of which will follow changing needs of the high frontier.

BEYOND 2001: TWENTY-FIRST CENTURY SPACESHIPS

Spaceliners for the People

A few decades from now, perhaps as early as 2020, powerful scramjet spaceliners may fly three hundred to five hundred tourists to an orbiting resort, the Hilton on High, that will offer zero-gravity swimming pools and earth-viewing picture portals in every suite. What will make such a fantasylike scene become reality is a new hypersonic flight technology based on an air-breathing engine called a scramjet. "Why carry all your oxygen," Arthur C. Clarke has asked, "when for the first fifty miles you're climbing through an ocean of it?" Clarke is not alone in his advocacy of this high-tech propulsion system that has the potential of opening up the space frontier to millions of people who will never be astronauts or mission specialists. The fact is that NASA and the United States Congress House Committee on Science and Technology has recommended an extensive research program to develop the space propulsion technology that will in all likelihood dominate the twenty-first century.

These future spaceliners, with wings as wide as a football field, will integrate three propulsion systems into one spaceplane that will be capable of flying up to orbit and returning to earth again in the manner of a traditional airplane: a horizontal takeoff and landing. Although it sounds deceptively like the space shuttle, with the exception of lift-off, nothing could be further from the truth. The shuttle is a rocket all the way to orbit, but not the scramjet, which would use rockets only for the final boost into orbit, once the spaceliner reached a speed above 9,000 miles (14,500 kilometers) per hour, about half the speed needed for orbital insertion. The scramjet, unlike the shuttle, would also be highly maneuverable within the atmosphere.

With 1,000 tons of fuel at takeoff, it would be powered by traditional turbojets for the begin-ning of the flight, until the spaceliner (let us name it the *Goddard*) reached cruising altitude and a velocity of several hundred miles (kilometers) an hour. The scramjet engines would then roar to life, their supersonic airflow fueled with hydrogen, accelerating the *Goddard* to twelve times the speed of sound, Mach 12, and an altitude of 130,000 feet (40,000 meters). Then the rockets would kick in to double its speed to the orbital velocity of Mach 24, more than 18,000 miles (28,000 kilometers) per hour. After a few orbital maneuvers, the pilot would announce to his passengers that the shining High Hilton is visible to the right of the spaceplane. Some of the older passengers will recall a scene from Stanley Kubrick's *2001*.

The design equations needed to study and test the scramjet engines at hypersonic speeds

The second-generation space shuttle may be capable of on-demand launch and hypersonic maneuvering to and from orbit, features the present-day space shuttle fleet does not have. Courtesy General Dynamics.

are so complex that it will take a new generation of supercomputers to do the work. NASA's name for their future supercomputer is the Numerical Aerodynamic Simulator. This advanced computer is also known as the Cray-2 supercomputer, and it was delivered to NASA's Ames Research Center in California in late 1985. At that time the Cray-2 was the world's largest and fastest supercomputer, although the next-generation machine, projected to be eight times faster than the Cray-2 and with a memory four times larger, is already in design.

The Cray-2, and the supercomputers that follow it, will design the rockets and the spaceplanes of the future. It is capable of performing 2 billion calculations a second and has a memory of 256 million words. Without it, a scramjet engine could never be built. We may ultimately thank the supercomputers for helping to create the economical spaceliners that will take us to our vacation resorts above the earth.

Sailing the Solar System

The twenty-first century may see sailing spaceships making their way across the vast reaches of the solar system, taking supplies to Mars before a landing party arrives or to bases on the moons of Jupiter or Saturn. As they begin their journeys above the earth, their great sails unfurled in orbit, they will at first appear as brilliant stars in the night skies, and then slowly fade week after week as they steadily leave the earth behind them.

These sailing spaceships will not have to carry hundreds of tons of oxygen or hydrogen to propel them through interplanetary space. Their fuel will be free for the taking and it will be available everywhere throughout the voyage. What is this magical power source? Sunlight, the power of millions of photons exerting a pressure on the thin, aluminized sails that could stretch out as much as a few square miles (kilometers). While the solar sails will accelerate slowly at first, their speed will constantly increase. A small mylar solar sail of 10,000 square feet (929 square meters) could reach a velocity

of about 25,000 miles (40,225 kilometers) per hour in about six weeks—close to the average speed of an Apollo spacecraft making its round-trip to the moon. On longer voyages to the outer planets, this constant thrust could accelerate a small solar sailing ship to speeds of up to 124,000 miles (200,000 kilometers) an hour. They could become the celestial clipper ships of the Space Age, offering inexpensive transport of materials around our solar system.

Solar sailing ships are being seriously studied by space engineers in the United States, France, and Czechoslovakia. A nonprofit corporation, the World Space Foundation, based in California, is directing the research in the United States, and its engineers have already built a prototype solar sail, which they hope can be tested on a future flight of the space shuttle or Europe's Ariane rocket.

Meanwhile, enthusiastic French engineers have challenged the world to a solar sailing race to the moon. Such a race may be sailed in the first decades of the next century, and it won't depend on the weather because no matter when the race begins, above the earth, the sun will always be shining.

Pickups in Orbit

Already there are several specially designed propulsion units to do service in orbit, and the next few decades will see an ever-increasing diversity of big and small workhorse spacecraft that perform transport, repair, observation, or pickup of satellites, as well as building and assembly work. Just as any street in Manhattan, London, or Paris has an amazing diversity of vehicles, so too will the above-earth orbitways have rocket vehicles of every shape and size, performing a myriad of tasks.

The so-called payload assist modules taken into orbit by the space shuttle have, with a few exceptions, guided satellites into their final, high, geosynchronous duty orbits. Called by several different names (assist modules, transfer stages, orbital transfer vehicles), these propulsion units extend the reach of shuttle activities.

Unmanned work rockets, what NASA calls orbiting transfer vehicles, will extend the reach of space shuttle operations by taking satellites to geosynchronous orbit. Courtesy Martin Marietta.

Each type has its own boost capability so that the satellite bound for geosynchronous orbit, be it a lightweight or a heavyweight, will be matched to the right booster.

There will also be the boosters for interplanetary missions that will send the *Galileo* and *Ulysses* space probes on their way. Several different types of orbital maneuvering vehicles—free-flying spacecraft that can be used again and again, and then returned to earth for repair if necessary—will be able to help build space stations and platforms, inspect and repair satellites, or deliver or pick them up. In orbit by the early 1990s, these free-flyers will be remotely controlled by either a ground control station, the space shuttle or the space station. They will be outfitted with a TV system that will allow them to be teleoperated with precision. Eventually they will have a robotic servicer kit attached that will enable them to perform mechanical repairs of satellites or erect portions of large space structures. One of the largest missions such a free-flyer might be required to undertake would be to retrieve the large Hubble Space Telescope, bring it back to the space station for servicing, take it back to its higher orbit, and then return to the shuttle, or space station—all without refueling. And sometimes the free-flyers will be sent on hit missions to deorbit satellites that have outlived their usefulness. By putting them into lower orbits, the free-flyers will have sentenced these worn-out satellites to a fiery death as they decay and they eventually plunge to their ends through the atmosphere.

For all their versatility and projected accomplishments, the remote-control free-flyers will be of a modest size and will conserve the valuable space in the shuttle cargo bay as they are launched into their work orbits. Their design dimensions are about 15 feet (4.6 meters) in diameter, but only 3 feet (1 meter) wide. A single shuttle mission could launch a dozen of these handy space robots, all stacked neatly in a row in the cargo bay.

Space Van Orbiters

While scramjet technology appears to be the state-of-the-art propulsion for the next century, offering a single-stage spaceplane that will take us and our payloads to orbit at substantial reduced costs, it will probably not be operational for another forty years. Does this mean our only choices are between the space shuttle and Europe's Ariane rockets?

There are several ideas for less expensive

In the 1990s, unmanned free-flying spacecraft called orbital maneuvering vehicles will be able to help build the space station and inspect, repair, retrieve, and redeploy satellites. These service spacecraft will be taken to orbit by the space shuttle and returned to earth for repair when necessary. Courtesy LTV Aerospace and Defense; and TRW, Inc.

ways of getting people and supplies off the-planet and into useful orbits. One such plan, offered by the U.S. firm, Transpace, Inc., is built around a compact alternative to the shuttle called the space van orbiter. These space vans, about half the space shuttle's size, would be piggybacked on 747s to an altitude somewhere above 40,000 feet (12,200 meters), at which time their rockets would kick in and take them into low earth orbit.

Lower payload costs and extremely simple launch operations compared to the shuttle make the space van concept an attractive one. Transpace, Inc. believes that their spaceplanes could make as many as 1,500 flights each year and cost only one-fifth of what the space shuttle does to operate.

Space Tugs and Space Tankers

When a moon base is built in the twenty-first century and when asteroid mining becomes feasible within the next hundred years, new forms of space transport will be built.

A lunar space tug will be among the first such ships to be built. Filled with hydrogen and oxygen in earth orbit, these fuel transports would be some 200 feet (61 meters) in length and capable of delivering 1,000 tons of payload to the moon's surface. A cluster of six huge tanks that hold the hydrogen and oxygen payload surround the central propulsion engine and its propellant.

Once the deep well of earth's gravity is left behind, space transport becomes much less costly and more practical, and the number of commercial space ventures can multiply. For many people, the idea of mining the asteroids whose orbits come close to the earth remains in the realm of science fiction. But several studies indicate that it may become a practical method of obtaining valuable resources for humankind's expansion into space during the next century. What makes such a venture practical or impractical is, as on earth, the cost of doing business. In the asteroid mining business, the main cost of doing business would be the cost of transporting materials. Take away the earth's

gravity well and add an efficient, low-cost propulsion system to move the materials around, and you are in a space business with a healthy bottom line.

What will that all-important low-cost propulsion system be? It is called the mass driver, and it was first discussed in scientific literature by the wonderful scientist and author Arthur C. Clarke, who also brought to the world the concept of geosynchronous communications satellites, back in the 1940s. Clarke discussed the so-called mass driver in the context of an electromagnetic launching device on the surface of the moon. His basic mechanics of the electromagnetic catapult have been worked out in much greater detail based on other technological advances in the last three decades, and the space colony movement in the 1970s incorporated the device into its vision as an economical way to get lunar materials launched off the weak-gravity surface of the moon, after which they would be transported and used to build large space structures in which thousands of men and women and children could work and live in space. (Read Chapter 11, "Settlement and Colonization of Space.")

The same electromagnetic principles that could launch material at a high velocity off the surface of the moon could also be used as a reaction engine of a rocket. As a means of propulsion, the mass-driver reaction engine would be an electromagnetic device that would produce pulsed magnetic fields, which would accelerate any solid material, such as particles of moon material, inside a long shaft for thrusting power. A mass-driver reaction rocket could easily deliver four tons of constant force, equal to the space shuttle's solid rocket engines, and transport large payloads through interplanetary space— including asteroids that could be moved from their orbits to high orbits around the earth and the moon. And if the size of the solid particles to be accelerated is reduced, the thrust power could probably be increased to 15,650 miles (25,200 kilometers) per hour.

The electric power for the mass-driver spaceship would come from the ever-shining sun, collected by solar-collector wings attached on either side of the acceleration shaft, which would be several miles (kilometers) in length. With free sunlight for power and inexpensive reaction material for the mass driver, asteroids weighing millions of tons, with diameters of thousands of feet (hundreds of meters), can be moved closer to the earth-moon system for mining operations that could provide an abundance of new resources to replace the dwindling supplies on planet Earth.

Fusion Spaceships

The era of chemical rocketry will be replaced by nuclear forms of propulsion in the first decades of the twenty-first century. Many experts agree that it will be some form of fusion spaceship that will dominate the next century, enabling rockets to travel to Mars in a few weeks and eventually head out of the solar system as humankind's first starships on missions to neighboring stars. The dream of reaching the stars is alive and well, although the starships that may someday travel the trillions of miles (kilometers) between the stars represent formidable challenges to the most brilliant engineering and scientific minds today.

Even the United States government pursued development of nuclear rockets for seventeen years and spent $1.7 billion on the programs. The idea of a fusion-propelled spacecraft was first described by scientists at Los Alamos in 1955. Referred to as a pulsed nuclear rocket or a pulsed fusion system, it depends on the continuous detonation of low-yield hydrogen bombs behind the vehicle. The extremely hot plasma jet and shock waves from these many microexplosions would hit a huge pusher plate to which giant shock absorbers would be attached, and together they would smooth out the separate detonations and provide constant thrust. This was the work of the famous Project Orion, which struggled along from 1958 to 1965, until it became a victim of the *Sputnik* response and the profound institutional changes that took place in the United States, including the formation of NASA and a new list of national priorities. Under the leadership of the von Braun

team, chemical rocketry, not nuclear, would take men to the moon.

Freeman Dyson, the well-known American physicist, was involved in Project Orion almost from the beginning, and he developed several nuclear-pulse designs, which included the giant pusher-plate concept. Once such design called for thousands of microhydrogen "bomb" explosions every few seconds, which would give the spaceship its thrust. Dyson estimated that this starship could reach 3 percent the speed of light, have a payload of 45,000 tons, and reach the nearest star to the sun, Alpha Centauri, in about 130 years. That the *Star Wars* films have given many of us false expectations about interstellar travel times is undeniable. Real space travel between the stars is still a tremendous challenge for some of the best minds the world has so far produced. The distances to be traveled are difficult to comprehend. Only one part in one hundred million of the volume of the universe is filled with stars, and if we represented our sun by a basketball atop the Pan Am Building in New York City, the next star would be another basketball some 5,000 miles (8,000 kilometers) away in Hawaii. On the same scale, the earth would be a tiny pea-size object just over 100 feet (30 meters) up Park Avenue from the Pan Am Building.

After Project Orion

In 1978 the British Interplanetary Society published *Project Daedalus*, a detailed feasibility study, including the mathematical calculations, of an unmanned, robot-monitored interstellar mission, based on available technology, that would fly to Barnard's Star in forty-seven years at 14 percent the speed of light. The Daedalus star probe study used a variation of the nuclear-pulse propulsion system, and the mission fuel requirements would be 20,000 tons of deuterium and 30,000 tons of helium-3. The deuterium would be obtained on earth from the oceans, but the all-important helium-3 would have to be mined from the atmosphere of Jupiter. From the time the starship started to be built to the time earth received data from its flyby of Barnard's Star, it would take some eighty years.

Without question, *Project Daedalus* was a seminal study in the pursuit of interstellar travel. Its tens of thousands of words, and hundreds of diagrams and calculations leave no doubt that dozens of brilliant people are seriously studying the problems of journeys to the stars.

But will *Daedalus* ever fly? Probably conceptual parts of it will, but the explosive rate of technological advances is likely to drastically alter the design, including the propulsion system, of any unmanned starship that leaves our solar system in the last half of the twenty-first century. But as an example of the nuclear-pulse propulsion system that has been integrated with all other necessary starship systems, it is an important first effort. And it has motivated some brilliant minds to pursue their drawing-board dreams of the first starship. (See Chapter 12, "Extraterrestrials and Star Trips" for more information.)

The Laser Fusion Rocket

An advanced propulsion system that will power a starship or a fast interplanetary spaceship needs to be at least one thousand times more powerful than today's state-of-the-art chemical rockets, and research done at the Lawrence Livermore National Laboratory in the 1970s produced one on paper that may become the rocket of the twenty-first century: a laser fusion rocketship powered by thousands of microexplosions. Instead of hitting a giant pusher plate as they did in the *Project Orion* concept, microexplosions would be contained and directed by a magnetic field that would act as a rocket nozzle. As the plasma fireball expanded from the explosion, the magnetic field would blow it out the back as rocket exhaust.

For the explosions, a high-energy laser would first create an implosion system that would compress hydrogen to more than ten thousand times its liquid density. This would make possible the efficient thermonuclear burn of small pellets of heavy hydrogen isotopes. A rotating mechanical accelerator in the rocket engine would inject five hundred of these pellets into the thrust chamber each second. As each pellet reached the fusion point, it would be struck by a laser pulse that would last for less than a billionth of a second. Optical mirrors would focus the laser pulses on the pellets.

When the fusion occurred, a fraction of the fuel mass would contain the same power that keeps the stars burning throughout the universe. Two-thirds of this fusion energy would then be converted into a moving stream of charged particles that would give the spaceship momentum and would be dispersed as exhaust. Such an advanced propulsion system could drive spaceships at a velocity that is 1/100th the speed of light, perhaps faster. But even at 1/10th the speed of light, the vast distances to the nearest moving stars would take fifty to one hundred years for a one-way trip. If the earth's polar diameter of 7,900 miles (12,700 kilometers) represents 1 inch (2.54 centimeters), then at the same scale, the nearest star from our solar system is some 50,000 miles (80,450 kilometers) away. Such distances challenge the ingenuity and creative genius of our species.

A fusion spaceship may set out from our solar system late in the twenty-first century, but if the technology is available in the middle of the next century, fast fusion spaceships will be traveling the trade routes between the planets first. With their speed, they will help to rein in the solar system and bring it closer to human scale. Commercial and scientific missions within the solar system will always have priority over a starship mission. Any spaceship capable of journeying to the planet Mars in a few weeks will be in high demand among the spacefaring nations and corporations of planet Earth. And if the laser fusion spaceship is built, it will owe its existence to the intensive fusion research of the last few decades of this century, especially to the top-secret work now being conducted on high-technology defensive space weapons. It seems appropriate that what may be the key to reaching the stars involves human ingenuity recreating billions of tiny stars inside the spaceships that may carry us to them across the vast interstellar void.

The Ultimate Fusion Rocket

If a temperature of 2 billion degrees can ever be achieved and sustained, the ultimate fusion rocket engine driven by hydrogen and boron becomes possible. Boron-11 is commonly found in nature, and whatever amounts were needed could be routinely extracted from seawater.

Such a hydrogen-boron reaction would produce only charged particles and practically no side reactions. It could eventually offer direct conversion of energy and would produce no neutrons or radioactivity. Such a boron-hydrogen fueled fusion spaceship would be so powerful that its design would not have to compromise with gravity. But its nuclear-pulse engine would have to withstand the incredible 2-billion-degree temperatures, and today's technology does not even begin to have the engineering solutions to produce materials that can withstand such temperatures, which are more than one hundred times hotter than the center of our sun and can be found nowhere in our solar system.

The Alpha Centauri Express?

Beyond a perfected fusion rocket sometime in the next century, what theoretical propulsion systems could help earth folk break out to the stars and open up the age of interstellar flight?

The matter-antimatter propelled spaceship (M-AM for short) is the one most often considered, but no expert is willing to predict when such a spaceship could be built—if one can ever be built. It must be emphasized that such a propulsion system remains highly theoretical in the last decades of the twentieth century.

In 1932 Paul Dirac, an English physicist, discovered the positron, which verified the existence in nature of a particle-antiparticle symmetry. All known particles have antiparticles, and when particles of matter and antimatter come together, their energy is released. The mass of both these particles is then converted into 100 percent energy. In theory, such a mass annihilation rocket would convert all—not just a fraction—of its fuel mass into energy.

If such a M-AM rocket system could be built to fully convert a pound or kilogram of fuel into an exhaust beam and reaction force, it would be *five billion* times the energy release from the equivalent fuel mass in the most advanced chemical rockets.

Although some two hundred to three hundred antiprotons have been stored for several days at the European Center for Nuclear Research, storage for any practical length of time is beyond the reach of present technology.

The challenges of producing practical quantities of antimatter, storing it, and directing its energy release are formidable. But if earthling minds and their creative technology ever solve these subatomic perplexities, this far-in-the-future rocketship may use the most common element in the universe—hydrogen—to speed it to new planets around distant stars. Pit antihydrogen, composed of antielectrons and antiprotons, against ordinary hydrogen's electrons and protons, and the reaction would in theory produce a 100-percent conversion to energy (mass annihilation) and an unimaginable performance billions of times more efficient than those baby rockets of the early Space Age—the ones we fly today.

Such a superrocket would harness the elemental forces of the universe to penetrate its time and distance. Humankind could begin its migration to other stellar neighborhoods in the Milky Way galaxy.

Another Way-Out, Exotic Rocket

If no scientist alive today has the slightest idea of how to design and build a photon rocket, why even consider it as a possibility for the far future? The only justification for thinking that a photon rocket may exist someday is that the idea of one exists, created by the human mind, and the historic record is filled with tens of thousands of ideas that have become reality.

What would a photon rocket be? It would be another form of mass-annihilation propulsion, like the matter-antimatter fueled rocket. Theoretically this rocket's exhaust beam of photons could thrust the spaceship to the speed of light, but the on-paper mathematical dreams indicate that the best designs would be limited to 60 percent the speed of light. The power requirements of such a rocket appear impossible today, however. Each *pound* of thrust would require 668 megawatts of energy, more than 2,200 times the energy produced by a small power plant. But if such speeds could ever be realized, they would put one-way trips to some of the nearest stars such as Alpha Centauri and Barnard's Star a decade away. People could travel to the stars and return within their own lifetimes; it would not be a journey requiring several generations. And the cosmic prize might be a "new" planet for humankind after Mars gets crowded.

8
Robots From Earth

SUN SHOTS

Forecasting the Sun's Weather

If life on earth is still ultimately a mystery, so too is our mighty yellow dwarf star—the sun—upon which the earth entirely depends. The interplanetary pioneers launched into solar orbits in the 1960s were sent to measure radiation from solar storms, the solar wind, magnetic fields, micrometeoroids, and other phenomena of interplanetary space. The big question: How dangerous was this deep-space weather to spacefaring astronauts who would be moon-bound before the end of the decade? The answer: Not as dangerous as some experts predicted. What follows are summaries of some of these early spacecraft pioneers.

Pioneer 4, launched on March 3, 1959, was intended as a moon probe and passed it at a distance of 37,300 miles (60,000 kilometers). The robot craft, which weighed a mere 13 pounds (5.9 kilograms), was the first U.S. spacecraft to go into orbit around the sun, and it yielded excellent radiation data.

Pioneer 5 was designed as an interplanetary probe. It was launched on March 11, 1960, and eventually went into orbit around the sun at a distance between Venus and Earth. Data on solar flares and the high-velocity solar wind were transmitted until June 26, 1960. The spacecraft's year (period to orbit the sun) was 312 days.

PRECEEDING PAGES: *Even though the* Challenger *disaster delayed the scheduled 1986 launch of* Galileo *to giant Jupiter, it will nevertheless reach the planet before 1990 and release its probe, which will slam into the atmosphere, encountering deceleration forces some four hundred times more powerful than the earth's surface gravity. This painting shows* Galileo *being launched from the space shuttle. Courtesy General Dynamics Corporation.*

Pioneer 6, 7, 8, and 9. All four of these interplanetary scientific spacecraft were shot into orbit around the sun—three between Venus and Earth and one between Earth and Mars—during the last half of the 1960s. These robots provided a tremendous volume of data on cosmic rays, the solar wind, and magnetic and electrical fields.

It was learned that:

• The sun is dominant in interplanetary space, and its highly charged solar wind particles reach out well beyond the orbit of Mars.

• The sun has an "atmosphere" (now referred to as the heliosphere) composed of interplanetary magnetic fields, whose lines of force, because of the sun's rotation, twist like streams of water from a spinning lawn sprinkler.

• The effects of the earth's magnetic field extend more than 3 million miles (4.8 million kilometers) outward from the night side of the earth.

• The sun's cosmic rays spiral around its force fields and therefore travel through space in well-defined streams.

Russia, too, launched a solar research spacecraft to study gamma particles and X-rays from the sun, radiation that influences the earth's weather and communications, and poses a danger to above-earth cosmonauts. Named *Prognoz I* and launched on April 14, 1972, the 1,862-pound (845-kilogram) spacecraft was not inserted into an orbit around the sun, but rather into a very elongated elliptical orbit around the earth, which took it from 590 to 124,000 miles (950 to 200,000 kilometers) above the earth—a full orbit every ninety-seven hours that stretched halfway to the moon.

These sun-orbiting pioneers provided a solar weather network, including data on what was happening on the far side of the sun before it rotated toward earth. They were, in part, sentries on the watch to warn us earthlings that a sun storm was brewing and heading in our direction. If so, space missions would be scrubbed, astronauts grounded.

Hail Helios

Two sunbound spacecraft, *Helios 1* and *2* (named for the sun god of ancient Greece), were launched into solar orbit in the mid-1970s, the first on December 10, 1974, and an identical twin on January 15, 1976. A joint space venture between West Germany and the United States, these sun-probing robots studied the sun's corona, wind, and magnetic field, and other aspects of interplanetary space.

The Helios spacecraft made closer approaches to the sun than any other man-made object ever had. The first such approach took *Helios 1* to within 28 million miles (45 million kilometers) on March 15, 1975, and *Helios 2* later beat the record, on April 17, 1976, with an approach to within 27 million miles (43 million kilometers) of our life-giving star, which emits more energy in one second than humankind has consumed during its entire history on earth.

The Sun-Earth Explorers

A series of three International Sun-Earth Explorers was launched in 1977 and 1978 to obtain data on sun-earth relationships at the outermost limits of the earth's magnetosphere. *ISEE 1* was a NASA project, and *ISEE 2* was a European Space Agency project. Both spacecraft were launched on October 22, 1977, and went into giant orbits the apogees of which were 137,000 miles (220,400 kilometers) above the earth.

Then the United States launched the third Sun-Earth Explorer into a very unusual orbit, on August 12, 1978. The so-called halo orbit was around the L-1 liberation point, located between the earth and the sun about 1 million miles (1.6 million kilometers) from earth. This *International Sun-Earth Explorer 3* therefore became the first spacecraft to orbit a point in space rather than a celestial body. It measured the solar wind and other deep-space phenomena, while its sister craft measured the same phenomena nearer to the earth.

Ground controllers changed the spacecraft's position in 1982 so that it could monitor the earth's magnetotail. It was repositioned again in 1984, through a complex of maneuvers around the moon, for a comet rendezvous, and in September 1985, this versatile robot, renamed ICE (International Comet Explorer), became the first spacecraft to fly through a comet's tail when it intersected the tail of comet Giacobini-Zinner and sent back first-of-a-kind data to earth before all the Halley's Comet probes homed in on their venerated celestial object in 1986.

Ulysses *to the Sun*

Sometimes between 1994 and 1996, depending on the final launch date and trajectory chosen—a spacecraft named *Ulysses* will arrive at its difficult-to-get-to destination orbit after a swing by giant Jupiter and a needed gravitational kick that it could get nowhere else in the solar system.

The *Ulysses* spacecraft is taking a long detour to receive a tad of Jupiter's gravitational power, because after its swing around the giant, it heads part of the way back into the inner solar system towards its target—our star, the sun.

But is there no shorter route to the sun? Why not blast the 814-pound (371 kilogram) spacecraft directly to the sun from the earth? What is important to know about the *Ulysses* mission (formerly known as the International Solar Polar Mission because both NASA and the European Space Agency are participating) is that it has to study the sun's polar regions and some of the interplanetary space between it and our star—all 205 million miles (330 million kilometers) of it. The plane of *Ulysses*'s orbit is therefore unlike that of any earlier spacecraft, and no combination of available rockets (above-earth space shuttle launch included) can produce enough energy to hurl *Ulysses* directly from earth over the sun's poles. Jupiter helps *Ulysses* by changing its trajectory so that it will climb out of the plane in which the earth orbits the sun (the ecliptic plane) and turn back again into an orbit that takes it above the sun's poles—regions of the sun and interstellar space that have never been explored and studied before.

In the late 1990s, *Ulysses* will spend about four months above one of the sun's polar regions, its nine scientific instruments measuring magnetic fields, composition of the solar wind, cosmic rays, X-ray and gamma-ray bursts, dust, and much more. Scientists expect some surprises as their instruments probe these never-before-studied regions of the solar system. Their expectations are based on the fact that such characteristics as the structure and shape of the sun's magnetic field are different in the polar regions—both outbound from the sun and inbound from interstellar space. And *Ulysses* will turn some of its sensors in the other direction, toward distant regions of our Milky Way galaxy and to galaxies beyond, to measure high-energy radiations.

After four months observing one pole, *Ulysses* will cross the plane of the solar system and climb toward the opposite pole, which it will then pass as it continues its journey. The primary mission will end before the end of the century, but scientists will be sifting through the billions and billions of data bits it sends back until at least the year 2008.

The Sun Grazer: Project Starprobe

After *Ulysses* has ended its mission in the 1990s, another sun spacecraft could fly by the end of the century. Only this time, instead of orbiting beyond the planet Mars, this state-of-the-art sunship would penetrate the plasma fire of our star's corona at a mere distance of 1.7 million miles (2.74 million kilometers) from the sun—about twice the Sun's diameter. This distance is sixteen times closer than that achieved by the current-record-holding Helios spacecraft of the mid-1970s.

Such a mission has been under consideration at NASA's Jet Propulsion Laboratory for several years, and scientists concluded in 1985 that Project Starprobe could fly—*if* given adequate congressional funding—by the mid-1990s.

Spoiled by the illusory media-made Apollo legacy, which convinced us that space travel is easy, and the Voyager spacecraft soaring across

the solar system, sending back a treasury of magnificent images, the general public does not appreciate the formidable engineering challenges that must be overcome to send a robot spacecraft through the intense 2,500-degree Kelvin heat of the sun's corona and expect it to sample the star stuff, take precise measurements, and unfailingly communicate the data from its dig with earth during its passage through the extreme temperatures. Even the heavy carbon shield cannot shed its surface material like the nose cones of the early space days because this would contaminate the all-important sampling of the sun stuff.

A scientific analysis of these virgin solar particles will allow scientists to better understand the solar wind at its point of acceleration in the corona before it blows throughout the solar system, influencing everything, including all the planets and the interplanetary space between. And such a star probe could further solve some of the remaining mysteries of gravity where it is most dominant in our little corner of the galaxy. Still, to create sun-grazing wings that will fly us closer to our life-giving star than ever before, one must start in Washington, D.C., Earth. It's a crowded and extremely busy place where many things get overlooked or lost—even, to us, the most important, celestial object in the universe.

THE SUN'S CLOSEST OFFSPRING

Catching Swift Mercury

Mercury, the fastest orbiting planet in the solar system and closest to the sun, whipping around the sun once every eighty-eight days, has been visited by only a single spacecraft—*Mariner 10*—since the Space Age began. But thanks to the sun's gravitational power, the spacecraft visited this planet not once, but three times, and took a look at Venus on its way in from a flyby distance of only 3,585 miles (5,768 kilometers).

Mariner 10 was launched on November 3, 1973, passed Venus on February 5, 1974, and made its first rendezvous with Mercury on March 29, 1974. After it entered a close orbit around the sun, it re-encountered Mercury twice, in September 1974 and March 1975.

The spacecraft, which had the benefit of a decade of technological experience and evolution of other Mariner spacecraft, weighed 1,175 pounds (533 kilograms) at launch and contained a scientific instrument payload of 172 pounds (78 kilograms). It was specially designed to withstand the high temperatures closer to the sun—four and half times those encountered by a near-earth spacecraft—and had attached a large sunshade, protective thermal blankets, and solar panels that could rotate to keep them at a constant, survivable temperature of some 239 degrees F. (115 degrees C.). *Mariner 10* could also handle up to 118 thousand bits of TV data each second; this design fact made it possible for the spacecraft to transmit some ten thousand photographs on its mission of four flybys—one of Venus and three to Mercury. It offered our first close-up of the first planet from the sun and added much to our knowledge of the planet: the gigantic metal-rich core that makes it the most dense planet; the most extreme temperature variations of any planet known; its great, unique cliffs; and the giant impact crater, Mare Caloris, some 810 miles (1,300 kilometers) in diameter. We knew much more about Mercury and Venus after *Mariner 10*. We also proved we could build durable robots for ever more challenging space missions; the best, beyond most hopes, was yet to come. Even today, after the historic Voyager flights throughout the solar system, the same can be said: The best is yet to come.

Return to Mercury?

While red-soiled Mars and the magnificent outer planets of Jupiter and Saturn, along with their retinues of satellites, beckon strongly and hold the attention or our technical ingenuity, there

may be other unmanned missions to hot and fast Mercury, neighbor to sun, in the twenty-first century. Before *Mariner 10* encountered the planet three times in 1974 and 1975, practically nothing was known about Mercury—a fuzzy sphere through earth-based telescopes. But then a man-made Mariner cast its electronic eyes upon it, and photographic resolution of Mercury increased five-thousand fold in just one year. Our knowledge of the planet took an astronomical leap, but it still holds its secrets.

Future missions could include:

• A Mercury orbiter spacecraft scanning the entire unknown surface with high-resolution cameras and other instruments
• A lander (an adaptation of the Viking Mars lander)

There are currently no plans for sending these robots to Mercury, but that could change. It all depends on the future wealth, self-esteem, and curiosity of the spacefaring nations. Without question, the most exciting mission would be a Mercury lander or rover that could sample the floors of the small, steep-walled craters in the polar regions. Here, in chilly darkness ever since their formation, may be found the planet's primordial substances trapped in the form of ice. On-site analysis and data transmission could give us physical evidence to show how Mercury was created when the sun was a young star—primordial ice, melted and analyzed inside the belly of a made-on-earth machine.

UNVEILING VENUS

Seething Sister

More unmanned spacecraft—sixteen—have been sent to Venus by the United States and the Soviet Union than to any other planet, although Mars is a very close second. The reason for such continued probing is obvious: Venus and Mars are the two most earth-like planets. Venus, traditionally called the earth's sister planet, is a virtual twin to the earth in terms of size and mass, but the sisterhood abruptly stops there.

If there is a hell planet in the solar system, Venus is it. Instruments flying by, orbiting, or landing on cloud-shrouded Venus tell us of a surface with inhuman temperatures reaching 878 degrees F. (470 degrees C.) that would turn lead molten and steel red hot. Breezes of fire move slowly across the seering surface. Gigantic lightning storms crack in the sulfuric sky. Atmospheric pressures are ninety times those on earth equal to diving down to an ocean depth of 0.5 mile (0.8 kilometer).

Venus is a greenhouse without a keeper. The so-called runaway greenhouse effect, whereby sunlight and heat become trapped and temperatures rise, has made it a giant inferno inhospitable to the life forms of earth. If we continue to call it our sister planet, an adjective must be added for accuracy: seething sister. Why continue to send automated spacecraft to such a hellhole, where no human is ever likely to walk?

Because finding out why the evolutionary paths of these two odd-sister planets are so different could actually have a bearing on the future survival of the healthy earth and its human race. Surely that is reason enough to continue to build robot craft and send them to probe the mysteries of hellish Venus.

West and East to Venus

American and Russia began their unmanned launches to Venus in the early 1960s. The first, *Venera 1*, was launched from the Baikonur space center at Tyuratam on February 12, 1961. This Russian mission was also the world's first space shot launched to another planet. Two weeks after launch, however, radio contact was lost, and the spacecraft missed Venus by some 60,000 miles (97,000 kilometers) and went into an orbit around the sun.

The United States had better luck with *Mariner 2*, launched August 27, 1962. This was

America's first successful planetary spacecraft. After a journey of three and a half months, *Mariner 2* flew by Venus at a distance of 21,594 miles (34,745 kilometers). Considering the planetary distance scale, this was admirable marksmanship.

Mariner's instruments took atmospheric readings and confirmed the intense surface temperatures (hotter than your hottest household oven) that earlier earth-based radar had indicated. During the flyby, *Mariner 2* also discovered that Venus had no magnetic field or radiation belt. The modest but sturdy 447-pound (203-kilogram) robot continued into an orbit around the sun and sent data back to earth for almost a year (348 days) from a distance of 53.9 million miles (86.7 million kilometers). Compared to the Russian fizzle that left earth first, *Mariner 2* demonstrated what "Made in the U.S.A." meant in the space robot business and pointed toward great missions yet to come—Viking and Voyager.

What follows is a brief listing of other successful U.S. and U.S.S.R. Venus missions, which came after *Venera 1* and *Mariner 2*.

Venera 4 *(launched June 12, 1967)* rendezvoused with Venus on October 18, 1967. A capsule descended through the Venusian atmosphere and sent back data for an hour and a half. This was the world's first landing on another planet, although transmission ceased during descent and no data was returned from the surface.

Venera 7 *(launched August 17, 1970)* encountered Venus on December 15, 1970, and released its spherical probe capsule to the surface. The world's first spacecraft to send data back from the surface of another planet, the capsule transmitted for twenty-three minutes from the hostile environment before the high heat destroyed the instruments. It proved beyond a doubt the hellish temperatures and extreme pressures.

Venera 8 *(launched March 27, 1972)* ejected its capsule, which made a soft-landing on July 22, 1972, and transmitted data for fifty minutes. Instruments measured the amounts of uranium, thorium, and potassium in the surface rock, and results indicated that it resembled the granite of planet Earth.

Mariner 10 *(launched November 3, 1973)* flew by Venus in February 1974 before going on to the planet Mercury. Its cameras were equipped with ultraviolet filters, and it took the first pictures of the cloud-veiled planet—some 3,500 in all—which showed the unusual global circulation of the upper atmosphere. A layering of Venuian clouds was also observed for the first time.

Russia's Venera 9 *(launched June 8, 1975)* was the world's first robot spacecraft to send back a photo of the Venusian surface, on October 22, 1975. *Venera 10*, launched less than a week later (June 14), also took a photo of its landing site, on October 25, 1975, and reported on surface conditions. *Venera 9* reported for fifty-three minutes, *Venera 10* for sixty-five minutes before hellish temperatures and crushing pressures destroyed them. Rocks with sharp edges, appearing broken and fractured, were seen in the first photo; some of the rocks in the second photo (*Venera 10*) appeared to be covered with a dark material that *looked* like lichen, but there is *no* evidence for even primitive life forms on Venus. These two Soviet spacecraft were more sophisticated than the earlier ones and put an orbiter spacecraft above the planet from which a landing craft was launched.

Two more Russian Venera spacecraft (*11* and *12*) were launched toward Venus on September 9 and 14, 1978, and arrived in December—*Venera 11* on Christmas Day and *Venera 12* on December 21. They landed successfully, but no photographs were returned.

The United States next sent *Pioneer Venus 1* (launched May 20, 1978) and *Pioneer Venus 2* (launched August 8, 1978) toward our seething neighbor planet. They represented the most sophisticated U.S. spacecraft to explore Venus up to that time. *Venus 1* orbited the planet, photographing the clouds and making a radar map of 90 percent of the surface; *Venus 2* sent four probes through the Venusian atmosphere to report on conditions. The large probe spacecraft

had a 13.5-carat diamond window for a clear view of the infrared spectrum. The diamond, which withstood the tremendous pressures of the descent, was found after an earthwide search. Three other windows of this one *Pioneer Venus 2* probe were made of sapphire.

Three smaller probes were also carried on the Multiprobe Bus, which was 8.3 feet (2.5 meters) in diameter. As *Venus 2* learned about Venus, the probes were launched and entered the atmosphere at 26,000 miles (41,834 kilometers) an hour—some forty-three times faster than the speed of a commercial jetliner. They glowed briefly like meteorites.

The large probe hit the surface at the day-side equator of Venus at a mere 20 miles (32 kilometers) an hour, after descending for fifty-five minutes through the Venusian atmosphere, all the while sending back data. One of the small probes survived impact and transmitted for sixty-eight minutes on the surface before falling silent. Data showed that this probe's internal temperatures steadily rose after impact to a high of 260 degrees F. (127 degrees C.) before its radio failed. During surface transmission time, there was no evidence of any leaks into the spacecraft chamber.

On all four *Pioneer Venus 2* probes, instrument doors and windows opened, booms deployed, delicate sensors began operating. It was a tremendously complex robot that had several offspring, all of which flew to glory and made their earthbound makers proud and happy.

The longer-lived *Pioneer Venus 1* orbiter unveiled 90 percent of the Venusian surface in less than two years and imaged a surface of great mountains and rift valleys, expansive plateaus and shallow basins. Most of Venus consists of upland rolling plains. The tremendous data bank, covering all aspects of the planet, supported the greenhouse effect at work and explained the horrible 878 degrees F. (470 degrees C.) temperature.

Venera 13 *and* 14 *(launched on* October 30 and November 4, 1981) were the only interplanetary launchings in that entire year. After attaining orbit around Venus in March 1982, they

The unmanned spacecraft Magellan, *also known as the Venus Radar Mapper, will unveil cloud-covered Venus in great detail in the late 1980s. The radar will have three hundred times the resolution of the earlier Pioneer Venus Orbiter, and it will produce photographlike images of areas down to .62 mile (1 kilometer). Courtesy Martin Marietta.*

each launched landing craft, which provided the first element-by-element analysis of surface rocks after rock samples were drawn into test chambers and were analyzed by X-ray fluorescence spectroscopy. *Venera 13* also took the first color photograph of the Venusian surface.

V*enera 15 and 16 (launched on* June 2 and June 7, 1983) both went into orbit around Venus in October of that year. They began mapping the unseen north polar region of the planet by synthetic-aperture radar. Lack of craters indicates an extremely active surface volcanism and the possibility that the region mapped by spacecraft may be geologically young—on the order of one billion years.

FUTURE SPACECRAFT TO VENUS

Magellan: The Venus Radar Mapper

The United States may return to Venus in the early 1990s with a sophisticated radar imaging spacecraft originally called the Venus Radar Mapper (VRM) and officially named *Magellan* in 1986. Placed in an elliptical orbit around the planet, it will complete a circuit once every 3.1 hours. The state-of-the-art radar system, called synthetic-aperture radar, will operate for forty minutes of each orbit, when the spacecraft is at an altitude ranging from 190 to 1,060 miles (300 to 1,700 kilometers). The radar data will be recorded on board the spacecraft and later transmitted to earth at a rate of 250 kilobits each second. (The Voyager spacecraft could send only 115 kilobits a second as they flew past Jupiter.)

The earlier Pioneer Venus Orbiter's radar system provided only general maps of most of the planet's surface because its resolution capability was limited and could not show features any smaller than 180 miles (290 kilometers) across. The *Magellan* spacecraft will have three hundred times the resolution power of the earlier spacecraft and will be able to return photographlike images of features as small as 3,300 feet (1 kilometer) across from more than 90 percent of Venus. The VRM will "see" through the thick, poisonous clouds that are transparent to microwave radar and return detailed images of features formed by the seething winds, large landslides, perhaps ancient channels cut by once-flowing water, dry ocean beds, and the first detailed look at what may be the largest known volcanolike feature in our solar system. Elevations as small as 330 feet (100 meters) will also be recorded and converted to images with new digital processing techniques.

Such detailed information on the surface geology of Venus will complete the detailed mapping of the two most earthlike planets in our solar system. Mars was mapped with the Viking orbiter spacecraft in the late 1970s. By understanding how Venus and Mars evolved to their present states, we may learn what the future holds for earth and perhaps influence that future for the benefit of generations to come.

The *Magellan* radar mapper will be assembled in large part from space parts from other spacecraft—*Viking, Voyager,* and *Galileo.* It probably will be launched from the space shuttle with an upper stage booster in the late 1980s and will arrive at Venus five or six months later, depending on its final trajectory. The primary mission of radar mapping will continue for 243 days. Ve-

nus will surely retain many of its mysteries even after this extensive radar study, but *Magellan's* unveiling will certainly solve some of them, and there will no doubt be some marvelous surprises in the multitrillion bit data streams flowing to earth at the speed of light.

Venus in the Twenty-First Century

The twentieth century marked the beginning of the Space Age, and the 1970s were the golden age of planetary exploration. But Venus continues to be probed by robots made on earth. After extensive radar mapping has been completed in the early 1990s, other automated probes will be fired to our hot celestial sister planet. Russia often flies by Venus on the way to other targets and drops off scientific modules. *Vega 1* and *2,* the Russian spacecraft sent to Halley's Comet for close looks, continued the Venus unveiling by this method of dropping off two descent craft. Russia and France plan a cooperative asteroid mission called Vesta before the turn of the century and on their way to Venus plan to launch two descent modules, which would photograph the planet's surface during descent.

The United States plans a Venus atmospheric probe in the future, which would be hatched by a mother flyby craft. It will measure trace-gas components of the atmosphere to an accuracy ten times greater than that of the Pioneer Venus spacecraft. Such precise analysis of the Venusian atmosphere will help toward a more complete understanding of conditions in the inner solar system at the time planets accreted, more than four billion years ago.

Other Venusian descent spacecraft of sophisticated design will land in regions of interest discovered by the orbiting radar-imaging robots of the United States and the Soviet Union. But there is no likely justification for sending a manned mission to this hellish place. Development cost would be horrendous, and robots can be built to survive the inferno and do whatever they are designed and programmed to do for long periods of time—be it in the atmosphere (in which case they would be easier and less costly to design) or on the furnacelike surface. It is therefore highly unlikely that earthlings will ever hear a human voice from the Venusian surface say, "That's one hot foot for a man . . . . "

The distant future could see, through planetary engineering (see Chapter 11), the redesign of Venus to make it more earthlike and habitable, but Mars would certainly come first. In any event, this possibility, by no means certain, is hundreds, perhaps thousands of years away. Meanwhile our robots will continue to lift the veils of Venus, discover its secrets, and help us keep our earth act within its delicate balance.

THE NEW MARS

Down on Mars

After eight years of intense efforts for dozens of scientists, and fifteen years of full-time involvement for other Viking team members, the moment of truth was near—the *Viking 1* lander craft was descending to the rust-colored surface of Mars. It would either soft-land and survive or crash-land and become expensive Martian junk. It was early in the morning on July 20, 1976, the seventh anniversary of the first manned landing on the moon. On earth, the bicentennial celebrations of the United States continued.

Observers watched as data points appeared one by one on TV monitors at the Jet Propulsion Laboratory (JPL) in Pasadena, California, indicating the curving descent path of the *Viking 1* lander. As the spacecraft reached an altitude of 78,720 feet (24,000 meters), its velocity had dropped to 4,070 miles (6,549 kilometers) per hour. "We're coming down," said Albert Hibbs, a senior missions planner. "It's a long period of glide . . . before the parachute comes out."

At 4,592 feet (1,400 meters), the terminal descent engines started. Then, at 5:12:07 A.M. Pacific Daylight Time, a voice in Mission Control called out, "Green for touchdown . . . Touch-

down, we have touchdown."

Cheering broke out in the control center and the crowded auditorium. Human discipline and control, usually dominant in great explorations, surrendered to a surge of human emotions. Earth folk had accomplished the first soft landing of a robot spacecraft on the mysterious and mythic Red Planet, given the name Mars by the Romans. But the excitement was far from over. Soon after landing, image data began to arrive at JPL after its 18 minute and 18 second journey to earth. A TV monitor traced out, line for line, pictures taken from the surface of Mars, the first of the lander's footpad, the second a panoramic view of the Martian landscape in front of the lander.

The robot Viking 1 *dug a small ditch* (LEFT MIDDLE) *in the red Martian soil soon after it touched down on July 20, 1976. "We are the Martins now," Ray Bradbury said happily when* Viking 1 *reached Mars and sent back its first picture.* Courtesy NASA.

A May day on Mars. This photo was taken by the Viking 2 lander on May 18, 1979. A thin coating of ice is seen on the rocks and soils. How thin is the Martian frost? It has been estimated at about 1/1,000th of an inch thick. Courtesy NASA.

"It's mind-boggling what has happened," said Noel W. Hinners, associate administrator for space science, at a news conference soon after the first photographs from Mars were shown. His eyes were moist as he spoke. "I had tears in my eyes this morning for the first time, I guess, since I got married. It's really an emotional experience."

Ray Bradbury, the famous author of *Fahrenheit 451* and *The Martian Chronicles*, witnessed the event as an invited guest.

"Well, *we* are the Martians now!" Bradbury said happily.

Early Mars Missions

Viking's marvelous triumph was the culmination of almost fourteen years of unmanned exploration of the Red Planet, beginning with Russia's first attempt to reach Mars on November 1, 1962, with the launch of *Mars 1*. The first

Mars mission, however, lost radio contact with earth on March 21, 1963, after flying some 66 million miles (106 million kilometers). Twelve other spacecraft would attempt the curving journey to Mars before the footpads of *Viking 1*'s lander came to rest on the red Martian soil.

As with all explorations of new frontiers, there are disappointments and surprises along the celestial road to Mars before the Viking missions showed us the reality of another world—a world that for centuries before had lived only in our imaginations.

Other missions to Mars during the 1960s and 1970s, and what was learned from flying them, unquestionably helped the Viking spacecraft—the two orbiters and two landers—toward their historic touchdowns.

Mariner 4. *Launched two years* after Russia's *Mars 1* attempt, on November 28, 1964, this was the United States's first mission to Mars. When the spacecraft flew by Mars on July 14, 1965, at a distance of only 6,118 miles (9,844 kilometers), the modern era of Mars exploration began. *Mariner 4* took twenty-two photographs of the Martian surface, which showed a moonlike, heavily cratered terrain. Considering what more extensive reconnaissance later showed of the Martian surface, these first photos were unrepresentative and misleading. Another boring moonlike celestial body? Spacecraft instruments also learned that the Martian atmosphere was very thin, composed mostly of carbon dioxide, and that the pressure was only 1 percent of the earth's. When humans first set foot on Mars, they will have to wear space suits.

Mariner 6 *and 7. These missions* were two more Mars flybys, launched within a few weeks of each other, on February 24 and March 27, 1969. Five months later, the two spacecraft flew past the planet, taking a total of 201 photographs. *Mariner 6* flew by at a distance of 2,120 miles (3,411 kilometers) above the equator on July 31; *Mariner 7* followed on August 5 with a close approach of 2,190 miles (3,524 kilometers) above the southern hemisphere. Their improved photography equipment delivered high-resolution images that homed in on surface ar-

eas 85 by 200 miles (137 by 322 kilometers), including a giant crater named Nix Olympia (Snows of Olympus). Mars was revealing itself as less moonlike: a polar ice cap of dry ice, haze and clouds, evidence of wind weathering and water erosion, a thin atmosphere of carbon dioxide, sometimes suspending dust particles. Mars was yielding its secrets up to the flyby robots from planet Earth. The Red Planet was looking more like a cold, dry desert.

Mars 2 *through 7. Russia took* up the Mars challenge again in May 1971, with the launching of two spacecraft, almost a decade after their first attempt. *Mars 2* arrived on November 27 and went into Martian orbit. A lander capsule was ejected, but it crashed on the surface—the first man-made object to reach the surface. *Mars 3* went into orbit on December 2 to survey the planet and also ejected a lander capsule. But data transmission from the surface lasted only twenty seconds, and only a useless, extremely narrow strip of photograph was received on earth.

The Russian Mars series continued in 1973, with the launches of *Mars 4, 5, 6,* and *7,* but there were no successful capsule landings. The *Mars 4* spacecraft failed to achieve Martian orbit and flew past; the capsule of *Mars 6* crashed; and the *Mars 7* capsule missed the planet entirely, although the orbiters, including *Mars 5,* sent back useful data.

Mariner 9. *Launched on May* 30, 1971, by the United States, *Mariner 9* proved to be the most successful Mars mission until the Viking spacecraft arrived some five years later, in 1976. Arriving at the planet on November 13, after a journey of 167 days, it beat Russia's *Mars 2* into Martian orbit and became the first spacecraft to orbit another planet. But what did its cameras see? Nothing at first! A gigantic dust storm, the largest ever recorded, was raging on Mars. One scientist suggested that perhaps they had visited cloud-shrouded Venus by mistake. His humor was not appreciated by everyone.

By the end of January 1972, the dust storm was beginning to clear, and the high-resolution cameras began to do their work. They would eventually take 7,329 photographs that covered about 90 percent of the Martian surface.

Mariner 9 revealed Mars for the first time in human history. It spent 349 days in orbit around Mars and mapped 100 percent of the planet, showing that its surface had two very distinct hemispheres—an ancient cratered surface in the southern hemisphere and a geologically younger surface in the northern hemisphere—with exciting features such as volcanoes larger than any on earth, immense canyons, dry river channels, large lava flows, and polar regions that indicated past glacial periods. *Mariner 9* also photographed two small Martian moons, Deimos and Phobos. In all, the spacecraft transmitted fifty-four billion bits of science data as compared with two billion bits from all previous Mars flights. Alone, without its failed sister craft *Mariner 8*, which crashed in the Atlantic because of a launch failure, *Mariner 9* paved the way for *Viking 1* and 2, those Space-Age art objects produced by the golden age of planetary exploration, robots that extended the human senses to another planetary world for the first time in human history.

Viking Robots on Mars

Viking 1 and 2 were highly sophisticated spacecraft, each consisting of an orbiter and a lander—four separate machines in all. After separation command from Mission Control, the landers were fully on their own, controlled by an onboard computer. The nearly twenty minutes' travel time for radio signals to reach earth made this portion of the mission impossible to control from earth. The landers had to navigate, enter the atmosphere, decelerate, and land on a surface that still was not known completely and could offer surprises such as boulders or small, steep-walled craters that could topple the landers. The computer brain was said to have an intelligence equal to a seven-year-old child—but only in limited directions.

The *Viking 1* lander settled on the Martian surface, on what scientists believe to be an ancient lava flow, at a speed of 8.2 feet per second (about 5.6 miles, 9 kilometers, an hour) after its

journey of 200 million miles (322 million kilometers). On earth, it was the summer month of July. The *Viking 2* lander touched down on the surface six weeks later, on September 3, 1976. The two landing sites were about 3,100 miles (5,000 kilometers) apart.

The Viking robots weighed about 1,960 pounds (889 kilograms) and measured 5 feet (1.5 meters) across and 1.5 feet (0.5 meters) high. Their soil-sample collecting arms stretched out about 10 feet (3.5 meters).

Each lander had one million separate parts and contained the equivalent of two power stations, two computer centers, a weather station, two chemical laboratories, three incubators to test for Martian life, a scoop and backhoe for digging and collecting soil samples, a miniature transport system for bringing samples inside for analysis, and a TV studio. It was a microfactory that would fill a building on earth. Landers 1 and 2 were the most complicated automated spacecraft ever launched from earth, but they operated well beyond all expectations and added reality to the romance of Mars created in the popular literature of the nineteenth and twentieth centuries.

The Good Vikings

The Viking orbiters surveyed potential landing sites before the landers were released for touchdown. With hundreds of images transmitted to earth during the first weeks in orbit, Viking scientists determined which landing sites would offer the best prospects for successful landings.

Both orbiters took some 52,000 photographs during their missions, and the two landers took 4,500 additional photos of the surface. About 97 percent of Mars's total surface was photographed at a resolution of 984 feet (300 meters), while another 2 percent was seen at a resolution of 82 feet (25 meters). By the summer of 1980, the two landers had returned more than three million weather reports from the surface of Mars. Infrared observations from orbit exceeded one hundred million. Scientists continued to analyze the mountain of data throughout the 1980s.

The Most Significant Viking Discoveries

- THE REASON the Red Planet is red is because of the oxidized iron that gives the soil its reddish color.
- MARS'S SOIL is like firm soil or sand on earth—cohesive and fine-grained.
- MOST OF the surface rocks resemble basalt lava on earth. The chemical makeup is like weathered, altered basalt, and water and sulfur compounds were discovered in the soil.
- THE SURFACE of Mars is similar to the deserts of earth, but it is much drier than any place on earth. Water is locked in the north polar ice cap in the form of subsurface permafrost.
- VIKING LANDER experiments in both locations on Mars gave puzzling results of the soil analyses, but they found no organic chemicals or existing life. Intriguing chemical properties of the red soil mimicked certain reactions of living systems, but the final scientific consensus is that no life was found at the two Viking landing sites.
- THE MARTIAN atmosphere contains nitrogen, oxygen, carbon, water vapor, argon, and traces of several other gasses—everything needed to support life—except that they are not in the right proportions for earth life forms.

- THE THIN Martian atmosphere is about 1/100th as dense as earth's.
- MARS PROBABLY had a much denser atmosphere in the past, but over time, nitrogen has escaped into space and has modified the atmosphere.
- THE MARTIAN sky is pink, not blue, because of suspended particles of fine red dust about 1/25,000 of an inch (0.001 millimeter) across.
- THE CLIMATE of Mars's northern and southern hemispheres is very different. Global dust storms originate in the south, and water vapor is fairly abundant in the far north during its summer.
- MARS TODAY is seismically much less active than earth. Even though great volcanic mountains were found, they are now inactive.
- THE TWO Viking landing craft continuously monitored weather at the landing sites. The winds on the surface were light, about 15 miles (24 kilometers) per hour. Surface temperatures ranged from -20 degrees F.(-29 degrees C.) in the afternoon to -120 degrees F. (-84 degrees C.) at night. Clouds and fog occur despite the fact that atmospheric water content is extremely low compared to earth.

The Viking spacecraft, orbiters and landers, presented the new Mars. Although the landers were designed for ninety-day missions on the surface, the *Viking 1* lander operated until late in November 1982, while the *Viking 2* lander sent back valuable data until April 12, 1980. In late 1980, the *Viking 1* lander was renamed the Mutch Memorial Station in memory of Dr. Thomas A. Mutch, the former Lander Imaging Team leader, who disappeared in the Himalayas while mountain climbing.

We learned more about the Red Planet in one year than we had about the earth in one thousand years. The old romance of Mars is gone; the dreams of canals and other artifacts of intelligent life have been replaced by scientific reality. But a new romance of the planet has replaced

it. Its future will be shaped by humankind, their manned ships, and their settlements. Mars is, after all, the next most friendly planet in the solar system, and we will someday walk in its red soil, see its pink sky with our own eyes, and live our lives there.

The Mars Album

If the fifty thousand-plus photos taken by all four Viking spacecraft were developed in an 8- by 10-inch format and spread out, each touching the other, they would give the entire Mars album an area of at least 50 square miles (130 square kilometers), equal to the area of a fair-sized city.

THE FUTURE EXPLORATION OF MARS

The Mars Observer

Before men and women journey to Mars sometime in the twenty-first century, several unmanned spacecraft will follow Viking's lead and further study the Red Planet.

In the early 1990s, the United States plans to launch a robot spacecraft toward Mars, which will go into Martian orbit about one year after launch. Called the Mars Observer (the shortened name for the Mars Geoscience/Climatology Orbiter), the Mariner Mark II spacecraft will conduct a detailed study of the surface and atmosphere of Mars to better understand the complex weather and climate of the planet, and the interaction between the soil and the atmosphere.

This mission will be the first of a new class of planetary observers that will use already developed, sophisticated earth-orbiting spacecraft that are adapted to planetary missions. The Mars Observer has a high scientific priority. By determining the planet's global surface composition and the role of water in the Martian climate, the spacecraft may answer many questions regarding Mars's evolution. Once the planetary evolutions of Mars, Venus, and Earth are more thoroughly understood, we stand a better chance of influencing our destiny as a species.

The Mars Observer mission, planned for one Mars year (equal to 687 Earth days), could be extended like the Viking missions if the spacecraft performance permits it. Scientific sensors will furnish, among other data, a mineralogical map of Mars; profiles of temperature, dust, and water vapor; and characteristics of the planet's magnetic field.

The spacecraft's initial orbit will be a low-altitude one so that its instruments can obtain higher quality data. At the end of the mission, however, a propulsion maneuver will boost the Mars Observer into a higher orbit so that the spacecraft doesn't crash on the surface. This will allow the mission to meet the planetary protection guidelines that prohibit unsterilized spacecraft from landing on the Martian surface before the year 2039.

Other Unmanned Missions to Mars

At least two other advanced reconnaissance missions to Mars will round out the Space Age's first century. NASA plans to send a Mars Aeronomy Orbiter before the year 2000, which will study the interaction of the upper atmosphere and ionosphere with radiation and particles of the solar wind. A Mars Surface Probe mission is also in the planning stages. It would establish seismic, weather, and geological stations on the Martian surface that will provide the detailed knowledge base necessary for the planning of a manned mission to Mars.

In the early years of the next century, a sample return mission may make the Earth-Mars round-trip. Mobile rovers would move about the Martian surface to collect rock and soil samples, and take measurements. Such a mission could also help develop and test some of the new technology needed for a manned flight to the Red Planet.

Red Star on Mars

If their plans are accomplished, the Russians will beat the Americans back to Mars. Although Russia has not flown a mission to Mars since 1973, Soviet scientists revealed some ambitious planetary mission plans, which included Mars, at a 1985 meeting at the Johnson Space Center. The Russians plan to have two spacecraft in Mars orbit by 1989, and the mission will include the first landing of a robot probe on a Martian moon—Phobos. U.S. scientists were surprised at the candor and detail provided by the Soviet scientists about their forthcoming mission plans, and there was renewed speculation about possible joint space ventures between the two countries.

The Russians plan to place two orbiters in elliptical paths around Mars, and they then will circularize the orbit of one spacecraft so that its speed will match that of the Martian moon Phobos. The spacecraft will hover over the moon's surface, perhaps as close as 164 feet (50 meters) above it, taking photographic images of details as small as individual rocks and soil areas. If all goes well up to this point, a landing probe will then be deployed to the surface of the moon—an extremely tricky maneuver because the moon has such weak gravity and the landing probe could easily bounce back into space and fail at landing.

Once the probe is on the surface of Phobos, a springlike device will allow it to jump from place to place on the surface to explore different areas. The hovering orbiter will also be equipped with what the Soviets call a laser mass spectrometer, which will zap portions of the surface below with a laser beam and then analyze the resulting vapor.

If this complex Phobos mission is successful, the second orbiter and its landing probe could carry out the same experiments on the other Martian moon, Deimos.

This Russian Mars mission is far more complex than the U.S. Mars Observer mission, and, if all goes well, it will probably arrive two years ahead of the American spacecraft. The Martian moons are important because so little is known about the many small bodies of the solar system. Some experts speculate that the moons of Mars could be captured asteroids composed of primordial solar system material. Because they are small celestial bodies, it is likely they have not gone through the kinds of major changes that are caused by the internal heat of larger bodies.

It is very encouraging that, as the Space Age advances into the next century, the major spacefaring nations of the world are renewing their efforts to send robot emissaries to the Red Planet. These machine scouts will pave the way for what will probably be the first life on Mars—humans from the planet Earth. (For details of the possible future manned mission, read Chapter 11, "Settlement and Colonization of Space.")

TO THE COMETS AND ASTEROIDS

The Halley Armada

The neglected members of our solar system, in terms of spacecraft rendezvous missions, were the asteroids and comets, but that changed dramatically toward the end of the twentieth century when man began to probe, sample, and analyze the wanderers at close range.

It was the return of the famous Halley's Comet in 1986 that sent forth a five-spacecraft armada from earth to fly by the great comet that has been returning to the inner solar system and the sun for at least 2,200 years. Besides these Russian, European, and Japanese spacecraft, earth satellites and in-place interplanetary spacecraft also intensively studied awe-inspiring Halley's Comet from their various locations throughout the solar system.

Now the most studied comet in history, Halley's opened the door with its 1986 passage through the solar system to unmanned spacecraft exploration of those "minor" objects of our cosmic neighborhood—comets and asteroids. Future missions include plans to scoop up some comet gas and launch into orbit around an asteroid.

The Comet Rendezvous and Asteroid Flyby

In the 1990s, a newly designed Mariner Mark II spacecraft, developed for NASA by the Jet Propulsion Laboratory, will begin its maiden voyage, probably from the payload bay of a space shuttle, atop an upper stage rocket. After six months, the spacecraft will fly by an asteroid named Hedwig and then continue toward its rendezvous with Comet Wild 2 (or an alternative such as Kopff or Honda-Mr´kos-Pajdu´sako´va) several years later.

Travelers Past Halley's Comet and Their Discoveries

- THE RUSSIAN and French spacecraft *Vega 1* and *Vega 2*, both launched in December 1984, first dropped off Venus probes and then flew on to encounter great Halley on March 6 and 9, 1986. Their wide-angle and narrow-angle visual cameras resolved details as small as 656 feet (200 meters). Even the United States played a part in the Vega spacecraft by developing and providing a sophisticated comet-dust collector.

- THE EUROPEAN Space Agency's British-built *Giotto* spacecraft, launched in the summer of 1985, encountered the comet on March 14, 1986, and flew closer to the nucleus than any other spacecraft. Its color and infrared cameras approached to within 376 miles (602 kilometers) of Halley's nucleus and survived the encounter even though *Giotto* was knocked out of commission by cometary debris for almost thirty minutes. Only two of its ten instruments escaped damage. *Giotto* determined that the nucleus of Halley's Comet was 7.5 miles (12 kilometers) long, was shaped like a peanut, and was dark as coal.

- JAPAN'S SMALL interplanetary spacecraft *Suisei* passed Halley's Comet on March 8, 1986. Its four instruments, including an ultraviolet television system, made good the Japanese tradition of quality and delivered images of Halley's huge ionized hydrogen coma. Another probe launched by Japan, *Sakigake*, passed 4.4 million miles (7 million kilometers) from Halley's on March 11, 1986, and measured the solar wind.

Mariner Mark II is NASA's economical answer to budgetary restraints. A simple modular spacecraft, it can be reconfigured for different missions that will fly beyond Mars to the asteroid belt, the outer planets, and deep-space regions of the solar system. The spacecraft in this series will take advantage of ongoing advances in technology and have the additional design freedom offered by a space shuttle/upper stage launch.

While the mission will be the Mariner Mark II's first flight, its two mission targets will also represent firsts. Its flyby of asteroid Hedwig at a distance of about 3,800 miles (6,100 kilometers) will be the closest a spacecraft has ever come to any of the more than 3,300 asteroids, which someday may be supplying the earth and other solar system settlements with raw materials.

After its instruments scrutinize the small asteroid, 70 miles (113 kilometers) in diameter, the Mark II spacecraft will journey out beyond Jupiter's orbit and then swing back toward the sun for its Comet Wild 2 rendezvous some four years after its launch from above the earth.

But this is hardly the end of the mission. Unlike the several spacecraft from various countries that cast quick, although important, glances at Halley's Comet in 1986, the Mariner Mark II will fly alongside the comet for nearly three years! All the while it will be taking data as this cosmic duo, one natural and one made on earth, make their closest approach to the sun in the late 1990s and then head outward again.

Scientists believe that Comet Wild 2, named after its discoverer, Dr. Paul Wild, is a marvelous target for the fly-along mission because it has never gone closer to the sun than Jupiter's orbit and probably has retained its pristine ices composed of the primordal stuff from which the solar system was formed. The spacecraft, still in the design and development stages, may also carry a surface penetrator device that could be fired into the comet's nucleus to measure quantities of as many as twenty chemical elements.

The information return from the six-year-plus mission will be astounding. If only the Mariner Mark II had been ready to accompany Halley's Comet on its outward journey to the distant solar system in 1986, we would have had the armchair ride of the century.

The First Asteroid Flyby

Project scientists for the *Galileo* mission to Jupiter decided in 1985 to change the trajectory of

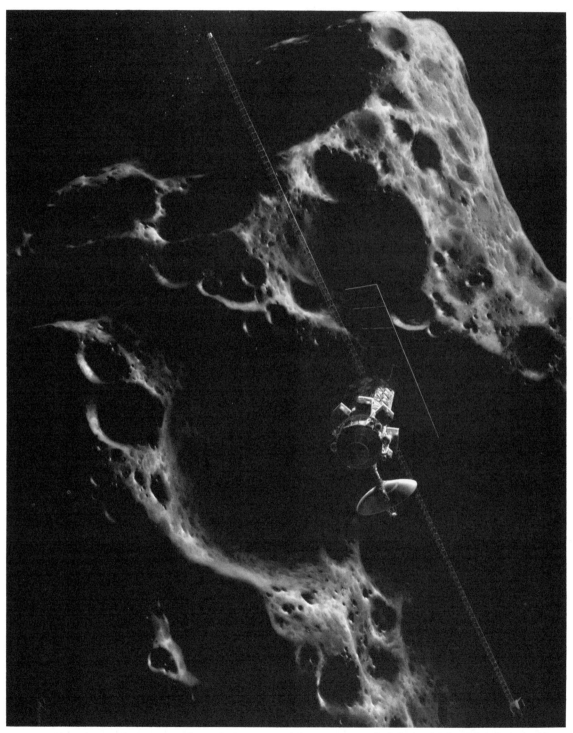

An asteroid rendezvous mission may fly in the last years of this century. Called the Multiple Mainbelt Asteroid Orbiter/Flyby mission by NASA, it would study several asteroids and orbit a single one for detailed study. Painting by Paul Hudson. Courtesy Jet Propulsion Laboratory.

the spacecraft and detour it past the rocky asteroid named 29 Amphitrite, with a diameter of 125 miles (201 kilometers). Amphitrite's surface was to be scanned by *Galileo*'s cameras and mapping spectrometer to give us the first close look at one of those cosmic chunks of rock that make their orbital homes between Mars and Jupiter. But the mission plan was scrubbed after the *Challenger* exploded, although a similar one could be flown in the early 1990s, when *Galileo* is finally launched.

Venus Plus Asteroids

The 1990s will also see the Russians going toward the asteroids, but it seems they go nowhere without first stopping at their planetary compulsion, Venus. This mission is no exception. Is it because much of the U.S.S.R. is so cold and Venus so hot?

Proposed for launch in 1991, the cooperative Soviet-French mission is called Vesta, named for the brightest known asteroid as seen from earth. Two spacecraft in one booster rocket will be powered into their respective trajectories— one to Venus and the other to Vesta. The Venus spacecraft would eject a lander above Venus and then continue on to yet another asteroid. The spacecraft heading for Vesta, the French contribution to the mission, could possibly drop an asteroid lander onto the surface of the asteroid as it flew by and analyze its bright, reflective surface, most probably composed of basaltic minerals.

Vesta is the brightest of any of the large asteroids and can be seen with the naked eye. As a target for the Russian and French spacecraft *Vesta*, however, it is a small celestial body, measuring only 342 miles (550 kilometers) in diameter.

A Comet Sample Return

In the late 1990s or the first decade of the twenty-first century, the United States may launch a comet sample return mission. Currently called the Comet Atomized Sample Return mission, its goal would be to visit the same short-period comet that would already have been analyzed earlier by the flyby mission. The spacecraft would collect samples of cometary dust and gas during the flyby and then return them to earth for detailed laboratory analysis. Such a study of pristine cometary material could answer many questions about the origin of our sun and solar system.

The technologically complex mission would require solar electric propulsion, robotics, and remote ice-coring equipment. But to catch a piece of a cometary vagabond, just one of the some one hundred billion that orbit around the sun and planets in the Oort cloud of comets, would be an amazing feat.

Two More Asteroid Missions

In the future, selected asteroids may be sites for in-space mining operations. These would provide raw materials for refining and final building materials for the larger space stations, lunar and planetary bases, and city-size space colonies of the future. NASA has plans, if not final funding, for two asteroid missions, one which will go to the asteroid belt and one which will study the great chunks whose orbits bring them relatively close to the earth. Knowing exactly what these asteroids are made of is an important goal for these missions.

The Multiple Mainbelt Asteroid Orbiter/Flyby mission, ready for launch in the last years of this century or the first years of the next, will study a variety of asteroid types as it flys by on its way to a target asteroid, which it will orbit and study in detail. A Mariner Mark II spacecraft will journey to the asteroid belt between the orbits of Mars and Jupiter.

Closer to home, the earth-approaching Asteroid Rendezvous mission will take a long-term, detailed look at one of the several dozen known asteroids whose orbits cross the earth's orbit around the sun. This group of celestial chunks may be geochemically different from the mainbelt asteroids, and their accessibility from earth makes them prime targets for the mining of extraterrestrial materials. These asteroid assay missions will no doubt have the full attention of several multinational and Fortune 500 corporations.

OUTWARD TO THE GIANTS

The Little Spacecraft that Could

June 13, 1983, is a historic date, but only the rare person could tell you why it should be remembered. That was the date that the hardy little spacecraft *Pioneer 10*, launched more than a decade earlier (on March 3, 1972), left our sun's solar system forever to wander throughout the immense Milky Way galaxy. It was the first man-made object to leave our local neighborhood of the universe.

At the historic crossover point, when the spacecraft went beyond the orbit of Neptune (now the most distant planet), it was 2.81 billion miles (4.52 billion kilometers) from earth. Round-trip communications with the amazing spacecraft take about nine hours at the speed of light, and scientists hope they can continue communicating with *Pioneer 10* until the early 1990s, when it is some 5 billion miles (8 billion kilometers) from its planet of origin. As the first human artifact to leave the solar system, *Pioneer 10* is a triumph of human ingenuity and engineering skill. But it did so much more, and so much more for the first time, that it deserves to be known as the earth's most famous spacecraft.

The main mission of *Pioneer 10* was to journey to giant Jupiter and its retinue of moons

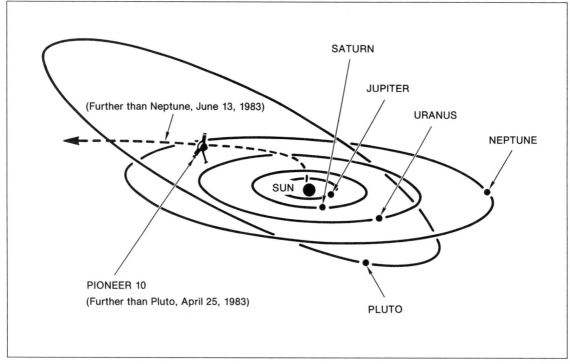

ABOVE AND FACING PAGE: *The little spacecraft that could. Why is June 13, 1983, a historic date? It is the first time a robot spacecraft (and human artifact) left the solar system forever. When* Pioneer 10 *was launched on March 3, 1972, scientists were not even sure it would survive its flight through the asteroid belt.* Courtesy TRW.

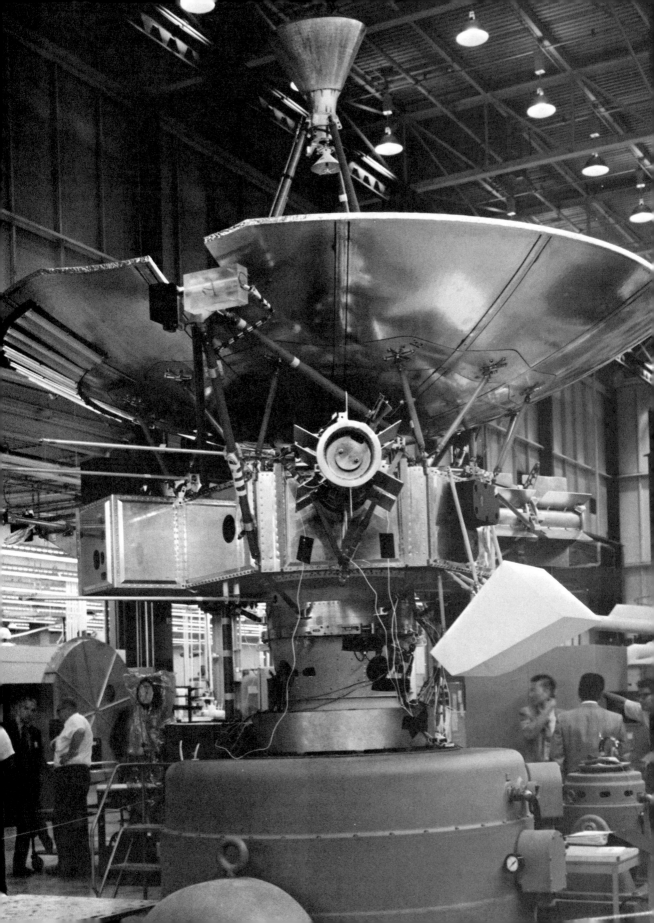

some 620 million miles (998 million kilometers) from earth and survey this mysterious planet up close for the first time. It accomplished this scientific feat, and attained many other firsts along the way. Because *Pioneer 10* was the first spacecraft to fly beyond Mars, scientists did not even know if it could survive its flight through the rocky debris of the asteroid belt. But it did, thus paving the way for the Voyager spacecraft that would follow and reveal marvels that rivaled the human imagination.

When this exemplary robot flew past its destination planet, Jupiter, its data streams of 1,024 bits a second produced some 500 photographs. The images were computer reproductions of digital data, which recorded 64 different intensities of red and blue light. In just one second, 1,000 intensities were measured, coded, and sent back as data to earth to be reconstituted into two-color images that showed a giant, turbulent planet without a solid surface. *Pioneer 10* confirmed, among many other discoveries, that Jupiter was composed of liquid hydrogen.

The 570-pound (259 kilogram) spacecraft carried twelve scientific instruments that could answer more than a dozen scientific questions. It could respond to commands from earth from as far away as 2 billion miles (3.2 billion kilometers); maneuver in space and face its high-gain dish antenna toward earth; keep the precise time and maintain temperature; and store over 50,000 bits of data for transmission to the home planet.

The spacecraft's flight stability was attained by a controlled spin of five revolutions per minute. Nuclear energy—in the form of four radio-isotope thermoelectric generators (two each at the end of two 9-foot, 2.7-meter, booms—powered *Pioneer 10* and its sister craft that followed. The modest 8-watt radio transmitter, with the help of powerful receivers on earth, has stood the test of more than a decade. But its signal is extremely weak when it reaches the earth, on the order of 20,000 trillionths of a watt. If this infinitesimal amount of energy were collected for nineteen million years, it still would only light up a Christmas tree for 1/1,000th of a second.

And then there is the famous plaque, a message sent with eternal hope to other intelligent life in the universe. This gold-anodized plaque, bolted to the mainframes of both *Pioneer 10* and *11*, uses as its key the most common element in the universe—the hydrogen atom. Along with other basic information about our solar system is a drawing of a human female and male. That the plaque was put on *Pioneer 10* and *11* at all says much about the human spirit.

Once every million years or so, *Pioneer 10* may come close to another star system, but the odds of this cosmic bottle ever being found by other intelligent life must be considered dim. But again, the act of placing it on the spacecraft perhaps tells more about us than the entire primitive machine that carries it. Farewell, good *Pioneer 10*; may the thousands of stars and billions of years be kind to you.

A Tough Flight to Follow

The sister spacecraft to *Pioneer 10*, *Pioneer 11*, had several firsts of its own. Identical in design and instrumentation, *11* was more than a back-up mission. Its historic significance is that it was the first spacecraft to fly past the majestic planet of Saturn after its swing around Jupiter.

Launched almost a year after its trajectory-blazing sister craft, on April 6, 1973, it too passed through the once feared asteroid belt without damage and on then on to Jupiter. It flew past the giant on December 3, 1974, at a much closer distance than did *Pioneer 10*—26,725 miles (43,000 kilometers) versus 81,000 miles (130,000 kilometers). And its trajectory by Jupiter was different also; it flew over the planet's north and south poles for the first time, regions not seen from earth.

On went *11* toward Saturn, where no spacecraft had ventured before. Even the experts did not know if it could survive its passage near the rings of Saturn. As *Pioneer 11* swept by Saturn and its moons on September 1, 1979, it provided the first close-up images of the planet shaped like a flattened sphere, the swirling atmosphere with a large white spot, and the wondrous rings. The on-board instruments found out that Saturn, like Jupiter, radiates about twice as much energy as it receives from the sun; that the planet has a high-altitude haze above the clouds, possibly composed of ammonia ice crystals; that the atmosphere has high-speed jet streams; and much more. A cloud of hydrogen was discovered around the rings, and two new rings, called F and G, were discovered. Most importantly, the flyby of Saturn by *Pioneer 11* all but proved that spacecraft would not crash upon the shores of the rings and could fly through the system without damage. It paved the way for the soon-to-come Voyager spacecraft, the jewels (along with Viking) of the golden age of unmanned planetary exploration.

Voyagers to the Stars

Launched in 1977 in a trajectory toward Saturn, *Voyager 1* first dazzled the minds of millions of us on earth with thousands of images of Jupiter taken during its flyby. One of the two most sophisticated space robots every built, *Voyager 1* neared its close encounter with the beautiful planet, Saturn, in early November 1980. Reflecting on the historic significance of the cosmic event, Carl Sagan said: "I believe we are at a moment that will be remembered for tens, hundreds, perhaps even thousands of years . . . . We are at a moment of extraordinary discovery. There are six or eight new worlds up there [moons of Saturn] that we are about to see for the first time . . . . The exploratory instinct," Sagan concluded, "which has taken us to the vicinity of Saturn is part of the reason for our success as a species."

After its Saturn flyby, *Voyager 1* headed toward interstellar space, while its sister ship, *Voyager 2*, launched only two weeks earlier, continued to fly toward its own unique trajectory through the Saturn system some ten months later, and then onward toward its encounters with Uranus and Neptune. These wonderful and remarkable machines had already been flying for three or four years when they reached Saturn. They were the most complex space machines ever built.

Their imaging system, which produced the greatest celestial show every seen on earth, was far more sophisticated than the earlier Pioneer spacecraft. Two vidicon TV tubes, a wide angle and a narrow angle, were designed for slow scan readout and required forty-eight seconds to produce each image. They were black and white cameras, but the images were taken through different color filters and the data were combined to produce color images through computer processing and enhancement. It took five million bits of information to produce one picture, and the data streams were one hundred times faster than those from *Pioneer 10* and *11*—115,200 bits a second from the distance of Jupiter. More than seventy thousand images were transmitted to earth by *Voyager 1* and *2*.

The Voyagers weigh 1,820 pounds (826 kilo) each, more than three times the weight of the Pioneer spacecraft. Their largest component is the high-gain radio antenna dish, which measures 12 feet (3.7 meters) in diameter. Their boom span, from tip to tip, is about 30 feet (9 meters). One boom carries the cameras and several other instruments; the other supports the three nuclear generators that produce a total of 400 watts of electrical power for the spacecraft. A single magnetometer boom extends even further out than the instrument and power booms. The beauty of *Voyager 1* and *2* is in what they have done, not how they appear. By looking at these machines, no person could imagine what beauties they were capable of sensing and sending back to planet earth.

Voyager Revelations . . . So Far

Both spacecraft were launched from earth in the summer of 1977, *Voyager 1* on September 5, and *Voyager 2* on August 20. Even though *Voyager 1* blasted off about two weeks later than its sister spacecraft, it nevertheless arrived at its closest point to Jupiter four months earlier, on March 5, 1979, because of a faster trajectory. *Voyager 1* spent 92 days observing Jupiter, its moons, and the interplanetary space around it. *Voyager 2*'s instruments, eleven in all, spent 133 days studying Jupiter, with its closest approach occurring on July 9, 1979. A combined total of 250 billion bits of data was returned by the spacecraft dur ing their Jupiter encounters. Some of these data became 33,000 images.

Because of the abundance and depth of knowledge gained by the Voyagers, our two giant planets have become the new Jupiter and the new Saturn. More was learned about them in seven and a half months than was learned during the rest of human history, and the Voyagers are still flying and sending back their data streams to us.

A sampling of Voyager discoveries about Jupiter include:

• An intensely hot, doughnut-shaped plasma sulfur cloud circling Jupiter encloses the orbit of the moon Io.

• A thin ring of dark particles circles Jupiter.

• Eight active volcanoes on Io, the first active volcanism found off the earth.

• Electrical current between Jupiter and Io measures at five million amperes.

• Three new moons, including odd-shaped Amalthea, which looks like a captured asteroid.

• Detailed images of Jupiter's moons Io, Europa, Ganymede, and Callisto.

• Gigantic cloud-top lightning bolts observed on Jupiter; thunderstorms that could engulf earth.

• A rotation period of about six days for the outer edge of the Great Red Spot, which moves in a counterclockwise direction.

And then, after the wondrous images of Jupiter and its moons, these precious spacecraft flew on to ringed Saturn to unveil its secrets. As *Voyager 1* and *2* flew by Saturn in November 1980 and August 1981, they discovered, among other things:

• Saturn's rotation period is 10 hours, 39 minutes.

• Winds near Saturn's equator blow at 1,100 miles (1,770 kilometers) per hour in an eastward direction.

• Most of the debris in the rings is composed of water ice. So too are several of its moons.

ABOVE: *The spacecraft* Voyager 2 *continues to whisper its data-bit images from the far reaches of our solar system. The painting shows the relationship between the spacecraft and Uranus during its closest encounter on January 24, 1986.* FACING PAGE: *These three images of Uranus were taken on January 23, 1986, before the closest encounter. Different filters were used for each image; the methane filter image on top shows the planet surrounded by high, bright clouds.* Courtesy Jet Propulsion Laboratory.

• Rings within rings were discovered, and they numbered in the thousands. Density waves, created by gravitational interactions, are partly responsible for the complicated structure of Saturn's rings.

• A doughnut-shaped hot plasma cloud circles Saturn. Its temperature is three hundred times hotter than the sun's corona and twice as hot as the torus around Jupiter.

• Saturn's five inner satellites—Mimas, Enceladus, Tethys, Dione, and Rhea—were imaged and each was shown to have distinctive features.

• Several new moons were discovered, bringing the count to twenty. The outermost moon, Phoebe, is composed of rock and may be a captured asteroid.

• Titan, Saturn's largest moon, lost its title as the largest moon in the solar system after Voyager's instruments measured it with radar. The title went to Jupiter's Ganymede.

• Titan has an atmosphere composed of nitrogen, methane, and several organic compounds. Methane may flow on Titan like water does on earth. Titan is the only moon in the solar system with an atmosphere and has planetary scientists very excited.

As *Voyager 2* left Saturn behind in September 1981, *Voyager 1* was already four months away from the planet, searching for the boundary within the solar system, where the solar wind fades away and the cosmic rays from interstellar space replace it. But because of *Voyager 2*'s trajectory and a rare alignment of the planets that occurs less than once every century, it had two other planets on its cosmic itinerary before it headed into interstellar space. Uranus and Neptune would also be scrutinized by this robot's eyes and ears before the end of the 1980s.

Voyager's Big Whisper

When *Voyager 1* was at its closest point to Saturn, on November 12, 1980, its data stream was sent back to earth with 21.3 watts of power, about one-third that of a 60-watt light bulb.

This radio signal, traveling at the speed of light, took 1 hour, 24 minutes, 46 seconds to travel the 947 million miles (1,524 million kilometers) to earth. When the signal intercepted earth, it had a diameter of 22.5 million miles (36.2 million kilometers). The signal was so faint when it was received by the large-dish receiving antenna that it amounted to only one million-trillionth of the energy needed to operate a 25-watt lightbulb.

Voyager 2's *Flyby of Uranus*

Four years after *Voyager 2* left Saturn behind, in late 1985, Jet Propulsion Laboratory scientists prepared for the spacecraft's encounter with the emerald planet of our solar system—green, ringed Uranus. The planet's green cast is the result of methane in its atmosphere, which absorbs red light. The reflected sunlight off the planet therefore shows mostly blue, green, and yellow light. This light, like *Voyager's* radio signals, takes 2 hours, 45 minutes to reach earth.

On January 24, 1986, the spacecraft made its closest approach to this faraway sphere of gas at a distance of about 50,000 miles (80,450 kilometers). It was some 1.8 billion miles (2.9 million kilometers) from earth and traveling at more than 45,000 miles (72,405 kilometers) per hour. Again *Voyager* performed its magic and gave humankind its first close-up views of the planet and its thin rings composed of dark material that reflects no more light than blackboard slate.

Just specks of light through telescopes on earth, the moons of Uranus (Miranda, Ariel, Umbriel, Titania, and Oberon) were seen in detail for the first time, and more than a dozen new moons and rings were discovered. Even the tremendous distance from earth did not degrade *Voyager's* wonderful images. The information and images sent back from Uranus represent an improvement in space communications since 1960—when *Pioneer 5* set a deep-space distance record of 22 million miles (35 million kilometers), but without any image data—by a factor of about 1 million.

Close-up images of the moons of Uranus taken by Voyager 2, including those of Miranda (the closest view) and Ariel, revealed dramatic fractures, cliffs, and craters for the first time. Courtesy Jey Propulsion Laboratory.

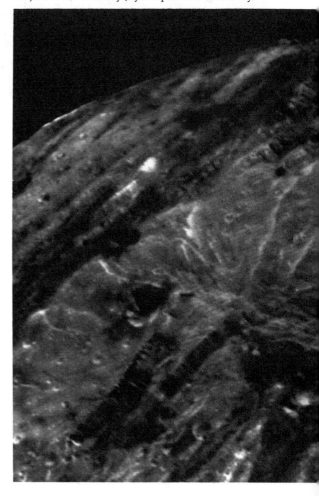

And Onward to Neptune

If *Voyager 2*'s superb performance continues through the 1980s, and there's every reason to believe it will, faraway Neptune will receive its robot glance in August 1989, as the spacecraft soars by it on its way out of the solar system. Twelve years after *Voyager 2* left earth, at a distance of some 3 billion miles (4.8 billion kilometers) from the sun, it will turn its cameras and instruments on Neptune, now the most distant plant, which takes 164.8 earth years to revolve once around the sun. Neptune's largest moon, Triton, retains a thin atmosphere of methane. *Voyager*'s close-up views may help determine

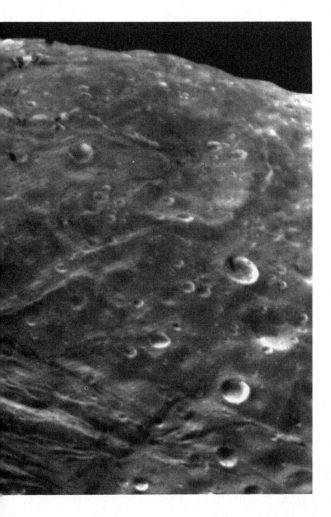

why Triton is the only large moon in the solar system to orbit in the opposite direction to the rotation of its planet. This and other eccentricities in its moon family suggest that a mysterious event may have taken place in Neptune's past. Some astronomers speculate that little Pluto, now considered a planet, may once have been a moon of Neptune. *Voyager 2*'s data streams may solve some of the puzzles of this last planet on the amazing grand tour of our solar system.

Good-bye Forever, Voyagers

By the end of the 1980s, all planets in the solar system will have been visited by robot spacecraft and their sensory instruments except for iceball Pluto, which may host its own robot spacecraft in the early part of the twenty-first century.

Voyager 1 and 2 will leave our solar system by the end of this century and curve through interstellar space for thousands, even millions, of years. During these millennia, during which civilizations rise and fall, they may never come close to another star or planetary system. Indeed, to traverse only one-sixth the distance between our solar system and the center of the Milky Way galaxy (about 5,000 light-years), it will take the Voyager spacecraft about one hundred million years! They're hardly traveling at the hyperspace speeds of our *Star Wars* friends.

But anything is possible, even an encounter with another spacefaring civilization in the galaxy. Although the distances between the stars are immense (only one part in one hundred million of the volume of space is filled with stars), the Voyagers have time on their side. The odds are astonishingly stacked against intelligent life ever plucking one of the two Voyagers out of the cosmic ocean, but if they do, we have sent them a message from earth. Affixed to each spacecraft is a gold-plated copper phonograph record that contains 118 photographs of the earth and the activities of its creatures, ninety minutes of music from all over our planet, greetings in almost sixty human languages and one whale language, and a message from the one-time Presi-

dent of the nation that built the spacecraft—Jimmy Carter of the United States of America. The last paragraph of the President's message to unknown intelligent life reads:

"This is a present from a small distant world, a token of our sounds, our science, our images, our music, our thoughts and our feeling. We are attempting to survive our time so we may live in yours. We hope someday, having solved the problems we face, to join a community of galactic civilizations. This record represents our hope and our determination, and our good will in a vast and awesome universe."

And the magnitude of the cosmos really hits home when we realize that the Voyagers may still be flying when human life no longer exists, when even the sun and planets have transformed again to dust and gas. The Voyagers may be, in some future, far or near, the only human artifacts left in the universe.

Galileo *to Jupiter*

Scheduled to be launched in May 1986 from the space shuttle, Project Galileo was delayed several years by the *Challenger* disaster. This interplanetary spacecraft, when it is finally launched, promises to be one of the most exciting interplanetary missions in the last decade of this century.

Great Jupiter, which holds 71 percent of the total mass of all the planets, still holds us in thrall, as does its unique retinue of moons. And the two-in-one *Galileo* spacecraft, consisting of an orbiter and an atmospheric probe, will be the first spacecraft from earth whose primary mission will be entirely devoted to the Jovian system, what some scientists have referred to as a "miniature solar system."

Not just a flyby mission, Project Galileo will study Jupiter and some of its large moons for about two years. *Galileo* will also be the first

In this painting by Don Davis, based on a computer simulation of the flight path, Voyager 2 *encounters Neptune in August 1989, thus ending its tour of the solar system.* Courtesy Jet Propulsion Laboratory.

spacecraft to go into orbit around one of the giant outer planets, and the first to deploy a probe to descend into Jupiter's atmosphere to measure its structure and chemical composition, and to perform other scientific measurements before the planet's intense pressures and temperatures destroy it.

Project Galileo was named for the famous Italian astronomer who discovered the four major moons of Jupiter some 350 years ago. In 1985, NASA decided to add an asteroid flyby mission to *Galileo*'s flight plan. The Jupiter-bound robot craft could still accomplish a flyby of an asteroid located in the middle of the asteroid belt before flying on toward Jupiter. This detour for *Galileo* could be the first asteroid flyby of the Space Age, and it would accomplish one of the primary objectives of unmanned spacecraft exploration years before it could otherwise be done.

About five months before *Galileo* arrives at Jupiter, the atmospheric probe craft will separate from the orbiter spacecraft, and each will follow an independent flight path to Jupiter. A few hours before the probe craft and its six instruments plunge into Jupiter's immense atmosphere, the orbiter will fly by the amazing moon Io, at a distance of only 600 miles (965 kilometers), to make close observations of the most volcanic surface in the solar system. At the same time, Io's gravity will slow the orbiter, helping to insert the spacecraft in its proper orbit around Jupiter.

The probe's descent module will strike the upper layers of Jupiter's atmosphere at about 112,000 miles (180,208 kilometers) an hour. Powerful deceleration forces will subject the probe craft to the equivalent of four hundred times the earth's surface gravity during this period. The probe will then deploy a parachute and begin to take atmospheric measurements—the first direct sampling of Jupiter's swirling, ever-changing, and multicolored atmosphere.

Forty minutes after the probe enters the atmosphere, scientists hope it will have descended below Jupiter's lowest water clouds, having measured all the important cloud layers during the descent, including the topmost clouds in the

This painting shows Galileo *releasing the probe to Jupiter.* Courtesy General Dynamics Corporation.

equatorial zone, which probably consist mostly of ammonia. During its expected hour of life during descent, the probe will transmit data to the orbiter, which will in turn transmit it to earth at rates up to 134,000 bits per second. If all systems perform well, it is expected that the probe's six instruments will also determine the structure and location of the clouds, as well as precisely determine the ratio of hydrogen and helium, and measure lightning, energy absorption, and radio emission.

The descent probe is designed to penetrate the Jovian atmosphere to a depth that equals between fifteen and twenty earth atmospheres. Then it will succumb to the unearthly increases in pressure and temperature that will crush and finally vaporize it. The orbiter will receive the probe's last data bits, transmit them to earth, and then continue its twenty-two month mission in orbit around Jupiter, its instruments sensing the ever-changing moods of the giant and some of its sixteen moons.

Galileo's *Orbits*

Laden with eleven sophisticated instruments, the orbiter will make ten elliptical circuits of Jupiter in twenty-two months. The entire surface of the giant planet will be surveyed. Its orbits will also take it closer than ever before to some of the larger moons, such as volcanic Io, where a near-infrared mapping spectrometer will measure and map the heat from this hot moon's volcanoes. The instrument payload, weighing 132 pounds (60 kilograms), will study the composition of the moons' surfaces, the structure of the upper atmosphere of Jupiter and some of its moons, atmosphere and cloud temperatures and composition, magnetic fields, the clouds of ionized gases in the magnetosphere, high-energy particles, plasma waves, and small particles such as micrometeorites.

But for the average earthling, the thrills will come from thousands of images taken by a new camera system that will deliver the highest resolution images ever produced by a spacecraft. Instead of the vidicon tubes flown on previous planetary missions, the orbiter will contain solid-state change-coupled devices, extremely sensitive and accurate light-gathering instruments. The moons of Io, Europa, Ganymede, and Callisto will be photographed with a resolution twenty to three hundred times better than what the marvelous Voyager spacecraft cameras delivered. During the orbiter's closest encounters with the moons, the best resolution will zoom us in for views between 98 and 165 feet (30 and 50 meters), making it possible to see lava flowing on Io or icy crust cracking on Europa. *Galileo* will deliver close-ups of these strange worlds, and we will sit in our living rooms or offices and marvel over what our robot senses have brought forth.

The Jupiter Future

As the trillions of data bits from *Galileo* are transformed into new knowledge about our giant planet, new mission plans will spring forth, compete with one another for limited pieces of the budgetary pie, and eventually fly forth with their state-of-the-art instruments in the first decades of the twenty-first century. Future robot craft to Jupiter will answer questions that the *Galileo* mission hasn't even asked yet; we have to wait at least until 1995.

A future orbiter with a few entirely new senses will no doubt please planetary scientists. It will probably see where *Galileo* was blind. And lander spacecraft will no doubt snoop and poke and dig on the strange terrain of sulfur–snow–covered Io or order electronic eyes into what some scientists believe to be a deep liquid water ocean under the ice crust of Europa. It is even possible that our faraway machines may discover life in the depths of Europa's underground oceans. But even if life is not found, there is still the water—a truly priceless treasure wherever it is found in our solar system.

Return to the Rings

Even after the thousands of beautiful images sent earthward by *Voyager 1* and *2*, Saturn still enchants us and beckons our ingenious machines and our future astronauts. A new Mariner Mark II spacecraft will be put into Saturn orbit near the turn of the century and will do for Saturn what *Galileo* did for Jupiter, but the probe portion of what is now called Project Cassini will go to Saturn's largest satellite, Titan, the mysterious moon cloaked in an atmosphere rich in nitrogen. Such a Titan probe mission has planetary scientists and space enthusiasts drooling with anticipation.

Titan is the only moon in the entire solar system to have a substantial atmosphere. Composed mostly of nitrogen, Titan's atmosphere also has substantial amounts of methane and probably some argon. Reddish smog cloaks the moon's cold surface, and scientists believe the atmospheric haze is produced by chemical reactions and made up of organic compounds. What has scientists so excited is that this moon's atmosphere could resemble that of primitive earth, long before life arose. The Voyager space craft discovered that the hazy atmosphere is at least 250 miles (402 kilometers) thick, and that its pressure on the surface is at least twice that of our planet's atmosphere. Space folk on earth want to know what's going on above and on the surface of Titan. Because of the very low temperatures (-280 degrees F., -173 degrees C.), it is possible that lakes and streams of liquid methane or nitrogen are present on the surface. This is why a Titan Probe/Radar Mapper mission has a high priority at NASA and JPL and will probably arrive at the Red Moon before the year 2005, with a launch date before 2000.

Project Cassini to Saturn will be a joint effort of NASA and the European Space Agency. Preliminary definition work on the Titan probe was completed in 1985, and it called for the probe descent craft to be mounted on the side of a Mariner Mark II orbiter spacecraft. On arrival at Saturn, the Cassini orbiter spacecraft would execute an orbital tour of Saturn and its moons, using Titan for gravity-assist maneuvers. The probe would be released from the Mariner Mark II by a spin-eject device, and then parachuted through Titan's red atmosphere to land on the mysterious surface that promises as many surprises as Jupiter's Io has volcanoes.

During the two-phase descent, lasting three hours, a heat-shield nose cap made of beryllium will bear the brunt of the atmospheric friction before being jettisoned at an altitude of about 155 miles (250 kilometers). The pilot and main chutes will then deploy, after which the instrument inlet ports will open so that the sensors can begin their work at an altitude of about 124 miles (200 kilometers). During this phase, the science atmosphere experimentation will take place for about two hours, and during the last phase of descent, an imaging system could provide views of the surface. Three hours after probe entry, the 90-pound (41-kilogram) Titan probe should impact on the surface at about 7 miles (11 kilometers) per hour. If the instruments survive the landing, we could look upon the first images of Titan's strange surface.

At that moment, if all goes well, there will be jubilation at Jet Propulsion Laboratory, recalling the past exultations of the Viking robot landings on Mars and the Voyager flybys of Jupiter and Saturn. More than likely it would be the first great interplanetary flight of the new century. The Space Age will not be even 50 years old.

Future Journeys to Saturn

Other orbiters, other probes, other flybys on their way to Uranus and Neptune will cast their electronic eyes at Saturn and its myriad moons during the next century. Titan may prove so exciting that we will visit it again bearing state-of-the-art machines with their artificial intelligence. Both Saturn and Jupiter offer several choices for manned bases on their moons, and this may come toward the end of the next century, if there are apparent practical reasons to make such manned journeys to the outer planets.

Saturn certainly deserves the same kind of intensive coverage of its atmosphere, and weather patterns, and survey of its moons that Jupiter will receive with *Galileo*. Another Saturn orbiter mission beyond Cassini will no doubt fly in the first decades of the twenty-first century and will present us with the first high-resolution images of the magnificent rings, close approaches to some of the understudied moons, and additional radar coverage of Titan's surface. And flybys to the more distant planets of Uranus and Neptune can point instruments to be used for the first time at Saturn and its moons to fill in certain knowledge gaps or answer elusive planetary questions. But no matter how much we learn about this second-largest planet in the solar system, whose enigmatic rings generate radio emissions with a power of some 3 million watts, Saturn will never lose its unique beauty to human or robot eyes.

Project Cassini will fly on a new Mariner Mark II spacecraft to Saturn and its atmosphere-cloaked moon, Titan, near the turn of the century. A probe will be sent through Titan's thick nitrogen atmosphere, which may be produced by the chemical reactions of organic compounds. If Cassini's probe survives its descent, it will send back images of the mysterious surface. Rivers or streams of liquid nitrogen could be present on the moon. Courtesy Jet Propulsion Laboratory.

Two Giants and a Midget

There will be other robot voyages to the three outermost planets—Uranus, Neptune, and Pluto—after *Voyager 2* casts its fleeting glances at the first two before 1990. The wealth of knowledge about Jupiter and its moons gained from the *Galileo* mission should be compared with similar information from the other three giant planets. Such comparative studies offer a much fuller scientific return from a mission. Once the next generation of data streams arrives from the *Galileo* and Cassini projects, Uranus and Neptune are next. Little Pluto and its big moon, Charon, will be last in the cosmic cue up.

Why go to the dark corners of our solar system? Unmanned spacecraft continue to create a detailed map of the solar system, and no one wants an unfinished map. Flyby, orbiter, and probe missions to the three outermost planets will complete this great task. In-depth studies of the atmospheres of Uranus, Neptune, and Pluto will discover the percentages of hydrogen, helium, and deuterium present. These data will allow scientists to test cosmological models to better understand what happened in the early seconds of the universe after the big bang. It may also improve their understanding of the origin of life in our solar system. But most important will be the completion of the reconnaissance of our solar system that, in the long run, may have direct bearing on our survival as a species. We are, after all, already shopping around for another planet or moon on which to settle.

To reach the planets beyond Saturn in less than ten years, these flyby and probe missions require Jupiter's gravitational help as the spacecraft swing by the giant. Without such swingby assists, a mission to Pluto could take some forty years given present propulsion systems.

But there will be no such swingby trajectory opportunities until the early in the twenty-first century. That's when the Uranus, Neptune, and Pluto missions will fly. Before 1990, *Voyager 2*'s Uranus and Neptune flybys will help define the goals of these future orbiter and probe missions. By 2020, even distant Pluto, probably a methane snowball, will have yielded its mysteries to the far-flung senses of our made on earth robots. And the robots may be microspacecraft by then, spacecraft weighing only a few pounds (kilograms) that can do everything a 1-ton or more space probe did at the end of the twentieth century. Space science is rapidly gaining the ability to dramatically miniaturize everything in a spacecraft. But size is not relevant as long as the data streams reach earth and are processed. If microspacecraft can do the work and cost much less to launch, there may be another golden age of interplanetary robots traveling to the planets and moons of our solar system. And in the even more distant future, our robot emissaries may gestate and metamorphose during their voyages, their designs incorporating man-made biological systems using advanced DNA techniques.

Robots to the Stars?

When *Pioneer 10* left the solar system behind in June 1983, it was traveling at a speed of about 25,000 miles (40,000 kilometers) an hour. Had it been heading in the direction of Alpha Centauri some 4.3 light-years away (now the closest star to our solar system), it would take some 80,000 years to reach it. But our first interstellar spacecraft was headed toward a point near the boundaries of Taurus and Orion, and it has been estimated that it will come "close" (within a few light-years?) to another star only once every million years or so. For *Pioneer 10* to enter the planetary system of another star, the human race would have to wait ten billion years!

While faster velocities for escape from the solar system are possible even with today's propulsion technology, the numbers don't change that much. The stars are not beyond our reach, but they certainly are beyond our time.

Voyager 1 and 2 will also pass beyond the orbit of Neptune by the year 2000. In about forty thousand years, it has been estimated, *Voyager 1* will pass within 1.7 light-years of an obscure star designated AC + 79 3888. Some scientists predict that, given advances in state-of-the-art, deep-space communications, we could listen to the Voyagers for up to one hundred years. Still, they would be at a distance equal to only 1/1,260th of the distance to Alpha Centauri. Such vast distances and time periods do not make us optimistic about sending robots to the nearby stars in the forseeable future, but breakthroughs in the first decades of the twenty-first century could change the interstellar flight numbers that now appear so discouraging.

Some of the future planetary missions will no doubt follow the Pioneers and the Voyagers out of our solar system and into the shallows of interstellar space. Designers and flight teams of future missions consider long-term missions and trajectories into interstellar space because the astounding lifetimes of our early space ro-

bots have made it essential to think in terms of decades and even centuries. It is possible a machine from earth will reach a close star in the twenty-first century. Spacecraft will certainly be testing the cosmic waters just off the shores of our solar system. If a technological breakthrough occurs, and the history of the Space Age says that is inevitable, we may launch our fastest ships, which conceivably could reach some of the nearest stars in centuries instead of millennia. At least the discoveries of the Hubble Space Telescope in the next decade will tell us where such interstellar robots should be sent. The stars may be closer than we think. We should have an answer before the Space Age celebrates its centennial.

FUTURE ROBOT SPACECRAFT

Mission	Launch Date	Encounter/ Data Return	Mission	Launch Date	Encounter/ Data Return
TO THE SUN			Mainbelt Asteroid Mission	2000	2002
Ulysses	1989	1992-1993	**TO JUPITER AND MOONS**		
Project Starprobe	1998	2000	Galileo	1991	1995
TO VENUS			Jupiter Moon Lander	2003	2005
Magellan (Venus Radar Mapper)	1988–1989	1989–1990	**TO SATURN AND MOONS**		
Soviet/French Vesta (Vesta encounters Venus on its way to the asteroid belt.)	1992	1993	Project Cassini (Titan Probe and Radar Mapper)	1995	1998–2000
TO MARS			Saturn Orbiter	2002	2005
Mars Observer	1990	1991	**TO URANUS AND NEPTUNE**		
Mars Aeronomy	1995	1996	Voyager 2 Uranus Flyby	1977	1986
Mars Sample Return	2000	2001	Voyager 2 Neptune Flyby	1977	1989
Soviet Mars Moon Mission (Phobus)	1988	1989	Uranus or Neptune Orbiter	2005	2014
TO THE ASTEROIDS AND COMETS			**TO PLUTO**		
Asteroid Flyby and Comet Rendezvous	1991	1994–2000	Pluto Flyby	2009	2019
Soviet/French Asteroid & Comet Mission (Vesta)	1992	1996–1998	**AN INTERSTELLAR FLIGHT**		
Comet Sample Return	1998	2001	Distant Planetary System Probe	2020?	2099?

NOTE: Launch and arrival dates are often revised, especially those that are more than two years away.

9

Commerce
& Science In
Orbit

INDUSTRY IN SPACE

Gold Rush Above Earth

By the year 2000, commercial space operations are predicted to generate at least $50 billion each year in gross revenues. The gold rush will soon be going on in the orbitways above us, and the lucky prospectors are likely to be Fortune 500 companies that are now involved in dozens of space-related technologies, including launch services, telecommunications, materials processing, and remote sensing services. Besides some of the giants such as John Deere & Company, McDonnell Douglas, and the 3M Corporation, which have been actively using the space shuttle for their research and development of new products, there are more than 300 companies that are seriously planning space ventures as part of their future growth. This does not include the more than 2,100 companies around the world that are already engaged in supporting space activities. Some experts (including some members of the United States Congress) believe the space activities as a worldwide industry will generate some $500 billion by the turn of the century and provide twenty thousand new jobs.

In 1984 NASA adopted a new commercial space policy that fundamentally changed its relationship with private enterprise. One of its stated goals is to stimulate private space investment by providing federal seed money to industry and by purchasing commercial space products and services on a selective basis. By assisting companies with new high-tech commercial space ventures, NASA will encourage private sector investment in the high frontier. Also, from time to time, an entire Spacelab mission will be made available to private industry for research and development.

Such incentives will send even more corporate prospectors packing their payloads for space. Within a few decades, many of them will be digging for their in-orbit profits in commercial areas totally unforeseen today. The growth of commercial space activities will be astonishingly rapid, and after years of R&D investments, the prospectors will hit their pay dirt in the orbital streams above earth.

Space-Based Telecommunications

If Arthur C. Clarke wrote nothing more than his now famous technical paper "Extraterrestrial Relays," on geostationary communication satellites, which was published in October 1945 in the journal *Wireless World*, he would deserve a place in the history books for having foreseen the revolution that these satellites have created. As the twentieth century comes to a close, the biggest in-space business is communications satellites. They have revolutionized worldwide communications, and over the next few decades they will change even more how the nations and people of planet Earth interact, in subtle ways that are not clearly or fully understood today. Without them there would be no live newscasts or sportscasts from just about any place in the world, nor would there be historical televised events such as the first manned landing on the moon or the 1985 Live Aid rock concerts that raised money for victims of the African famine. Communications satellites (comsats for short) are the biggest business in space today, and this big space business has just begun its first real growth spurt.

Communications satellites generate some $3 billion in revenues each year, and forecasts predict that more than four hundred comsats will be launched by the year 2000. It has also been

PRECEDING PAGES: *As business on the space frontier grows, more companies will want to rent an orbit. Space Industries, Inc., under an agreement with NASA, will provide an orbiting processing facility that will be available in the early 1990s, before the space station is built.* Courtesy NASA.

estimated that by the turn of the century, twenty shuttle and expendable rocket flights each year will be devoted to carrying comsats into orbit. More than sixty different services and products of Space-Age telecommunications have been identified, and the international competition to sell these services and products will be hot. Specially designed data-transfer satellites will link the worldwide offices of multinational corporations. Schools, libraries, and eventually homes may someday have access, at reasonable rates, to the entire contents of the Library of Congress, the British Museum, France's Bibliothèque Nationale, or the other great libraries around the world. Educational programs at all levels in many disciplines can be beamed to institutions or homes. Wristwatch-size radios will be able to send and receive messages from any two points on earth at a reasonable cost. Electronic mail and international phone calls will become dramatically less expensive and more widely used than they are today, and there is tremendous growth potential for direct-broadcast satellites that will beam their programming to home or business dish antennas. One study projects that twenty-four countries and five regional organizations will be owners or operators of communications satellites by 1991. Even small countries such as Ireland and Luxembourg will have their own geostationary satellites serving them from high above the earth. All of this from the world's first commercial satellite, *Early Bird*, launched by the Communications Satellite Corporation (Comsat) in 1965, which could handle a maximum of 240 voice channels, or TV or high-speed data.

Comsat's Silver Anniversary

Since April 1965, when Comsat launched *Early Bird 1* into geostationary orbit for transatlantic coverage, more than ninety communications satellites have been launched at a cost of about $3.5 billion. The first global coverage was established in 1969, in time for the first manned moon landing mission, portions of which were telecast around the world. But while the monop-

oly of Comsat and its international parent consortium, Intelsat, is rapidly crumbling under the pressure of competition from dozens of countries and other organizations, some of Comsat's recent agreements demonstrate that this corporation, which in 1970 paid out the first dividends ever paid to stockholders from a commercial space venture, will not be squeezed out of the space market it created. One deal involves providing 1,500 Holiday Inns with teleconferencing capabilities. Another ten-year contract with the National Broadcasting Company directs Comsat to distribute programming to 182 NBC affiliate stations. In the educational field, Comsat has an agreement with the American Bar Association and the American Law Institute to transmit educational law programs. Comsat is also cautiously moving into direct satellite broadcasting to homes—a venture that involves a merger and $500 million in capital to put the operational system in place. Even though Comsat will face even stiffer competition in the 1990s from new companies, such as Antares Satellite Corporation, which should be operational in the new decade, this first commercial, money-making space corporation will still have much to celebrate on its silver anniversary.

Corporate Space Divisions

Besides the traditional aerospace companies moving into space telecommunications ventures, as well as launch services, there will be many other corporations moving aggressively into this profitable space frontier. In the United States, for example, Federal Express plans to design its own satellite and earth stations that will connect to its central telehub. Its plans go beyond document transmissions, and may include transmission of video images, voice, and data on a global scale—nothing less than a worldwide network of high-speed communications. Federal Express hopes to go beyond telecommunication products and become a commercial leader in space transportation and space services. So do a lot of other companies. But they must move cautiously because there are still some gaps in

A Sampling of What the New Generation of Commercial Satellites Might Do for Us

Search and Rescue. The in-place search and rescue program, SARSAT-COSPAS, is an international cooperative effort involving Canada, France, the Soviet Union, and the United States. Since the program began in September 1982, it has been credited with saving more than five hundred lives through its ability to quickly pinpoint downed aircraft, boats or ships in distress, or land vehicles in trouble.* The life of a Belgian race driver was saved, for example, when his all-terrain vehicle crashed in a remote area of Somalia, in Africa, and a satellite picked up the distress beacon and fixed his position, allowing a doctor to fly to the crash location. The lives of two Canadians were also saved by the satellites when they were stranded in the wilderness after their canoe overturned in the rapids of the Winisk River in a remote area of the Hudson Bay region.

This particular search and rescue system is projected to remain operational through the early 1990s, but private enterprise will no doubt be involved in the new generation of such satellites.

Personal Security. Geostar Corporation, headed by Princeton's Gerard O'Neill, is planning to launch its own satellite communications system offering several services to private individuals at reasonable rates. One of these services would provide search and rescue capabilities, but the proposed Geostar plan would be a two-way system, whereby subscribers could transmit their positions to the satellite. A driver in trouble, for example, could call for help without knowing his location. The Geostar system would also allow the transmission and reception of short text with a limit of thirty-six characters between any two points on earth. It would also have the ability to aid people threatened by attack—on the street or at home. By pushing a button on the transceiver, a potential victim would signal the satellite, which would locate the nearest police car and give the officer the exact location and identity of the person. Within seconds, help would be on its way, and the person would be so informed. The Geostar system could handle several million subscribers, and it could give them a number of services including those already mentioned.

Geostar Corporation had twelve thousand orders for its position-fixing transceivers at the beginning of 1986, and it expects to have three dedicated Geostar satellites in orbit by the early 1990s.

Personal or Business Tracking and Navigation Services. Future satellite systems in geo-

* *The emergency transmitter for land, air, and sea costs between $150 and $300 and can be bought at most marine and aviation stores.*

the market studies of commercial space ventures. Even the demand for launch services up to the turn of the century is not certain, and the large investments involved flash caution signals to corporate decision-makers. But that is not stopping dozens of companies around the world from forming their own space divisions to go after those profits in orbit, because if they wait much longer to initiate their R&D and their options planning, they will be too late—earthbound, with all the orbits covered by their competitors.

A Satellite for All Occasions

Commercial satellite capabilities are expanding rapidly. The coming generation of satellites will be larger, more sophisticated, and more versatile than those flying a decade ago. Private enterprise will be creating systems in all areas of telecommunications. Some will provide services never before available, and others will improve on, and compete directly with, services offered

A Sampling of What the New Generation of Commercial Satellites Might Do for Us *(cont'd)*

synchronous orbit some 23,500 miles (37,811 kilometers) above the earth will be able to track and locate anything on the earth's surface: cars, boats, packages, missing children or adults, pets, works of art—virtually anything to which a miniature radio transmitter can be attached. Ten million packages could be tracked every hour by such large satellites. The same is true of road vehicles. Chrysler Corporation has a system on the drawing board that uses satellite and computer technology to exactly pinpoint a driver's location on a color video map inside the car or truck. A driver would know instantaneously where his vehicle was relative to his destination. Called the Chrysler Laser Atlas and Satellite System (CLASS), it would use satellites similar to the government's Navstar system, which will have eighteen satellites in orbit by the early 1990s. Each satellite transmits a unique signal, and ground-station computers can interpret the different signals and get a fix on any object to within 300 feet (91 meters). The CLASS system would depend on a similar comsat capability, which is integrated with computer and laser video-disc technology.

Personal Communications. There are some six hundred million telephones in the world, but they are unevenly distributed, with 75 percent of them concentrated in only nine coun-

tries. Half of the earth's population lives in countries where there is less than one telephone for every one hundred persons. The city of Tokyo, for example, has more telephones than the entire continent of Africa.

The revolution in commercial communications satellites may change this imbalance in the next few decades. Besides Geostar Corporation's plans to have worldwide, person-to-person transmission and reception of brief messages in just seconds included as one of many services to subscribers, several other companies are planning to enter the satellite telephone competition as well as offer other services. Mobile Satellite Corporation of the United States seeks Federal Communications Commission approval of a long-distance telephone system via satellite that it projects will cost its customers only 15 cents a minute. Another firm, Omninet, hopes to offer paging and emergency location-finding services on a commercial basis.

There will be additional innovative services generated from the state-of-the-art satellites, and these services will not be offered just to governments and large corporations; in the 1990s, these services will be offered to millions of people around the world. Because of international and corporate competition, the prices of these services should decrease dramatically over the first few years they are in place.

by a government space programs or international cooperative agreements. A growing variety of unique satellite systems will be able to directly benefit most of the 4.8 billion people on planet Earth by the year 2000.

Commercial Launch Services

Now that free enterprise has been given the incentives to enter space for profit, many services

will be available for hire in the space marketplace. Just as any earthbound business can go into the marketplace and lease anything needed, the same will be true for businesses that must enter space to produce their products, launch their satellites, or repair their scientific instruments.

As a result of the grounding of the space shuttle and Ariane in 1986, several companies, including the traditional aerospace corporations around the world, are preparing—depending on the demand—to market their own launch services. Their success or failure will depend heavily on the outcome of the price war between Ariane and the space shuttle once they begin flying again.

As is the case with any new space venture, some of the undercapitalized companies may fall by the wayside, especially if they are priced out of the space marketplace by competitive pricing of the launch packages offered by Europe, the United States, and other countries. Nevertheless, some of these space entrepreneurs will cetainly survive, and the result will be that space industry customers will have a larger selection of launch services from which to choose, some of which will be tailored to certain orbits and payload weights.

One such privately owned launch service in the United States, Space Services, Inc., has a promising future because of its unique payload and because it has the approval of the Department of Transportation. Space Services has developed its own rocket, the Conestoga, with which it plans to launch cremated human remains in special burial satellites that will orbit the earth. Working with the Celestis Group, a Florida-based consortium of morticians and former Kennedy Space Center engineers, Space Services will offer a space burial for about $3,000, which is competitive with the costs of traditional burials. In the future, the company hopes to offer burials in deep space, beyond earth orbit, with escape trajectories that would take the burial satellite beyond the moon and into a deep-space orbit. A new cremation technique of ash reduction will allow an individual's remains to fit inside a small, thin capsule no more than 2 inches (5 centimeters) in length. Several thousand of these capsules would be placed inside the burial spacecraft on each mission.

Space Services projects that their Conestoga rocket would launch about ten space burial missions a year by the mid-1990s. The burial satellite will be designed to have a highly polished and reflective surface so that the relatives and friends of the deceased will be able to see the bright satellite pass overhead in the night sky when the weather is clear.

Rent-an-Orbit

As industry in orbit develops beyond its embryonic stage of experimentation on space shuttle flights and private companies seek profits in the frontier above earth, the variety of support services for their payloads will increase. American, European, and Japanese aerospace companies will be building and launching unmanned space platforms that will be rented to companies for research or materials processing. If a firm wants to fly a research payload for six months, they will rent the space on one of these modular free-flying platforms. Each of these for-hire space platforms will probably be designed to accommodate a certain class of payload weights.

Space Industries, Inc., of Houston, Texas, plans to build and launch a processing facility in space before the space station is in orbit. Commercial paying customers may be using the first of the automatic facility's two modules by 1990. The space shuttle will service the unmanned laboratory and collect the processed materials during scheduled missions.

This project was made possible by an agreement with NASA which allows the corporation to defer its launch and other costs until it is making money in orbit. Then it would pay NASA 12 percent of its yearly revenues to pay off the launch costs of about $75 million dollars. Space Industries projects that the fully operational facility, with its two modules and four solar array panels, will be flying by the mid-1990s.

Unmanned satellite servicing vehicles, remotely controlled from the space shuttle, a ground station, or the space station in the 1990s, will also be leased to perform all kinds of in-space work: satellite deployment or retrieval, inspection (including testing and check out), maintenance and resupply (including cleaning or resurfacing of the spacecraft), major repair and parts replacement, and so forth. If the free-flying platform to be serviced is a materials processing facility, the servicing vehicle could deliver new materials for processing and pick up the products or materials already processed. As with any business, once the demand for these services increases, the customer costs will go

SELECTED RESCUES BY SATELLITE USING THE SARSAT-COSPAS SYSTEM

Date	Location	Type of Ship/ Plane (Name & Ident.)	Number of Persons Rescued	Details
11/7/82	Bahamas	36-foot sailboat, "Blue Jeans" on way from Tortola to San Salvador Island	5	Probably would not have been found without COSPAS. Helicopter rescue by CG chopper from Grand Turk.
3/7/83	Sept Isle, Quebec	Cessna 183	2	France, Canada and U.S. picked up signal from COSPAS. First time all three in one emergency.
4/3/84	Northwest Alaska 68° 07'N 165° 40'N	Dog Sled Driver	1	Woman dog sled driver suffering from flu and cold. SARSAT/COSPAS provided notification and location.
6/23/84	Aleutian Islands, SW Alaska 51° 38'N 176° 48'N	Kayaks	4	Four men in 4 seagoing kayaks circumnavigating Adak Island encountered strong winds and high seas; 1 man injured. COSPAS/SARSAT provided only alert and position.
12/31/84	Somalia, Africa 7° 01'N 47° 29'E	Automobile	1	SARSAT/COSPAS provided first alert and position of car racer seriously injured in crash.
3/23/85	Western British Columbia, Canada 53° 29'N 129° 26'W	—	1	SARSAT/COSPAS provided alert and location of trapper out of food, wet, and with incipient hypothermia.
7/2/85	North Atlantic Ocean 45° 39'N 5° 40'N	Sailboat "Arcara"	2	SARSAT/COSPAS provided location of survivors in life raft. Sailboat had sunk after collision with unknown floating object.
8/5/85	Greenland 64° 50'N 49° 55'W	Persons	3	SARSAT/COSPAS provided only alert and location of persons lost on icecap and without food for five days.
10/3/85	Atlantic Ocean 44° 13'N 11° 49'W	Sailboat "Shuffle"	2	SARSAT/COSPAS provided position of sailboat in danger of breaking up in severe gales and mountainous seas.

Selected from the *Cospas/Sarsat Record: List of Events.* Courtesy Goddard Space Flight Center.

down. Such services will become available for the first time in the 1990s. It will therefore be the new century before any budget rent-an-orbit deals are offered.

"Made-in-Space"

The first "made-in-space" product went into the marketplace in 1985, but buyers must have complete faith in NASA's quality control because they literally cannot see what they are buying—it is invisible to the human eye.

This first space product, distinguished from the thousands of space technology spin-off products that we find in our homes or at the office, will not end up on the shelves of your neighborhood store. If you are employed by a high-tech corporation, a medical research center, or any other state-of-the-art research center, this invisible product is likely to be found in the laboratory.

What is the mystery product? Sold in lots of thirty million, the product consists of tiny microscopic spheres, made of polystyrene, that are only 1/2.500th of an inch (10 micrometers) in diameter. The latex spheres will be used to help calibrate and focus electron microscopes and to improve microscopic measurements in electronics, medicine, environmental pollution research, and other high-technology areas.

The eighteen billion spheres, made in space during the flight of the sixth space shuttle in April 1983, were packaged by the National Bureau of Standards into six hundred vials, each with a capacity of 5 milliliters, which contained the thirty million spheres. Each vial is filled mostly with water; the latex spheres are in a 0.4 percent concentration by weight.

The spheres were manufactured in special reactors aboard the space shuttle, and the lack of gravity (more precisely, microgravity) allowed these spheres, about the size of a human red blood cell, to grow uniformly in size and shape. On earth it was impossible to create spheres that measured up to such exacting specifications. This first made-in-space product proves that some manufacturing processes that cannot occur on earth can take place in space.

The price for a vial of thirty million spheres is $394. The first customers include seven companies, one university, and one government agency—the Food and Drug Administration. The sale proceeds will be shared equally by NASA and the Bureau of Standards. The market potential for these invisible-to-the-eye latex spheres has been estimated at between $200 and $300 million a year. If actual sales fall into this range, the first product made in space will be *the* success story for the first history written of the industrial age in space.

The Promise of Space Products

The unique environment of space, its vacuum and microgravity, will help to produce rare alloys, perfect ball bearings, new forms of crystals and crystalline semiconductor materials, and a new class of space-made pharmaceuticals that alone may become a $23 billion annual industry by the year 2000. The first five years of space shuttle flights, before the fleet was grounded, offered opportunities for private enterprise to do research and development in these state-of-the-art manufacturing processes, many of which will become full-fledged manufacturing operations during the 1990s, after the space station is orbiting the earth and the future space-based industrial park begins taking shape. There are very good reasons to expect some of the corporate giants such as McDonnell Douglas and 3M to have their own processing laboratories in space, either attached to their space stations or in separate modules nearby, by the end of the century. The main reason, of course, is profit. Within the next few decades, the new industrial environment of space will do no less than revolutionize some of the earthbound industrial processes as it creates products that affect our everyday lives and health. As branches of the knowledge tree experience a growth spurt in orbit, more and more products and processes will be discovered that cannot even be predicted today. Here are some of the areas that experts agree will be important in the infant years of the space industry.

Space-Made Alloys

Totally new, high-purity alloys manufactured in zero gravity will become some of the most enduring products of the Space Age. Research on Spacelab has shown that complex metal alloys can be made flawlessly in space, where large differences in the density of the material do not restrict uniform mixing as they do under the gravity of earth. The new mixes of materials that are possible in space will produce exotic alloys. An international research effort, which includes several European countries, Japan, and the United States, is underway and will also result in improved foundry production methods for companies that produce large quantities of metals here on earth.

Swedish research has developed a new aluminum under zero-gravity conditions that has a lower density and is lighter than regular aluminum. This "sponge" aluminum is produced by melting an aluminum alloy in a pressurized hydrogen atmosphere and then cooling it in zero gravity. This traps the hydrogen in the aluminum and makes it porous, giving it the lighter weight without any loss of structural strength.

John Deere & Company, a world leader in the production of farm tractors and other equipment, has committed a significant share of its research budget to experiments on processing iron in space. By observing the cooling and solidification of molten iron in a specially designed zero-gravity furnace aboard the space shuttle, Deere hopes to obtain new information that can be applied to its earthbound foundry operations. If its research results in reduced production costs and an improved metal, the corporation will have a competitive edge in cast-iron production. Cast iron is a very complex alloy, and each year some three million tons are produced. By processing the iron in zero gravity, the manufacturer will be able to control the shape and distribution of the graphite nodules throughout the iron. If the graphite nodules are spherical and evenly distributed, the strength of the iron will be higher. Research is also planned on the processing of iron-carbon composite materials, which promise to have new and useful properties such as higher heat conductivity. Managers at John Deere are pleased with their ongoing relationship with NASA. In fact, in the reception area of their technical center, space shuttle pictures are displayed instead of farm tractors and combines.

Alloy research on space shuttle flights may lead to a full research module on the space station in the next decade. If harder alloys can be manufactured in space, and most experts agree that they can be, then a dramatic improvement in the life of perishable cutting tools used in machining operations throughout industry will be the result. Such an improvement in hardness and cutting speed of these industry tools could have an annual market value of $500 million by the turn of the century. And someday, space-made alloys may be fashioned into rare and unique jewelry that will sparkle like the stars.

Zero-Gravity Crystals

Ongoing crystal growth experiments on Spacelab space shuttle missions prove that there is great potential in the zero-gravity manufacture of crystals. Crystals grown in space have fewer imperfections than those produced under gravity, and this improves their electronic characteristics. In the conditions of space, they can also be grown larger. Microgravity Research Associates (MRA) of the United States, a pioneer in microgravity crystal growth, believes that this research and development will lead to a new generation of higher-speed microelectronic components for computers, radar, and communications systems. These advanced space-produced crystal products could fundamentally change the entire electronics and computer industry.

On the flight of space shuttle *Challenger* in May 1985, a mercury iodide crystal, about the size of a sugar cube, was grown from vapor, and it had fewer imperfections than those made on earth under gravity. Such higher-quality crystals of this type could be used to improve X-ray and gamma-ray detectors. When used as components of some medical imaging devices, they will reduce the radiation exposure of patients. Even better space telescopes can be built with these higher-quality crystals.

Electronics corporations are also showing intense interest in the manufacture of gallium arsenide crystals in space, and a special furnace is being designed for this purpose. These crystals are preferred over silicon for solid-state electronic devices because they can achieve higher electron speeds in the microwave frequencies used in radar and high-priority communications systems used by the military. They are also more resistant to heat and radiation. The computer industry will be happy to have them for their ever-changing state-of-the-art computers.

The 3M Corporation has made a major commitment to NASA for space research and may use as many as seventy-two space shuttle flights through 1990s for new-product research. Two experiments studied productions of thin crystalline films during shuttle flights. One studied

The world's first astronaut from private industry, Charles Walker of McDonnell Douglas, is working with the continuous-flow electrophoresis unit. Biological substances, including many breakthrough drugs and medicines, are purified in space with this process. Some seven hundred times more material can be separated in space than on earth during the same period, and purity levels are better. Courtesy McDonnell Douglas Corporation.

films formed from a vapor created by organic solids. The other experimented with growing microcrystals on very thin organic film that were exposed to a vacuum and to zero gravity. The 3M Corporation also plans to develop a chemical laboratory for the space station in the 1990s. 3M's vice president for research and development, Lester C. Krogh, believes that research in space is essential for his company to keep the competitive edge into the twenty-first century.

"There already has been a transformation in the way our researchers think about science," Krogh says. "As we learn more about the properties and behavior of organic materials in space, we will apply that knowledge to what we do on earth."

Made-in-Space Pharmaceuticals

Complex and delicate processes for the separation of biological substances such as cells, enzymes, hormones, or proteins from the human body have been found to work significantly better in the microgravity of space than under the gravity of earth. This fact alone means that large pharmaceutical corporations such as Johnson & Johnson, SmithKline Beckman, Schering, and Upjohn will eventually build mini-factories in orbit to manufacture new and improved Space-Age drugs that either could not be produced on earth at all or could be made only in quantities insufficient to make their production economically feasible. These space-made drugs and biological substances have the potential of saving tens of thousands of lives and offering new and much improved treatments to millions of people suffering from diseases such as diabetes and hemophilia.

A cooperative relationship between McDonnell Douglas and 3M's Riker Laboratories Division has proved on various space shuttle flights that the separation process known as continuous-flow electrophoresis can separate over seven hundred times more material in space than on earth during the same time, and it can achieve purity levels that are four times higher

than when the very same equipment operates under the influence of gravity on Earth. The biological substances move through an electrical field of the electrophoresis unit and are separated out by their own unique surface electrical charges.

The first in-orbit test of the McDonnell Douglas electrophoresis unit took place during the flight of space shuttle *Columbia* in late June and early July of 1982. This was also the final test flight for *Columbia*. The separation chamber, about 6 feet (1.8 meters) high, was located on the orbiter's mid-deck. In the test, a mixture of albumins and a mixture of proteins were separated. The unit separated these substances at varying concentrations, and even the most dense mixture was separated and had a high degree of purity. The amount of material processed was even better than predicted. Compared to the on-earth unit, the spaceborne electrophoresis unit processed 463 times more material, and the quality remained high. The next test, conducted on shuttle *Challenger* in April 1983, separated 700 times more material than its earthbound counterpart and also attained much higher purity levels.

Then on the late August, early September 1983 flight of *Challenger*, the equipment separated live pancreatic, kidney, and pituitary cells for the first time. From the pancreatic cells came beta cells that produce insulin. This area of research promises new treatments, and perhaps even a cure, for diabetes. From the kidney cells came various hormones needed to treat disorders such as hemophilia. The semiautomatic equipment also isolated from the pituitary cells certain cells that were capable of producing substantial amounts of growth hormone. With each shuttle test, the future for an in-orbit drug industry looked bright.

McDonnell Douglas and Johnson & Johnson, its partner at that time, were enthusiastic about the results. There was no longer any doubt that such in-space processing of pharmaceuticals would be an extremely beneficial enterprise in the long run—both to medicine and the bottom lines of those future-looking commercial companies that led the way into orbit.

The Hush-Hush Hormone

Charles Walker (his flight suit name tag reads "Charlie") became America's first industry employee to fly on the space shuttle. He was McDonnell Douglas's mission specialist to oversee the continuous-flow electrophoresis processing of a medical hormone—a hush-hush hormone that, following Federal Drug Administration approval, will be in the marketplace before 1990.

What was the mysterious hormone that McDonnell Douglas and its last partner, Johnson & Johnson, referred to as a "hormone of interest"? The companies were very secretive about it because of the keen competition between the pharmaceutical giants. Certain space-produced hormones and drugs valuable to the treatment of cancer or diabetes will have multibillion-dollar sales potential in the marketplace. Perhaps it was the hormone that regulates the blood-clotting mechanism of the human body—that would certainly have tremendous value for all surgical operations. But no one was talking; even NASA was tongue-tied. But then, late in 1985, *Aviation Week & Space Technology* reported that the hush-hush drug was erythropoietin, which can stimulate the growth of red blood cells in the body. The drug could treat millions of people who have lost the ability to produce red blood cells, and its use before surgery could replace transfusions and their complications. After this space-produced drug is tested on people and receives FDA approval, it will go for sale in the marketplace, perhaps as early as 1988.

Charlie Walker's first flight aboard the space shuttle *Discovery* was in late August 1984. He flew on the *Discovery* again as mission specialist for McDonnell Douglas in April 1985. On his first flight, 83 percent of the material was separated, but when the automated system developed problems, he had to operate the system manually, proving that his ticket was worth the price. During his second shuttle mission, he processed 100 percent of the sample by the fifth day in orbit, and during his third mission, in 1985, he processed the batch of hush-hush hormone that was actually used for clinical testing. McDonnell Douglas has also approached other pharmaceutical companies interested in producing other medical products that can benefit large numbers of patients, and they will likely have their own laboratory aboard the space station or private enterprise facilities in the 1990s.

In the category of products under active study are space-processed monoclonal antibodies for blood and cancer diagnostic work and a material that could locate and tag cancer tumors soon after they form and a potential treatment for emphysema. The company predicts that as many as fifteen breakthrough pharmaceuticals will be on sale by the year 2000.

Cancer Drugs from Above

Industry's first astronaut, Charlie Walker, accomplished some other important research on his April 1985 *Discovery* flight: He activated thirty-six crystal-growth experiments that may eventually result in breakthrough cancer drugs.

Today's bioengineers are more and more often designing new drugs by tailoring their molecular structures to work with or against the atomic structures of the body's protein molecules. To accomplish such complex work, the scientists must first thoroughly understand the atomic structures of the protein molecules that the new drugs will assist or attack. But the bioengineers often find it extremely difficult, and sometimes impossible to grow protein crystals on earth large enough to allow them to understand their atomic structures. This is where crystal-growing in space becomes all important. During a Spacelab 1 flight, for example, one type of protein crystal grew one thousand times larger than the same type did on earth. Protein crystal–growing in orbit therefore has the intense interest of commercial firms and universities. The molecular models derived from such tests will be the foundations for the new miracle drugs of the twenty-first century. There are some 250,000 proteins critical to life and life's diseases that can be studied in space. Within some of these atomic structures, scientists may find the recipes for powerful new drugs, including those that can arrest the growth of cancer and perhaps render it impotent.

A Space Spin-off Sampler

When *Apollo 17*, the last lunar landing mission, returned to earth in December 1972, it had been just over a decade since John Glenn became the first American to orbit our blue planet. In those ten years some thirty thousand spin-offs of space technology—both products and processes—that found their way to commercial use had been identified. Another two thousand or more of these spin-offs are going into the commercial marketplace every year.

The technology spin-offs from the first thirty years of the Space Age have done nothing less than profoundly change our planet and the way we spend our daily lives on it. No citizen of an industrialized nation has remained untouched by the influence of space technology. And even today the continuing human endeavor in space dramatically changes how we lead our lives on the surface of the earth. Developments are occurring so rapidly in space industry that the state of space commerce fifty years from now remains difficult to predict.

The microelectronics revolution alone has given tens of millions of consumers around the world products that they use every day: pocket calculators, personal computers, advanced TVs, radios, and other small and major home appliances that offer more functions, and often better reliability, than the earlier designs. Beyond the home, we find Space-Age technology giving us quieter aircraft and automotive engines, advanced hydrofoils, better built bridges, new construction materials, and innovative management techniques. Even the restoration of the Statue of Liberty in New York Harbor was helped significantly by a special corrosion-protection primer paint known as K-Zinc 531 that was originally developed to protect NASA's primary launch facilities at Kennedy Space Center.

The Future Fortune Companies of Space

Because of their research and development programs on space shuttle flights, several large cor-
porations have already made long-term commitments to manufacturing and processing space products and providing space services in orbiting industrial parks. Any such Fortune list of companies involved in space industrialization would have to include General Electric, John Deere, McDonnell Douglas, and 3M. And the giant corporations of the telecommunications industry have already followed their hardware into space with new services. No doubt the majority of top-ten aerospace firms will be committed to the new orbiting technology: United Technologies, Boeing, McDonnell Douglas, Rockwell International, General Dynamics, Lockheed, Signal Companies, Northrop, and Grumman. The same international companies that were involved in certain aspects of the space shuttle program will also be heavily involved in the design and creation of the space station, which has become an international effort. As the space station becomes operational in the 1990s, there will be several dozen companies that will have forged the hardware and skills necessary to make a profit in the high frontier. Their names will appear on the Fortune list of space companies. So will the name of a leading corporation that builds industrial robots and a major construction engineering firm that will oversee the building of large space structures. In the United States, the American Stock Exchange and the senior executives of its member companies will be paying close attention to the in-space investment opportunities. So too will the chief executive officers of investment houses around the world.

SCIENCE IN SPACE

New Perspectives

As soon as rockets could carry scientific payloads into orbit, several scientific disciplines suddenly had global views, and their new per

Space Age Spinoffs

Wherever you are right now—home, office, or on the road—you will very probably see something that owes its existence to the technological impact of the Space Age. What follows are a few lists, broken into general categories, that give just a meager sampling of beneficial spinoffs from humankind's thrust into space.

THE ARTS

- COMPUTER AND video artwork

- STATE-OF-the-art film special effects

- RARE BOOK and document preservation (vapor-phased deacidification) and document enhancement techniques

- ART WORK separation (image enhancement techniques to separate X-ray images of painting where two or more exist on the same surface)

BUSINESS AND INDUSTRY

- ARTIFICIAL INTELLIGENCE (synthesized speech)

- STATE-OF-the-art computer systems

- HIGH-SPEED telemetry and data transfer (bank-teller terminals through multiplexing)

- COMPUTER-ENHANCED graphics for advertising and broadcasting

- ROBOTICS

- NASA SOFTWARE programs, over 1,500 available, for commercial use (Shell Oil Company adapted one program for use in production of chemicals for plastic products; another NASA software program designed construction cranes.)

- HEAT-RESISTANT paint

- NEW LUBRICANTS, including dry lubricants that are bonded to a variety of metals

- SENSORS OF all kinds, including those that detect radiation

- LASER CUTTERS and strippers for electrical wire

- FLAT POWER cable (first used for equipment on the surface of the moon)

- PORTABLE HIGH-intensity light systems, fifty times brighter than high-beam headlights on automobiles

- ULTRA-EFFICIENT industrial insulation (storage wells on tuna boats)

- DEEP-SEA fishing nets that increase productivity by 30 percent (developed as safety nets for building of shuttle)

CLOTHING AND APPAREL

- SPECIAL BRA for female athletes to prevent shoulder and back strains

- COOLING SPORTSWEAR for all athletes, which contains heat-absorbing gel packets

- COOLING VESTS for quadriplegics and for workers in hot climates

- SCRATCH-RESISTANT sunglasses (from NASA research on space helmet visors)

CONSTRUCTION

- FABRIC STRUCTURES made of spun-glass fiber and coated with Teflon® (examples: The Silverdome, home of the NFL's Detroit Lions in Pontiac, Michigan; and Sea World's Festival in Orlando, Florida)

- AUTOMATIC WELDING systems

- IMPROVED ANTI-corrosion coatings and heat-resistant paints

Space Age Spinoffs *(cont'd)*

- FIREPROOF BUILDING materials
- INNOVATIVE METHODS of preparing building specifications

ENERGY

- IMPROVED SOLAR cell collector systems for home and business
- SUPERINSULATION MATERIAL for energy conservation
- CONCEPT OF power satellites in geosynchronous orbit (solar energy collected by large solar cell arrays is converted to electrical energy, which is then transmitted as microwave energy to earth, where it is again converted into electrical power for the consumer)

HOME CONSUMER PRODUCTS

- ALL HOME appliances, small and large, from TVs to toasters, have been influenced by the microelectronics revolution
- HOME ANTENNAE to pick up direct satellite broadcasting
- NEW WALLPAPER materials made from metalized plastic fabric
- FOOD PACKAGING using metalized plastic film
- FIRE-RESISTANT materials for furnishings and clothing
- ULTRA-RELIABLE five-year flashlights; battery breakthroughs for home and garden appliances and tools; high-intensity, hand-held lights
- SENSORS FOR smoke alarms and security alarms; ultraviolet devices for home security systems
- NEW LUBRICANT technology to protect records

- IMPROVED WATER filters; a device to increase light bulb life

MEDICAL

- LOW-DOSE medical X-ray systems; portable X-ray devices for athletic events and other field situations
- ULTRASOUND EQUIPMENT and other medical imaging devices that depend on state-of-the-art computer systems
- NICKEL-CADMIUM batteries for heart pacemakers; programmable pacemakers; implantable heart-assist devices
- AUTOMATED BLOOD-pressure measurement devices
- MEDICAL TELEMETRY between emergency vehicles and hospital
- PORTABLE EMERGENCY medical systems
- TALKING WHEELCHAIRS (highly intelligible synthesized speech)
- EXCIMER LASER research for medical use in unblocking clogged coronary arteries of heart
- FLUID-CONTROL technology (artificial sphincter)
- IMPLANTABLE, AUTOMATIC drug delivery systems

TRANSPORTATION

- ADVANCED HYDROFOIL design
- BATTERY-POWERED cars and other vehicles
- MICROELECTRONIC SENSORS for cars
- ADVANCED SHOCK absorbers for highway crash barriers (This came out of research on astronauts' couches.)
- COMPUTER DESIGN of automobiles and other vehicles
- ADAPTATION OF lunar Rover drive control for handicapped drivers

spective from above the earth acted as a catalyst for new discoveries and new directions in research. In the past three decades, hundreds of sensor-laden satellites have sent their data streams filled with millions of bits of information to the receiving stations on the earth's surface for processing and interpretation. Whatever the science—archaeology, astronomy, geodynamics, meteorology, oceanology—the scientific instrumentation observed its subject from the advantage of a considerable height. New patterns emerged from the study of either the cosmos or the earth's surface. Astronomers could avoid the distorting and concealing atmosphere of the earth by getting above it; meteorologists could actually see large regions of the planet's weather for the first time; land use planners could accurately inventory crops around the world. As satellite instrumentation became more sophisticated, the entire earth's resources could be surveyed in about eighteen days. Crop harvests could be predicted, water resources better managed. It was another part of the continuing revolution we call the Space Age.

The Landsat Revolution

It was significant that the first of five Landsat satellites was launched into orbit in 1972 to test its capability to gather data on earth resources. Why? Because it was the same year that the last Apollo manned mission went to the moon and returned safely to earth. Congress canceled the other Apollo missions. We had done it; we had put twelve Americans on the moon. Now it was time to turn our full attention back to earth and use our vast resources and skills to improve humankind's lot on the home planet.

Looking back on the years from 1972 to 1984, when the Landsat series of satellites was launched, no one has been disappointed. The Landsat earth resources satellites have left their valuable legacy to humankind. They have offered something to nearly everybody because their sensors scrutinized the entire surface of the planet every few weeks, and the data was there for scientists in many disciplines. Landsat's

wealth of data has been made available to users in more than one hundred countries. The millions of images and the trillions of data bits the Landsat satellites have transmitted to earth have been invaluable tools for scientists keeping a current inventory of the earth's resources and monitoring the quality of our environment.

As the imaging systems became more sophisticated in *Landsat 4* and *5*, so did the data and image return on agriculture, rangeland management, oceanography, water management, flood damage assessment, beach erosion, urban planning, and geology. State-of-the-art thematic mappers, which supplemented the multispectral scanner of the earlier satellites, had 2.5 times the resolution of the earlier scanner and could see in seven spectral bands in the visible and infrared regions. *Landsat 5*, launched in March 1984, circles the earth every ninety-nine minutes. Its resolution is about 98 feet (30 meters) square. This last Landsat earth resources satellite has set the standards for the new generation of remote-sensing satellites.

A Satellite for All Seasons

The versatility of the Landsat satellites was truly amazing. Their sensors could prospect for mineral and oil deposits by imaging fault lines or help determine the best sites for dams or nuclear energy plants using the same information for all these tasks. They could discover new timber resources ready for harvest. By studying the imagery, experts would know what kind of spring runoff there would be because of the snow cover and what the water levels in the reservoirs would be. They would also be able to determine what amounts of forage there were on the rangelands for the grazing herds. Even the Alaskan Indians, Eskimos, and Aleuts used Landsat images to help them select thousands of acres of timberland and mineral exploration areas from the vast wilderness tracts offered to them to settle claims that went back to the purchase of Alaska from Russia in 1867.

Landsat 5 has also helped archaeologists discover Mayan cities, villages, and farmlands hid-

The Landsat series of satellites began in 1972, a significant date because that was the year we ended our Apollo moon program and brought our space program closer to earth. Imagery technology has advanced dramatically since the program began. These two photos, both of Detroit, show how image quality has improved significantly from Landsat 3 (top) *to the thematic mapper image of* Landsat 4 (bottom). Courtesy Eros Data Center.

SPACE SCIENCE CALENDAR

Date	Event
1988-1989	Hubble Space Telescope launched.
1989	Upper Atmospheric Research Satellite launched.
1989	First of new generation of weather satellites (GOES, Geostationary Operational Environmental Satellite) launched.
1990	Extreme Ultraviolet Explorer astronomical satellite launched.
1990	Solar Optical Telescope launched.
1991	Space Infrared Telescope Facility launched.
1991	Advanced X-Ray Astrophysics Facility launched.
1992	Europe's Infrared Space Observatory launched.
2007	Golden anniversary of the Space Age.
2008	Second-generation space telescope deployed.

den for centuries in the jungles in Mexico's Yucatan peninsula. More than 110 possible ancient sites were found by studying computer-enhanced, false-color images from *Landsat*. An archaeological team then confirmed several of the sites on a helicopter expedition that included discovery of a previously unknown Mayan site, complete with its own pyramid.

But it was crop assessment that demonstrated the greatest potential for the Landsat satellites. A crop inventory experiment over large areas of the earth proved that global forecasts of global wheat crop production could be done faster and more accurately than with other methods. The study involved the wheat fields in eight countries around the world, including Russia and China. This Landsat experiment led to a more advanced program begun in 1980 called the Agricultural and Resource Surveys through Aerospace Remote Sensing (AGRISTARS). The goal of this program is to thoroughly evaluate the development and use of remote sensing satellites for global resources information.

In August 1984, NASA turned over Landsat operations to the National Oceanic and Atmospheric Administration, the same organization that operates the U.S. weather satellite system. During the same year, the U.S. House of Representatives approved a bill to govern commercialization of land remote-sensing satellites. In the decades to come, remote-sensing satellites will be able to make increasingly fine distinctions and better predictions about the earth below. And like the people who are spread across the planet's land surface and have their own work specialties, these resource eyes in the sky will be specialists in detection. One may concentrate on the glaciers of Antarctica, another may collect data on the tropical waters near the equator, and still another may focus on the crops of the world, including the vineyards of California and France, or the wheat harvest of Russia.

The Future of Earth Resources Satellites

Scientists the world over hope to have a complete and continuous monitoring of the earth's dynamic processes by the end of the twentieth century. Some satellite programs already in place will continue with launches of replacement and upgraded satellites.

The Crustal Dynamics Project. This international program will continue to measure tectonic plate motion and earth rotation in the future. This will require a second Lageos satellite to be built by Italy and launched by the space shuttle. (See also Chapter 6, "Space International.")

Atmospheric Research. One future project will be the building of the Upper Atmospheric Research Satellite (UARS), which will probably be launched from the space shuttle in the early 1990s. The satellite will measure the extent of ozone depletion and possible effects this could have on the earth's climate.

Another earth-atmosphere research satellite is the Tethered Satellite System (TSS) to be built by Italy and trailed off the space shuttle by the United States. (See also Chapter 6, "Space International.")

Solar Radiation. The ongoing Earth Radiation Budget Experiment (ERBE), which involves three satellites, is precisely measuring the amount of solar energy absorbed in different regions of the earth and the amount of heat energy emitted back to space. How the earth interacts with the sun's energy is a critical factor in climate predictions. Besides the United States, France, West Germany, and the United Kingdom are also participating.

Oceanography from Space. The U.S. Navy's *Geosat A* ocean survey satellite, launched in 1985, is capable of detecting topographical ocean features such as eddies in the Gulf Stream, wave height and winds at the ocean surface, and the exact shape of the earth (called the geoid) to improve the precision of navigation. Future oceanography projects from space include the Navy Remote Ocean Sensing System (NROSS), which will carry the NASA Scatterometer, and a joint satellite project between the United States and France called Topex (for Ocean Topography experiment). Data from earlier satellites such as *Seasat* and *Geos-3* have demonstrated that altimetry satellite data provide a global view of the ocean's surface topography that cannot be equaled from land or ocean-based studies. The Topex satellite would be capable of revealing details of ocean circulation such as currents and eddies, as well as seafloor structures.

Better Eyes in Orbit. The sensor technology that evolved during the Landsat series of satellites will continue to be pushed to more precise, state-of-the-art applications of remote sensing of the earth's surface, especially in regions of the electromagnetic spectrum that are not now covered. Biogeochemical cycles such as the influence of vegetation upon the planet's carbon dioxide budget are being studied, and it is hoped that by the turn of the century, a more complete understanding will be gained of how land processes, strongly mediated by vegetation, interact with the global processes of weather and other cycles. The principal sensors that Landsat used were: the Thermic Mapper; the Multispectral Scanner; and the Advanced Very High Resolution Radiometer. The space shuttle has also contributed to the new knowledge of our planet's surface with the sophisticated space shuttle imaging radar called SIR-A (Synthetic-aperture Imaging Radar A), which has discovered, among other things, a vast array of dry riverbeds and tributaries under the sands of the Sahara Desert.

Agsat. One satellite system, named Agsat, designed by graduate students at Stanford University in the United States, would be capable of detecting changes in the chlorophyll levels of crops. With four satellites in place, Agsat could help farmers determine the health of their crops and the potential yield of their fields. The system could even schedule irrigation and fertilization and reveal clogged irrigation sprinklers. Farmers equipped with microcomputers could analyze their own data, or it could be distributed over cable television or radio stations.

Planet Earth will be under the constant scrutiny of dozens of satellite eyes from all the spacefaring nations. If there ever are found in the data streaming to earth some serious warning signs of ecological shifts requiring concerted international efforts to help the planet with its complex balancing act, let us hope the good of all mankind will be considered.

The Weather Satellites

In the spring of 1985, weather satellites celebrated their silver anniversary. The first experimental weather satellite, *TIROS 1* (an acronym for Television and Infrared Observation Satellite), was launched on April 1, 1960, and transmitted its first weather picture of clouds over the Gulf of St. Lawrence.

Shaped like a hatbox, *TIROS 1* weighed only 270 pounds (122.5 kilograms), about one-eighth of what the sophisticated NOAA (National Oceanic and Atmospheric Administration) weather satellites of today weigh. This weather satellite, the first of over forty launched by the spacefaring nations around the world since 1960, pro-

Many different types of imaging from the space shuttle, including traditional photographic images shown here, are advancing our knowledge of earth resources, archaeology, and geology. These two photos show, ABOVE, *a fossilized drainage pattern in South Yemen, which reveals evidence of ancient water in the desert; and,* FACING PAGE, *one of the most active volcanic regions on earth, the Kamchatka peninsula in the eastern USSR, where at least twenty volcanoes have erupted in recent times.* Courtesy Lunar and *Planetary Institute.*

This spectacular scene of an aurora in the southern hemisphere was taken during the Spacelab 3 mission that flew in late April and early May 1985. The crew commander, Robert Overmyer, took this photograph with a 35-millimeter camera. Courtesy NASA.

duced 22,500 photographs of the earth's weather before it failed.

The orbits above us were kept busy in the early 1960s with more weather satellites and the Mercury manned spaceflights of Shepard, Grissom, Glenn, Carpenter, Schirra, and Cooper. By 1965, nine more TIROS satellites were launched, flying with improved sensors and in polar orbits to increase picture coverage. Before these first weather satellites flew, less than 20 percent of the earth's surface could be monitored for weather conditions. Today we are looking at 100 percent of planet Earth in ever more complex ways, which soon will include

knowing the precise conditions of all the oceans on earth—temperatures, winds, storms, and so forth—at any given moment and having a complete update every six hours.

The thirty-plus weather satellites the United States has flown in the last three decades have returned more than five million television images of the earth's weather conditions, from which meteorologists can make their forecasts. Because of the satellite revolution in weather forecasting, today's five-day forecasts are as reliable as two-day forecasts were more than a decade ago.

All together, the weather satellites have had

more than eighty years of observing time in orbit and have traveled over 10 billion miles (16 billion kilometers). They have been responsible for moving the science of meteorology dramatically ahead and have saved thousands of lives around the world.

Weather Forecasting in the Twenty-First Century

Weather satellites have continued to evolve into extremely complex in-orbit sensing systems. The Nimbus series of seven satellites was placed into orbit between 1964 and 1978. More complex than TIROS, they were second-generation research satellites, and they carried advance cameras including a TV cloud-mapping camera system. The Nimbus also contained an infrared radiometer that could take pictures at night for the first time. These satellites tested the state-of-the-art meteorological equipment that led to the sophisticated twenty-four hour satellite weather coverage we have today.

Weather satellites have been launched by the European Space Agency (*Meteosat*), India (*Insat*), Japan (*GMS*), Russia (*METEOR*), and the United States (NOAA and GOES). Their advanced technology is capable of measuring atmospheric and ocean temperatures at different altitudes and depths, charting ice in the shipping lanes, discovering forest fires, measuring rainfall and forecasting next year's drought or harvest (as ESA's *Meteosat 1* is doing for Africa), and even detecting the health (by measuring the chlorophyll content) of crops as they grow hundreds of miles below the satellite sensors. Such satellites today do much more than detect weather conditions, and they will soon be monitoring and reporting on the total global environment.

As of late 1986, the United States has two different satellite designs in two different types of orbits. One of the latest, most advanced TIROS-N-type weather satellites is *NOAA-9*, launched in December, 1984. In a polar orbit, from an altitude of some 470 miles (756 kilometers), it covers the complete earth once a day, and its infra-

red sensors take vertical profiles of temperatures and water vapor in the atmosphere from the surface to the fringes of the atmosphere. The other type of satellite, GOES (Geostationary Operational Environmental Satellite), takes in the big picture of the western hemisphere from its high geosynchronous orbit. The National Oceanic and Atmospheric Administration has two GOES operating, under ideal circumstances, to cover the Atlantic and Pacific regions off North and South America.

What does the future hold for weather satellites? The next generation of GOES will be designed to last five years, and the first one will be launched by the space shuttle when flights resume. The same will hold true for the polar-orbiting NOAA series that must be replaced every few years. After the weather satellites of this new generation live out their extended lives, they may be replaced by a permanent orbiting module, filled with remote sensing equipment, that would be attached to the operational space station in the 1990s. Your TV weather reports may feature live reports from your space weatherman, who will be floating next to the weightless weather globe and pointing to the recent track and present whereabouts of a current severe storm.

The Hubble Space Telescope

The most eagerly awaited scientific instrument in history will soon begin its life in orbit above the earth. Finally, after years of delay because of budgetary problems, technical challenges, and the *Challenger* disaster, the space shuttle *Atlantis* is scheduled to launch the Hubble Space Telescope by 1990. Earthbound astronomers and other scientists will soon operate this great observatory in the sky, put it through its first year of rigorous testing and make some startling discoveries during the shakedown orbits.

It will be, as many people have been saying for many years, the most important event in the science of astronomy since Galileo pointed his small telescope toward the moon and the planets more than 375 years ago. Everyone agrees: The Hubble Space Telescope will revolutionize

The silver anniversary of weather satellites was celebrated in 1985. The first was RCA's TIROS 1 *launched in April 1960. Today the entire surface of the earth is covered, and more than thirty satellites have returned over five billion TV images and saved thousands of lives around the world.* FACING PAGE: *The image of a clear North America was taken in 1978.* ABOVE: *The storm image, taken by the NOAA-7 satellite on December 20, 1984, is of a rare winter hurricane named Lili.* Courtesy Goddard Space Flight Center; and RCA.

astronomy. This state-of-the-art telescope promises to do nothing less than herald a new era of scientific discovery that will last into the first decade of the twenty-first century.

Named after the famous American astronomer Edwin P. Hubble (1889–1953), who discovered beyond our own Milky Way distant receding galaxies that indicated an expanding universe, the Hubble Space Telescope (HST) is the most powerful, complicated, and precise telescope ever built. During its projected operational lifetime of fifteen to twenty years, it will open up the universe, even peer back in time some

fourteen billion years and view the universe when it was young, as the galaxies were forming after the big bang. It will detect and gather light from countless galaxies never before seen by the eyes or instruments of humankind. The light from these cosmic objects will be billions of years old, having traveled from the young universe to our present universe, when humankind is reaching out to understand its place in the cosmos. And during its first decade above the earth, it will focus its powerful technology on those cosmic puzzles—the black holes, exploding galaxies, and quasars, and solve, or at least shed light on, their mysteries. Here are some of the reasons that the Hubble Space Telescope will be the greatest telescope on or off the planet Earth. It can:

• Observe objects that are fifty times fainter than those visible to the most powerful earth-based telescopes.
• Peer seven times further into the universe than ever before.
• Observe a volume of the universe that is 350 times more than could be seen from earth.
• Concentrate its light 400 times more efficiently than the earthbound Hale reflector, which is twice as large.
• Produce images with at least 10 times (and perhaps as much as 100 times) better resolution than ground-based telescopes.
• Observe some 4,500 hours each year as compared to the best possible case of 2,000 hours a year for on-earth telescopes.
• Measure accurately, with its high-speed photometer, varying intensities of light, as many as one hundred thousand variations each second.
• Lock on to and track for hours a celestial object so small and faint that it is equivalent to resolving a small coin from a distance of 750 miles (1,207 kilometers).
• Spot faint objects so far away that it is equivalent to spotting a firefly from 10,000 miles (16,090 kilometers) away, which makes the Hubble Space Telescope some one billion (1,000,000,000) times more sensitive to light than the human eye.

With such capabilities, the Hubble Space Telescope will learn more about the universe in its first few years than astronomers and stargazers have learned throughout human history.

A Delicate Giant

Big, powerful, yet extremely delicate in its movements, which are guided by some of the stars in its software program of twenty million,

the Hubble Space Telescope was an engineering feat that pushed the state of the art in astronomical technology to the edge. This, the best cosmic tool yet, which will chip away at billions of light-years, had its share of engineering challenges and obstacles that at times appeared overwhelming to those deeply involved.

The basic dimensions of the HST are 43.1 by 14.1 feet (13.1 by 4.3 meters), excluding any antenna booms. It is a good size for its launching berth in the space shuttle's cargo bay. Its weight at launch will be 25,500 pounds (11,567 kilograms), but it will weigh next to nothing in the microgravity of orbit.

The primary mirror is 94 inches (2.4 meters) in diameter, and it is the smoothest mirror ever created. If the surface of planet Earth were proportionately as smooth as this mirror, then Mount Everest would be only 5 inches (12.7 centimeters) high. In order to get the mirror this smooth, engineers had to polish it twenty times to smooth out the microscopic blemishes that were found by laser detectors. The polishing and grinding alone took twenty-eight months. After the ultrasmoothness was attained, the mirror was coated with a pure aluminum reflective layer, which was only 2½-millionths of an inch (6-millionths of a centimeter) thick, and this coating was protected by a microthin layer of magnesium fluoride.

It is also the largest telescope mirror ever to be transported into space, and its location in orbit, some 310 miles (499 kilometers) above earth, will make it immune to the distorting atmosphere with which all ground telescopes must contend. Still, as it orbits the earth at almost 18,000 miles (28,962 kilometers) an hour, this boxcar-size instrument will have to make delicate and precise movements to keep locked on distant cosmic objects billions of light-years away. This was the major headache-maker for the engineers and scientists. How will the Hubble Space Telescope accomplish its delicate cosmic dance as it courts the universe?

The HST will use a combination of rate gyros, reaction wheels, star trackers, and very sensitive white-light interferometer sensors to carry out the ultraprecise pointing and stabilizing mo-

SPACE COMMERCE CALENDAR

Date	Event
1985	First "made-in-space" product, latex microspheres used in calibrating high-tech instruments, went on sale.
1988-89	First space-made drug, which stimulates growth of red blood cells, available in marketplace.
1992	First dedicated materials processing facility put into orbit by Space Industries, Inc., under agreement with NASA.
1995	Burial launch services available at competitive prices; about ten launches a year.
1997	Space station operational, with second-generation materials processing modules attached or on information free-flying platforms.
2000	Many private companies and several nations compete to manufacture and market "made-in-space" products.

tions. The gyros first will give the HST a reference frame—its current orientation and the position to which it will be moved. Ground control will then command the reaction wheels (23 inches, 58.4 centimeters in diameter) to either accelerate or decelerate to move the telescope toward the cosmic object. Next the star trackers will come into play, using preselected bright stars from the total program of twenty million stars, to point the HST to within about 1 arcminute (one-sixtieth of a degree) of its target object. Finally the fine-guidance sensors will take over, when the guide star is within an area equal to one-tenth of the moon's apparent area. These sensitive white-light sensors will act through the rate gyros and reaction wheels to point the telescope at the target to within 0.007 arc-second, which is analogous to having a laser beam directed at a coin-size target some 600 miles (965 kilometers) away and not having it move more than 1.2 inches (3 centimeters) off center. And this tracking must remain absolutely accurate for up to ten hours, as light is collected from distant and mysterious quasars never seen before, objects sending out their light from an early universe that no longer exists.

Inside the Space Telescope

Besides the primary and secondary mirror systems, there are five major scientific instruments inside the Hubble Space Telescope: two cameras, one photometer, and 2 spectrometers. Together, over the next twenty years, these instruments will collect thousands of billions of data bits that will be shaped into a new view and understanding of the universe.

Faint-Object Camera. This is the only instrument to be built and supplied under the direction of the European Space Agency. It was built by Dornier, a West German company. The Faint Object Camera photographs dim objects billions of light-years away by taking cumulative long exposures lasting for several orbits around the earth—up to ten hours. Its light detectors can recognize individual photons of light, and it has the highest spatial resolution of any optical system on the Hubble Space Telescope. A single image consists of some 250,000 pixels (picture elements).

With its dozens of prisms and filters, this is the camera that will discover planets around other stars and perhaps find quasars at the centers of faint galaxies billions of light-years from earth. It may eventually show us the early universe shortly after the big bang.

The Wide-Field and Planetary Camera. This camera examines large areas of space and plots the spatial relationships between distant galaxies and quasars to determine distance scales and test cosmological models of the universe. It also

The Hubble Space Telescope, its launch delayed because of the Challenger *disaster, will peer seven times farther into the universe than any instrument before it and concentrate its light four hundred times more efficiently than the large Hale reflector. The resolution is so good, it can be compared to resolving a coin at a distance of 750 miles (1,207 kilometers). The telescope is shown here without its solar panel wings. What wonders will it soon unveil?* Courtesy Lockheed Missiles & Space Company, Inc.; NASA; and Perkin-Elmer.

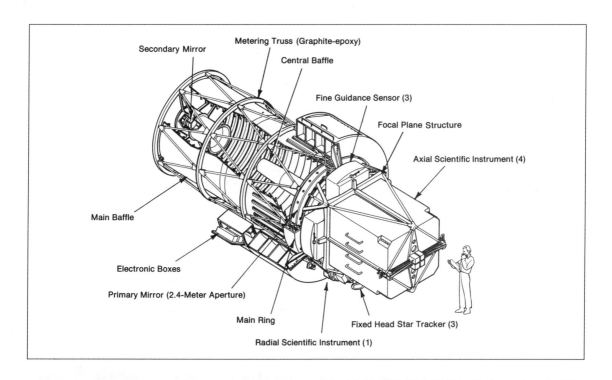

Secondary Mirror

Metering Truss (Graphite-epoxy)

Central Baffle

Fine Guidance Sensor (3)

Focal Plane Structure

Axial Scientific Instrument (4)

Main Baffle

Electronic Boxes

Primary Mirror (2.4-Meter Aperture)

Main Ring

Fixed Head Star Tracker (3)

Radial Scientific Instrument (1)

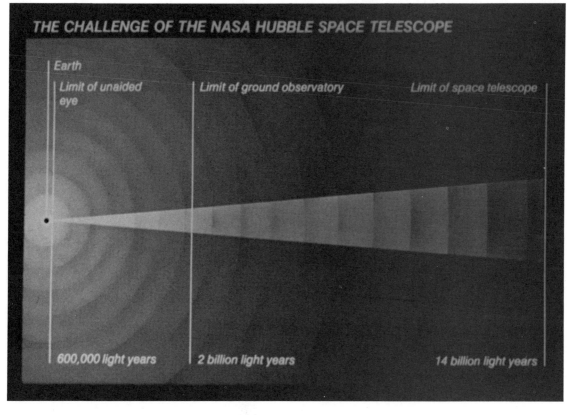

THE CHALLENGE OF THE NASA HUBBLE SPACE TELESCOPE

Earth

Limit of unaided eye

Limit of ground observatory

Limit of space telescope

600,000 light years

2 billion light years

14 billion light years

produces the highest-resolution images of quasars ever produced. At a higher focal ratio of f/30, it can also study the planets of our solar system (except Mercury, which is too close to the sun) and provide images equal to, and sometimes better than, the Voyager flybys of Jupiter and Saturn. The planetary camera's resolution is equivalent to resolving a cosmic object the size of a baseball from a distance of 200 miles (322 kilometers).

The Wide-Field and Planetary Camera consists of eight CCD (charge-coupled device) cameras. These eight CCDs are clustered in two groups of four, each with a telescope. A final combined image has 2.5 million pixels. The light intensities of each pixel are recorded on the onboard computer, transmitted to earth via telemetry through a data-relay satellite at the rate of one million bits a second, and assembled into images of the far universe never seen before.

The High-Speed Photometer. This instrument measures light over a certain wavelength range as often as one hundred thousand times each second, thus detecting any minute fluctuations in brightness. Detection of rapidly spinning celestial objects such as neutron stars or other compact stars occurs more frequently than ever before. Such precision will help calibrate faint stellar objects and pin down their cosmic distances from earth.

Faint-Object Spectrograph. Spectrographs are optical devices that divide light from a source into separate beams according to wavelength. The study of the spectrum of a light obtained from any cosmic source can yield valuable information about the star or galaxy or quasar—whether it is hot or cold, dense or rarefied, what its chemical composition is, its distance from earth, and its velocity. So much to be learned from light!

The Faint-Object Spectrograph obtains the spectra of extremely faint astronomical objects in the ultraviolet and visible wave bands that could not be obtained before. Study of these spectra helps astronomers solve such mysteries

The Advanced X-Ray Astrophysics Facility will be launched by the space shuttle in the 1990s and will probe the distant universe of black holes, quasars, stellar birth, and other X-ray sources. It will be one hundred times more sensitive than the earlier X-ray satellites such as the Einstein Observatory (also know as HEAO-2). Courtesy Lockheed Missiles & Space company, Inc.

as the violent centers of distant galaxies, the physical properties of quasars, or the chemical composition of pristine comets before they are altered by the sun. By studying the spectra of distant quasars, for instance, astronomers can reveal the properties of the universe as it was more than ten billion years ago, before many of today's chemical elements, forged in the centers of second-generation stars, even existed.

The High-Resolution Spectrograph. This precision instrument looks only at the ultraviolet region of the spectrum and sees light that cannot reach the surface of the earth. It is very accurate because it uses more light than the other spectrograph and resolves it into finer increments, thus providing the most detailed information yet obtained on a cosmic object's chemical composition. This spectrograph will study the composition of interstellar gas and dust as well as what makes up our Milky Way galaxy's halo. It will gather information that was impossible to obtain before the Hubble Space Telescope reached orbit.

These instruments were designed and built to be maintained, repaired, and upgraded by astronauts on future space shuttle flights, so they may represent only the first generation of instruments that are seeking—a common search among many exciting searches—the ultimate fate of the universe. Will it expand forever, or will it reverse its expansion and fall back again into a big crunch—an ultimate black hole to end our present universe? The Hubble Space Telescope may find the answers to these universal questions.

The Hubble Helpers

While the Hubble Space Telescope will dominate astronomy for the rest of the twentieth century, there will be other sophisticated astronomical satellites that will join it in orbit above the earth, each one with its own special window on the universe. For more than three thousand years, humankind was restricted to only one window—visual astronomy using human eyes and then optical telescopes to aid the eyes. But all that changed about fifty years ago, when radio astronomy was born and widened our view of the universe. In the Space Age, now that astronomy can view and record from orbit above the earth the high-energy radiations (radiations that cannot penetrate the earth's atmosphere and have therefore remained invisible to us until the later half of this century), entire new specialties of astronomy have been created: infrared, gamma-ray, ultraviolet, X-ray. Here are some of the astronomical satellites that will fly before the end of the century. They do not include the dozens of astronomical instruments that will do their work from the space shuttle's open cargo bay during future flights.

Advanced X-Ray Astrophysics Facility (AXAF). Launched by the space shuttle in the 1990s, this advanced imaging telescope will explore X-ray sources in space with about one hundred times the sensitivity of the High Energy Astronomy Observatories of the 1970s. *HEAO-2*, for example, known as the Einstein Observatory, recorded X-ray emissions from every known quasar, some of which are believed to be as far away as 14 billion light-years. The AXAF observatory will begin where these earlier observatories left off, probing the mysteries of star birth, star death, black holes, and quasars by catching and recording the X-ray emissions from these cosmic sources. It will create a new and complete X-ray map of the heavens, observe the coronal X-rays from almost all the one hundred thousand stars closest to our sun, and discover unknown quasars at the very edge of our expanding universe.

Extreme Ultraviolet Explorer (EUVE). This 1,200 pound (544-kilogram) astronomical satellite is scheduled to be launched in the early 1990s and will carry five scientific instruments—four telescopes and one spectrometer. The spacecraft is designed to probe the last unexplored region of electromagnetic radiation—the extreme ultraviolet—which falls between X-rays and ultraviolet. Like many other high-energy radiations, the extreme ultraviolet is invisible from the earth's surface because our planet's atmosphere blocks it out. Fewer than two dozen sources of the extreme ultraviolet have been discovered, and four of them are in our solar system: the sun, Jupiter, Saturn, and Saturn's mystery moon Titan. The other ones are white-dwarf stars and double stars that are under going gravitational tugs of war.

The Extreme Ultraviolet Explorer is expected to complete an all-sky survey of extreme ultraviolet radiation sources just as the highly successful *Infrared Astronomy Satellite (IRAS)* did for the opposite end of the spectrum. Scientists believe that the EUVE may discover one hundred sources or more. The white-dwarf star population may expand rapidly, and there will no doubt be some other surprise discoveries as there were with *IRAS*.

Gamma Ray Observatory (GRO). Scheduled for launch from the space shuttle in the early 1990s the GRO weighs more than 30,000 pounds (13,608 kilograms) and has been designed for in-orbit refueling. This observatory for the gamma-ray window on the universe will observe in detail what is occurring near neutron stars, black holes, supernovas, and other violent phenomena in the universe. It may also provide a fuller understanding of what is happening at the center of our Milky Way galaxy, the origin and characteristics of unidentified gamma-ray bursts, and the formation of all heavy elements—including those that compose the earth and our solar system—in supernova explosions.

Space Infrared Telescope Facility (SIRTF). After the success of the *Infrared Astronomical Satellite (IRAS)* in 1983, this satellite has a very tough act to follow. The *IRAS* opened a new window on the universe for the first time and observed infrared energy-emitting sources some one thousand times dimmer than can be seen from earth-based observatories. This included a distant galaxy (ARP-220) that emits an amount of infrared energy equivalent to two trillion suns, a possible solar system in the making around the star Vega, several new comets, and the locations and intensities of more than two hundred thousand celestial objects. *IRAS* carried out the first complete survey of the infrared sky and made many discoveries.

The Space Infrared Telescope Facility is a follow-up to *IRAS* that will fly in the 1990s. Originally designed to be mounted inside the space shuttle's cargo bay, it was redesigned as a free-flying observatory when mounting evidence indicated that the space shuttle is not always the best place to carry out astronomy.

The SIRTF's mirror will be just under 33.5 inches (85 centimeters) in diameter, and it will be cooled to very low temperatures, close to absolute zero, by liquid helium to eliminate background noise and increase sensitivity. It is designed to detect sources that are as much as one

thousand times weaker than what IRAS could detect. With such increased sensitivity, SIRTF will no doubt reveal some cosmic surprises of its own.

Infrared Space Observatory (ISO). This is the European Space Agency's contribution to infrared astronomy, and it is scheduled for launch in 1992 aboard the Ariane rocket. It will be available to scientists in Europe and the United States and will be operated around the clock. The infrared window to the universe will be carefully watched by the ISO and the SIRTF orbiting observatories, and the details of star gestation and birth, as well as other cosmic processes, will be there for all to see clearly.

Solar Optical Telescope (SOT). Another space shuttle flight in the 1990s will carry the Solar Optical Telescope into orbit. This optical-ultraviolet telescope will be able to obtain detailed spectra of solar gases. In the far-ultraviolet range, the SOT will be capable of resolving features of the sun as small as 12.4 miles (20 kilometers). The SOT will evolve over its scientific lifetime and will carry more sophisticated instruments to examine the properties of solar plasmas, including their temperature, density, velocity, and magnetic fields, more closely than ever before. Because the earth and humankind are totally dependent on the sun, its study will intensify as we cross the divide of the centuries and new interrelationships between the sun and earth become known.

Beyond the Hubble Space Telescope

Even before the Hubble Space Telescope has gone into orbit, NASA has funded Perkin-Elmer, manufacturer of the advanced optics system, to study technology requirements for the next generation of space telescopes. Lead times of a decade are common for major Space-Age projects that require innovative approaches and the development of new technology.

Assuming that the Hubble Space Telescope operates for twenty years, with some equipment updating during its lifetime, it will have to be replaced with a state-of-the-art space telescope by 2008. A larger space telescope of the twenty-first century, perhaps a multiple-mirror telescope with new technology, could be capable of resolving and studying cosmic objects that are one hundred times fainter than those seen by the Hubble. This means that it could study in detail any one of the two hundred billion-plus stars in the Andromeda galaxy, which is 2.2 million light-years away. Such a space telescope could get even closer in time and space to the big bang beginning of the universe.

While the new-generation space telescope would still be free-flying, other instruments could fly aboard a space station astronomy module. Already there are plans for the development of an extremely accurate astrometric telescope, which would detect planets around other stars. The space station could also have a servicing module for the advanced space telescope and all other astronomical satellites.

If large space colonies are built in the twenty-first century, they could have their in-space astronomical observatories flying near them, and large radio telescopes could be built on the far side of the moon to avoid radio interference from the earth.

A large, deployable reflector telescope has also been envisioned, which would carry out far-infrared observations. If its mirror was 33 feet (10 meters) in diameter, it would be capable of penetrating the cores of dark interstellar clouds and observing new stars being born.

Because new discoveries in astronomy are so unpredictable, the science of astronomy above the earth will remain exciting and full of surprises even after the Space Age celebrates its golden anniversary in 2007. And there is always that mind-boggling longshot—discovering *another* extraterrestrial intelligence to keep us company in the universe.

Be it a commercial or scientific payload, the costs to get it into orbit have dropped significantly since the early days of the Space Age. While today's cost to get 1 pound (0.45 kilogram) into orbit is about $8,000 (estimates vary depending on what costs are included), it cost a staggering $1 million per pound during the first few years of manned spaceflight. If significant reductions continue to occur in payload-to-orbit costs over the next twenty-five years, this would guarantee both industry and science permanent, productive, and inspiring places in outer space. And their presence and economic impact could eventually make above-earth tourism a reality for many of us.

10

Space Weapons
& Wars

Star Words

The debate over Ronald Reagan's Strategic Defense Initiative has been raging ever since the spring of 1983, when the President gave his famous "Star Wars" speech on March 23. Scientists, politicians, military and industrial leaders, writers and many citizens from all walks of life have expressed their opinions on the complex of issues involved in what is now a mammoth, multibillion dollar research program to see if the United States can develop a high-technology defense shield, deployed in orbits above earth, that will render attacking nuclear missiles and their warheads "impotent and obsolete."

A year after the speech, on receiving the Robert Goddard Award at the National Space Club, President Reagan told his audience that he considered the media's use of the term "Star Wars" for his Strategic Defense Initiative program very unfortunate.

"It isn't about war. It's about peace. It isn't about retaliation, it's about prevention. It isn't about fear; its about hope . . . . Let history record that in our day America's best scientific minds sought to develop technology that helped mankind ease away from the nuclear parapet."

The debate continues; round after round is fought among well-known heavyweights (Nobel Prize–winning scientists, best-selling scientist-writers, and presidential advisers and program directors) in the print or electronic media. It would seem that every possible viewpoint has been expressed, but new variations continue to be voiced.

That all of us agree with the president of the United States in his noble goal of ridding the

PRECEEDING PAGES: *If an SDI system is ever deployed, immediate access to space is essential to support the system. The Air Force has studied several advanced systems that could provide on-demand orbital missions. The concept shown here is an air-launched sortie vehicle that would climb to low earth orbit, conduct its mission, and glide back to earth like the space shuttle. Courtesy Boeing Aerospace Company.*

world of nuclear weapons is beyond dispute. But many serious questions have been and continue to be asked. How will the program influence arms control talks? Can such defensive space weapons be easily countered by the enemy? Will the tremendous costs strain even a healthy economy of the United States? Who decides where research ends and development begins during the first years of the program? These and many other debate topics must be thoroughly discussed during the research phase of the Strategic Defense Initiative—the initial phase that alone may cost some $26 billion and take us to 1990. Then, as we enter the last decade of the twentieth century, the next president and Congress must decide whether or not to develop and launch this future umbrella of high-tech lasers, beam guns, and other defensive weaponry that could once and for all free us from the nuclear nightmare. Or could it?

Towards the Limits of Technology

Will such a complex system of strategic defense weapons work? Some say yes and some say no; but it is really too early to do anything more than guess during these early years of research. No determination has been made as to which weapons systems will actually go into final design. And, of course, whether they work or not is only part of the answer. How well would they work? Would they be 100 percent reliable? Would they be vulnerable to attack? Would they be built and deployed at the expense of conventional forces? Many of us simply want to shake our heads and ask: Why make a dangerously complex world even more complex?

During a seven-hour Senate debate in the summer of 1985, Senator J. Bennett Johnston, a Louisiana Democrat who usually supports Pentagon programs, spoke up: "The truth . . . is we do not know what Star Wars is. It is a whole collection of technologies that we're going to be chasing out there with the mighty American dollar . . . . "

As the leading proponent of the space defense strategy, Lieutenant General James A. Abrahamson, Jr., director of the Strategic Defense

Initiative, strongly believes that a highly effective defense against ballistic missiles can be built someday. "The question is how soon and how affordable and what degree of effectiveness can initial steps allow us." This position was backed up by Dr. Greg Canavan of Los Alamos National Laboratories, who spoke for General Abrahamson at a symposium sponsored by The Planetary Society, "The Potential Effects of Space Weapons on the Civilian Uses of Space." In early 1985, Canavan stated the SDI position: "We believe that the technologies under consideration are promising enough that, with continuing effort, a future administration and Congress, if they choose, will have a very real option to design, build, and deploy an effective defense against ballistic missiles, a defense that will not threaten people as nuclear ballistic missiles do today."

Senator Carl Levin, Democrat of Michigan, questions the basic Reagan premise: "Star Wars is based on the fantasy that we can gain absolute security by developing new technology not vulnerable to countermeasures."

Carl Sagan, one of the most vocal opponents of the Strategic Defense Initiative, argues that "even scientists and engineers who support SDI doubt that a system much more than 50 percent effective can be deployed in the next few decades." The prime mover of the program, Edward Teller, the famous physicist who played a key role in developing the hydrogen bomb, has debated Sagan several times and sees it differently. "Even a moderately efficient defensive system," he believes, "will have a strong deterrent effect if combined with some retaliatory ability." Teller met with President Reagan four times, over a period of about a year, before the "Star Wars" speech in 1983 and discussed new ways of destroying and defending against enemy missiles and warheads. Without question, Teller is the leading scientist advocate of the Strategic Defense Initiative.

Wolfgang Panofsky, director emeritus and professor at the Stanford Linear Accelerator Center, sides with the opponents of the program and presents some persuasive arguments.

"The Strategic Defense Initiative is truly impressive; one can base so much political and strategic posturing on such limited technical and military potential." Panofsky goes on to argue that there is no justification whatsoever, neither technical nor strategic, to pursue anything beyond a limited research program that will examine the possibility of such defensive weapons and insure the United States against any technological surprises the Russians could discover.

"There were no studies preceding that speech even remotely indicating that nuclear weapons could be made impotent and obsolete' through defense," Panofsky states, "let alone indicating a stable path of how to get there from here. I object strongly to the politicization and 'high technology' salesmanship of SDI."

Counterpointing Panofsky's views is Gerold Yonas, chief scientist of the Strategic Defense Initiative Organization. He admits that "no single concept or technology has been identified as best or most appropriate," but argues that an aggressive research program must go forward because of Soviet advances in laser and particle beam research. "In contrast to our neglect of defense, the Soviets were spending as much on strategic defense as on strategic offense." Yonas is, at least, a realistic proponent and believes it is important "to approach the overall program at all times with a healthy skepticism as well as with the creativity associated with the exploration of a new field. With the current pace of technological advancement, and the strong motivation provided by the hope of a more secure and stable future, we believe we are well on our way." Chief Scientist Yonas believes that President Reagan's Strategic Defense Initiative is also important because it has sparked such a vigorous debate and brought sustained attention to the dilemma of human survival in the nuclear age. The debate continues.

The Manhattan Project Boys: Bethe versus Teller

Hans Bethe, the Nobel Prize–winning physicist, has been squaring off with Dr. Edward Teller, his old colleague from Manhattan project days, and Teller has not been above responding with some personal jabs, saying that Bethe has just been looking at the new weaponry in a superficial way: "What he has looked at, he agrees will work. What he hasn't looked at, he thinks won't work," Teller said on "Meet the Press" in July 1985. "Bethe is not a dreamer. Bethe is not a man with imagination. He is a wonderful man with facts and evaluating what's been done." In yet another interview, Teller gave his opinion that, "Bethe sees the future in a too-easy manner . . . . He was there when we constructed the first atomic bomb. Now he says there won't be anything new under the sun. Hasn't he seen enough new things?"

"Star Wars," Bethe responds, "will not work. There are many proposed devices; half of them may work, half of them won't work, and I won't enumerate them. But even those which may work, I think, don't give us security, because when they are up there in the skies, they are exposed to enemy countermeasures. And the worst enemy countermeasure, which is very effective, is for them to build more offensive missiles. And therefore, in my opinion, Star Wars is just an invitation to an arms race."

Bethe also opposes the SDI program on the basis of what would be its operational complexity. "The degree of perfection required of such a defensive shield is truly staggering," he says. "It is difficult to imagine a system more likely to induce catastrophe than one that requires critical decisions by the second, is itself untested and fragile, and yet is threatening to the other side's retaliatory capability." More recently Bethe has argued that a battle-management station would have to track some one hundred thousand objects—real and dummy warheads, boosters, and so forth—from launch to destruction or impact. "The SDI is a program with a goal that is not attainable unless and until there are technical breakthroughs that have not yet been conceived of." This in contrast to the comment of SDI director Lieutenant General Abrahamson that "there is very little question that we can build a highly effective defense against ballistic missiles someday." Perhaps the general knows something that Hans Bethe does not. "None of us has all the answers," said Edward Teller in 1985. Everyone is agreed that we must have a good

many more before the expensive Strategic Defense Initiative leaves its research phase and enters its second, more costly phase of systems development and space weapon prototypes. By the early 1990s, we probably all will know if the orbits above earth will be crowded with high-tech defensive weapons.

Costs on High

If advanced weaponry and systems of the strategic defense program ever become operational, some estimates put its price tag at $1 trillion ($1,000,000,000,000). Every man, woman, and child on planet Earth—all 4.8 billion of us—would have to contribute over 200 U.S. dollars to reach such a great total of money. Holding SDI costs to $1 trillion may become a challenge in itself. If the program is developed along the lines that President Reagan outlined, it could be double that figure, ending up to be more than double the entire annual federal budget—as much as $2 trillion.

The cost of the five SDI research years, which will approach $30 billion, causes a debate of its own, and that is only about one-fortieth of the often-projected cost of $1 trillion for an operational system.

"We must keep in mind the cost in opportunities foregone," said Senator Edward Kennedy in 1985, "—to strengthen our conventional forces, to achieve biomedical breakthroughs, or to keep U.S. industry competitive—when we divert such a significant share of our research budget to this single effort." Dr. Isaac Asimov, one of the most prolific writers in the world, who happens to write science and science fiction, is very skeptical about the whole Star Wars concept: "It's just a device to make the Russians go broke. But we'll go broke, too. It's very much a John Wayne standoff."

Carl Sagan comments sarcastically: "I can't think of anyone in the aerospace industry that has been critical of Star Wars; I wonder why? I think it's clear that if you wave a trillion dollars at the U.S. aerospace industry, you will get striking results, whether or not the argument for deployment is valid, and whether or not the net result will be a dramatic increase or, as I believe, a dramatic decrease in national security."

The proposed Strategic Defense Initiative already dwarfs the Manhattan atom-bomb project and the Apollo moon-landing program. The cost of researching and perhaps building such a defensive space shield against enemy nuclear missiles will remain controversial, as well it should. Such a ubiquitous program would affect the entire economy, which, in turn, plays an important role in national security. There could be a futuristic defense line in place on the high frontier but no money left to operate or repair the system. As with so many weapons systems, it could turn obsolete before it is even partly paid for. Edward Teller has remarked that "a defense is useless if it can be overwhelmed at a lesser cost by mere duplication of attack instruments. It will also be ineffectual if the defense itself can be destroyed at less cost than its replacement." The high costs of SDI and their effects on the national economy must be studied as vigorously and as closely as the designs, reliability, and effectiveness of the proposed hardware—the lasers, the space-based radars, and the electromagnetic hypervelocity guns that may never be called upon to flash their powerful energies or fling their "smart rocks" toward attacking enemy missiles.

Space Arms: Control or Race?

What the actual Soviet response to the concept of defensive weapons in space will be, not their frequent propaganda pronouncements, is another area of debate in the strategic defense controversy. Even during the research years of the program, the Russian position has evolved as a multitude of factors influence their policy—including the debate in the United States. There is little doubt that the Soviet research program has also been considerably stepped up in response to SDI, and some experts claim they are ahead in lasers.

Proponents of SDI claim it will motivate the superpowers toward arms control. George A.

Keyworth, President Reagan's first science adviser, believes that "the investigation of strategic defense options [is] an absolutely vital catalyst to real arms control." Princeton physicist Freeman Dyson believes that if the United States experiments with killer satellite systems, as has Russia, there could be a proliferation of advanced weapons and spacecraft in orbit.

"This possible future I call the technical-follies future," Dyson writes in his book *Weapons and Hope*, in which "space becomes the arena of a technological arms race pursued without much regard for military utility . . . . Real military advantage in space will lie with numbers and concealment."

Dyson contends that such a buildup of in-orbit technologies will have very little influence on "the realities of military and political power on the ground." Ridding planet Earth of nuclear weapons cannot be done with technology alone, Dyson is convinced, and he offers an alternative to his so-called technical follies future. It is the "defense-dominated future, [where] weapons of mass destruction are disarmed, not by defensive technology alone, but by legal and political restraints strengthened by the active intervention of technology."

This position emphasizing arms control is a tempered one compared to that of many opponents of defensive weapons in space. Hans Bethe believes that Russia and the United States must agree to the common danger and act. "The solution can only be political,"he says. "It would be terribly comfortable for the President and the Secretary of Defense if there was a technical solution, but there isn't any." Arthur C. Clarke agrees that negotiation between the superpowers is the only way. Labeling the entire SDI program as "technological obscenities," Clarke has no illusions about the challenges of bringing the East and West together. "I am not so naive as to imagine that this could be achieved without excruciating difficulty and major changes in the present political climate. But those changes have to be made, sooner or later."

What is the Russian response? A few days after the Reagan speech in 1983, the Soviet leader at the time, Yuri V. Andropov, said that if the concept ever went into development stages, "it would in fact open the floodgates to a runaway race of all types, both offensive and defensive."

Some Western experts believe that the Soviet Union and the United States are roughly equal in directed-energy weapons research, but that the United States is ahead in many of the other technologies, such as computers and automated control, that are essential for a space-based nuclear missile defense.

"It's not going to be a race between our 'Star Wars' and their 'Star Wars,' " says Stephen Meyer, a government consultant on the military policy of the Soviet Union, "but a race against our system and their efforts to overwhelm or neutralize it."

Proponents of strategic defense, including its director, General Abrahamson, often voice the view that the United States will gain bargaining leverage from the Reagan defense initiative. The Soviets, fully aware of the technological prowess of the United States, may fear an all-out race to build and deploy such space weapons and may attempt to head it off at the bargaining table. This view, proponents qualify, depends on the degree of support from the American people and Congress.

What happens or does not happen at the bargaining table during the research years of the Strategic Defense Initiative will certainly indicate in which direction the superpowers will go with defensive space-based weapons. The general belief among experts on the Soviet Union is that the Russians may also develop and build a defensive space umbrella against nuclear missiles. Meanwhile, the top-secret research laboratories in both countries are busy round the clock.

Today's Weightless Pentagon

Space has been militarized for decades, since the 1960s, when the first military satellites were launched into orbit. Even opponents of the Strategic Defense Initiative are in favor of maintaining an active launch program of military satellites that provide imaging reconnaissance, intelligence gathering, satellite communications among military units, and reference signals for navigation of ships, planes, and other military vehicles. All of today's military space hardware is concerned with the collection and transmission of intelligence information, and everyone concurs that such activity is a stabilizing influence on both sides. Imaging reconnaissance satellites, for example, are important in verifying arms control treaties already in place and in negotiating future ones. Sophisticated sensors in orbit are able to identify any rocket launched on the face of the earth and track its trajectory, and this gives a degree of security to both superpowers by lessening the threat of surprise attack. Each side knows, to some degree, what the other side is doing and what its military capabilities are, and this benefits all nations by calming fears and reducing tensions. Advocates of arms control believe that this is a healthy state of affairs and should continue.

"Militarization of space is not the issue," says Carl Sagan. "The introduction of weapons into space is. I am concerned that space weaponry, once given the go-ahead, will expand outward without limit."

Military satellites in orbit are becoming increasingly sophisticated, and their state-of-the-art capabilities are classified. There are four types of military satellites that serve today's weightless Pentagon.

Reconnaissance Satellites. Hundreds of reconnaissance satellites have been launched by the United States and the Soviet Union since the early 1960s. In fact, the United States alone has launched more than two thousand military satellites of all types since the Space Age began. Most of these spies in the sky are placed in polar orbits and contain high-resolution optical or infrared cameras. Although the specifics of these military satellites are classified, it is generally known that ground resolution of such cameras has advanced to the point of resolving objects of no more than a few inches (centimeters) across from a low earth orbit. It would be nothing for such a satellite to photograph the license plate of a Soviet military truck, for example, or have another camera system capture a larger perspective and show Russian soldiers during a training exercise. Thousands of specialized analysts in the West and the East intently study these images on a routine basis to gather and compile intelligence.

Since 1970, Vandenberg Air Force Base in California has launched more than six large imaging reconnaissance satellites. This number included three basic types: high-resolution film satellites (which periodically eject reentry pods containing exposed film); broad coverage film satellites (also with the ejection pod feature), referred to as "Big Bird" satellites; and the image transmission satellites, known as the KH-11 satellites, which return their imagery by digital radio transmissions. Developed by the Central Intelligence Agency, the KH-11 satellites are said to be the most expensive satellite hardware ever launched, valued at several hundred million dollars each.

The launch mix of these military satellites will be changing in the near future, now that the space station is ending its design phase and entering development. Future shuttle missions can also provide satellite retrieval, repair, or refueling missions for reconnaissance satellites.

A radar imaging reconnaissance satellite will soon be operational, and it will be able to photograph targets through the earth's cloud cover. It will have the capability of tracking Warsaw Pact armor, for example, in Eastern Europe. A new generation of the KH-11 image transmission satellites is also believed to be under development and will be launched by the space shuttle in the future. The United States Congress has also approved the development of a future spy satellite that will be able to detect Soviet ground-based laser weapons capable of attacking U.S. satellites.

Present- and future-generation early-warning satellites rely on state-of-the-art infrared telescopes placed in geosynchronous orbits to scan large land and ocean areas for the rocket plumes of Soviet boosters. This allows the United States to know about all Soviet satellite launchings and missile test activities and at the same time,

of course, provides a global alert system for any possible launch of offensive ICBMs. The current early warning satellite system observed about one hundred Russian space booster launches and four hundred strategic and tactical missile tests during 1984. In all cases, the infrared plume data is verified as a launch, as opposed to a nonrocket thermal event on earth, within three minutes of the Russian lift-off. Within five minutes, NORAD has characterized the type of rocket flight the Russians are conducting.

As these satellites wear out, they will be replaced by more sensitive advanced versions that will have a higher power capability and modifications of the large infrared telescope to prevent laser jamming from the Russians. They are also designed to overcome Soviet interference with their data flow to ground stations around the world by using satellite-to-satellite communications cross-links as well as a data rebroadcast capability. They will be launched by the space shuttle.

Electronic Intelligence Satellites. A space shuttle mission in early 1985 carried into orbit a large signal intelligence satellite for the National Security Agency. From a geosynchronous orbit over the equator, the satellite monitors Soviet radio transmissions. The capabilities of such electronic snooping satellites are highly classified, but it can be stated conservatively that their sensitivity increases substantially every few years.

The Russians also have an active launch program for electronic intelligence satellites. Since 1980 the Soviet Union has put four such satellites into orbit each year. In addition, they have put into orbit a number of ocean surveillance spacecraft that use the same radio intercept techniques.

In September 1984 the USSR launched *Cosmos 1,603* on their powerful Proton rocket, which is able to place up to 50,000 pounds (22,680 kilograms) into low earth orbit. This electronic intelligence satellite was the largest single military satellite ever launched, and it frequently flys over the United States (including ground tracks over Texas and the central United

The most advanced satellite navigation system in history will be in place by 1990. The Navstar Global Positioning System, with a total of eighteen satellites in three orbital planes, will provide highly accurate positioning information for land, sea, and aerospace forces. Courtesy U.S. Air Force.

States, off the coast of Florida, and off the coast of California at Vandenberg Air Force Base), intercepting radio communications and data. After launch and orbit insertion, the spacecraft maneuvered extensively in space in such an unusual way that the Air Force's Space Defense Operations Center had to marshal its full forces to track and determine this huge satellite's nature. In the beginning of its unusual maneuvering, it was thought possible that the mission was an antisatellite demonstration, but this was soon ruled out.

Military Navigation Satellites. The most advanced satellite navigation system in history will be in place by the end of the 1980s. Called the Navstar Global Positioning System, it will have a total of eighteen satellites in three orbital planes that will provide precise navigational data for the Department of Defense and commercial users. While Navstar is primarily an all-weather military navigation system, with a coded channel that will provide data to the armed forces and intelligence community, its commercial users will include shipping, air traffic, and search and rescue parties.

The space shuttle will launch this constellation of satellites into low earth orbit, and then specially designed rockets, payload assist modules, will deliver them into 10,900-mile (17,500-

kilometer) high transfer orbits. The system is capable of providing accurate location and velocity data for ships, vehicles, and aircraft. It will be able to provide instant three-dimensional navigation information from any point on earth. Ground troops can also be tracked or aided in their movements. In addition to its navigation equipment, each satellite will contain sensors to detect and report nuclear weapons tests in the earth's atmosphere.

Military Satellite Communications. The military uses communications satellites for more than two-thirds of its long-distance communications traffic—data, voice, image—and the increasing sophistication of the hardware makes it even more attractive for the global military network on land, at sea, in the air, and in space. As the armed forces become more dependent on them, ingenious technology for making satellites less vulnerable to enemy interception and attack are designed and built into them.

The U.S. Department of Defense's Defense Satellite Communications Systems (DSCS) now has jurisdiction over all satellite surveillance activities, including radio signal intelligence and radar intelligence. The ultimate goal of DSCS is to integrate all the communications systems of the armed forces of the United States into one command system. The Army, Air Force, and Navy will have access to all the data of the system—from weather, navigation, early-warning,

The weightless Pentagon. Since 1970 Vandenberg Air Force Base has launched more than one hundred military satellites of all kinds. This painting depicts one of the newest Defense Satellite Communications Systems satellites, the DSCS-3, which has built-in antijamming protection. Courtesy U.S. Air Force.

and other satellites—that comes under military jurisdiction. The data base will include every form of intelligence product, from enhanced imagery at many different wavelengths to raw data. When this sophisticated global system is completely in place, it will become the global military nerve network for the United States and the Western Alliance, and the primary resource from which important strategic decisions are made.

The first flight of space shuttle *Atlantis* in late 1985 delivered two advanced Defense Satellite Communications Systems satellites known as DSCS-3s. These were much improved over the operational DSCS-2 series. The new-generation satellites offer secure-voice and high-data-rate transmissions over super-high frequency channels. A new antenna design, with electronically steerable beams, is an antijamming feature, and state-of-the-art materials make them the first satellites protected ("hardened") against the electromagnetic effects of nuclear space weapons. These new satellites also contain a special transponder, which permits the transmission of emergency messages to nuclear forces by the president of the United States. A total of six DSCS-3 satellites will make up the system in geosynchronous orbit—four of them will be operational, and two will be spares.

Antisatellite (Asat) Weapons

As the superpowers and their allies grow more dependent on satellites for their intelligence and military decision-making, these spacecraft become important targets in time of war. Both the Soviet Union and the United States realize the importance of protecting their own space hardware and of having the capability to destroy that of the another.

The Russians were the first to test antisatellite weapons. In the fall of 1968, when America was racing toward its first manned landing on the moon, the Soviets launched two Cosmos satellites that could maneuver in orbit. Western experts believe these satellites were exploded

ABOVE: *An F-15 launches an antisatellite missile, a two-stage homing rocket, toward its target satellite in orbit. A kinetic-energy weapon, it will destroy its target by force and speed alone.* Courtesy LTV Corporation, Department of Defense photo. FACING PAGE TOP: *The interceptor portion, called the miniature vehicle, homes in on its target in this artist's conception by Richard Flores.* LTV Corporation. FACING PAGE BOTTOM: *This telescopic video image shows the impact explosion in orbit and the expanding debris.* U.S. Army photo, courtesy Department of Defense.

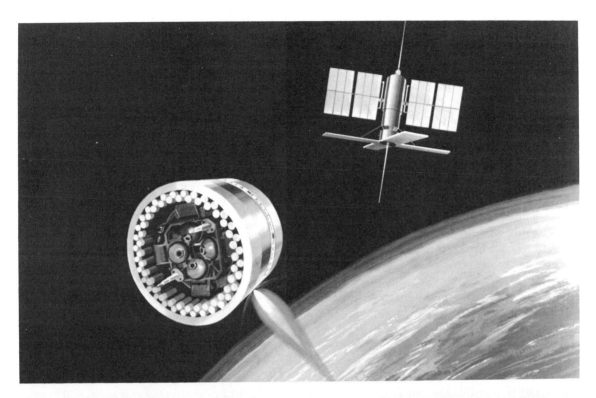

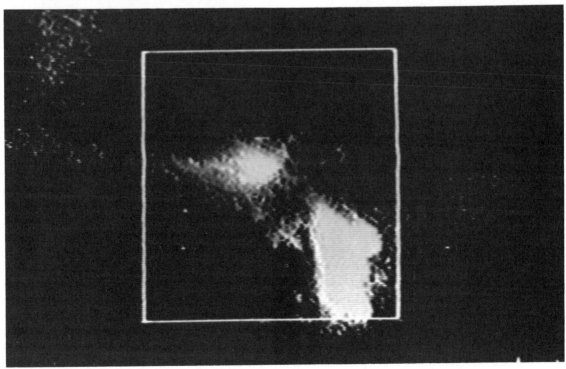

some 300 miles (483 kilometers) above the earth as they passed a target satellite. Then, in the late 1970s and early 1980s, the Russians tested a new-generation Asat that tracked down and destroyed its target.

The United States has also developed an antisatellite capability and publicly demonstrated its older-generation hardware in the mid-1980s by successfully firing Asat missiles from airborne F-15 fighter planes at targets in orbit. Once the F-15 reaches an altitude of 60,000 feet (18,288 meters), it launches a two-stage homing rocket into space. This system, which locates its target with an infrared guidance system and then destroys it by direct impact, without explosives, is faster and more versatile then the Soviet Union's direct-ascent antisatellite system. Both these systems, however, can reach only low-orbit satellites and cannot knock out the early-warning satellites in the higher geosynchronous orbits.

The fact is that any nation that possesses intercontinental missiles has an antisatellite capability, but antisatellite weapons can be based on a variety of technologies, including ground- or air-based interceptor systems, space mines, and many of the weapons being studied in the research phase of the Strategic Defense Initiative, such as the directed laser. An American reconnaissance satellite, capable of detecting Russian ground-based laser weapons, will be launched by the space shuttle before 1990.

Although efforts are underway to negotiate an antisatellite treaty between the superpowers, there is serious doubt that such a treaty could be verifiable. Some experts believe that the growing strategic importance of military satellites may make the negotiation of an Asat treaty impossible. The Reagan administration, believing that the Soviets are well ahead of the United States in antisatellite testing, ordered continued testing during the 1980s. The administration's position was that a comprehensive antisatellite ban was neither verifiable nor in the national security interest of the United States.

A study by the Office of Technology Assessment in late 1985 confirmed that the Soviet Asat program was a threat to some U.S. military satellites and warned that the threat could become much greater if controls and limits were not negotiated between the Soviet Union and the United States. Because of its far-flung military outposts, the United States is more dependent on satellites for its military communications network. The study suggested that the United States could possibly protect its strategic satellites without necessarily building new antisatellite systems. Two alternatives were given: build and place in orbit backup satellites that would be used on a standby basis; and electronically jam the Soviet Asat attack vehicles. The future of antisatellites, their technology and their deployment and use, is dependent on the research years of the Strategic Defense Initiative and its Russian counterpart, and the negotiations between the superpowers.

Future Space Weapons

For tens of millions of moviegoers around the world, future space weapons seem real because of the high-technology skills of special effects wizards. Dozens of Hollywood productions, especially the great *Star Wars* trilogy, have dazzled our eyes and vibrated our eardrums as intensely bright lasers flash across the screen delivering their split-second destruction to the enemy. These images remain embedded in our brains, and we all think we know what laser weapons are after seeing them so many times in the media.

But wait. Operational laser weapons do not even exist now. At least that is what the general public is told, and there seems to be little reason to speculate that some people with top-secret clearances know differently. This seems especially true in the face of the Strategic Defense Initiative, which is charged with the responsibility of researching the weapons of the future. In the research years of this program alone, the United States will spend close to $30 billion to find out if lasers, other directed-energy systems, the so-called smart-bullet weapons, and the complex, beyond the state-of-the-art computer, guidance, and power systems to support them can actually defend the United States and its allies against attacking nuclear missiles. What are the most likely weapons systems to be produced if President Reagan's strategic defense is approved by Congress to enter the development phase? Many different weapons are being discussed, researched, and tested, but no one knows for certain which weapons concepts will fall by the high-tech wayside and which may eventually go into development. The weapons that have been openly discussed in the media, by researchers, and by the opponents and proponents of a strategic defense program are space-based, air-based, and earth-based. It is not where such future weapons are located that is important in the overall strategic defensive scheme, but rather which combination of weapons systems most effectively and economically destroys the attacking ICBMs of the enemy. What weapons offer the best and most perfect defense against the thousands of real missiles and decoys that would form the so-called threat cloud of an all-out offensive attack? Billions of dollars are now being spent to answer this question.

Aiming at Trajectories

Three chances to hit the target are better than one, and this basic concept has been incorporated in a three-tier defensive system. In theory, enemy missiles can be destroyed during any one of three phases of their trajectories:

• During the boost phase, when the missile is launched and climbs out of the atmosphere (lasting three to five minutes)

• During the mid-course phase of flight, when the warheads separate from the missile (lasting about twenty minutes)

• During the terminal phase, when the warheads reenter the atmosphere (lasting about two minutes)

For the most part, the weapons technologies undergoing intense research are directed at the

first two phases of an ICBM's trajectory—the boost and mid-course phases. No military planner wants to run the risk of being at bat with two strikes, especially when a strikeout means that a nuclear missile gets through to its intended target and us—the living "soft targets."

Various types of lasers, space-based and ground-based, electromagnetic rail guns that can shoot computer-guided projectiles (smart rockets), and subatomic particle beam weapons are often mentioned as the weapons under research that would be capable of destroying missiles during these first two phases of the trajectory. Because the reaction time to initiate the defensive weapons is measured in seconds, some experts believe that the necessary speed would remove direct human decision-making and rely on a new generation of computers and their software created before the actual event—software that, because of its complexity, would be created by other computers and that would never be tested except at the time of the ultimate showdown. The trajectory flight times of nuclear missiles do not allow for software glitches or debugging procedures. It would take only thirty minutes to have the nuclear nightmare come true.

The Laser Weapons

Corporate and military research centers have been developing lasers and their potential as weapons for well over a decade. In the late 1970s, the Army Ballistic Missile Defense System Command contracted with Lockheed Missiles & Space Company to research and develop a killer laser system. At about the same time, the Army gave TRW a contract to develop a high-intensity antiarmor laser.

Today, during the research years of the SDI program, laser research has greatly intensified, and a number of systems are the focus of detailed development and testing. From the technological viewpoint only—not the all-important political one—there would appear to be no doubt that in the near future powerful and accurate space-based laser beams will be able to streak earthward toward enemy missiles at the speed of light, lock their hot beams onto the outer rocket skins, and burn holes in the fuel tanks, causing the rockets to explode and leaving the debris to fall to earth. In theory, some of the enemy rockets would be destroyed before they could complete their climb through the atmosphere and release their warheads. The accuracy of such a laser system must be to within a few feet (less than a meter) over a distance of several thousand miles (kilometers).

Such lasers could be based in space, where they would need guardian satellites called porcupines to defend them, or they could be based on the ground, with mirrors deployed in orbit to reflect and redirect their beams to enemy targets. Even before laser weapons have reached the point in development when they will be considered reliable antiballistic missile weapons, they could be used as antisatellite weapons at ranges as great as 22,000 miles (35,400 kilometers), reaching all the way to geosynchronous orbit and the military satellites stationed there.

As with everything else connected with the Strategic Defense Initiative, the experts disagree about whether or not such weapons conceived as defensive in nature would make efficient offensive weapons. One side claims that it would be too costly and that many other alternative and already available forms of weapons could perform any offensive task better; others claim that lasers would almost inevitably become offensive weapons capable of shooting down airplanes, setting tens of thousands of fires simultaneously in enemy territory, or even assassinating groups of leaders gathered together in one location.

The lasers under study during the research phase of SDI include various types: chemical, X-ray, free-electron, and excimer.

Chemical Lasers. These would be space-based systems capable of knocking out enemy missiles during the few minutes of their boost phase. The chemical laser depends on highly reactive elements, fluorine and hydrogen, which create a combustion fuel. It is fed into a mirrored resonator cavity, where intense beams of coherent

Artist's concept of a space-based laser, a directed-energy weapon that is under intensive research. In theory, such a system could intercept enemy ballistic missiles in their boost phase before they deploy warheads and decoys. Courtesy Strategic Defense Initiative Organization.

light are extracted from the chemical reactions. A new technology called adaptive optics helps to aim and amplify the laser light. The amplification process of the mirrors and resonant chambers has been compared to a violin body in the way it picks up and amplifies the string vibrations.

The U.S. Air Force began secret chemical laser testing under the code name Sigma Tau in 1976, but the program has since been declassified. This laser produced low-energy infrared light beams that were not powerful enough to destroy a missile. MIRACL and Alpha are two other chemical laser test programs currently underway. In late 1985, the Navy's MIRACL (an acronym for mid-infrared advanced chemical laser) high-energy laser was successfully tested against a Titan 1 missile casing in a ground test at a laser test facility at White Sands Missile Range in New Mexico. The pressurized second stage of the Titan 1 exploded after the laser burned a hole through it in several seconds. A

carbon dioxide laser at Los Alamos also has vaporized a tiny gold target with its 25 trillion watts of power. It is also likely that SDI research will discover entirely new, powerful lasers based on different chemical reactions and pumping mechanisms.

X-ray Lasers. This laser concept, the brainchild of Edward Teller, so-called father of the hydrogen bomb, did more than any other proposed technology to bring about the Strategic Defense Initiative. It is currently out of favor, however, because the strategic defense plans emphasize all nonnuclear weapons, and the X-ray laser depends on the explosion of a "small" nuclear de-

Sentries on high. These chemical laser systems might be able to destroy enemy missiles with their intense beams of coherent light. The beams are extracted from chemical reactions. Courtesy Lockheed Missiles & Space Company, Inc.

Before and after a laser test. The Navy's chemical laser test program for its MIRACL (mid-infrared advanced chemical laser) was test-fired successfully against a pressurized second stage of a Titan 1 in 1985. Courtesy Department of Defense.

vice. The weapon was developed at Lawrence Livermore Laboratories and was also referred to as the bomb-pumped laser. It was first test-fired beneath the Nevada desert in the fall of 1980 under the code name Dauphin, and it has been tested several times since. The latest test, code-named Goldstone, took place 1,800 feet (549 meters) beneath the Nevada desert on December 28, 1985.

Teller considered the physics of the weapon "technically sweet" and reviewed the technology with Hans Bethe, whom he hoped he could win over as an advocate of the weapon and its defensive capabilities. Bethe admitted that the technology would work, but he later became one of the most outspoken opponents of SDI.

Although the X-ray laser is no longer the centerpiece weapon of the new defense strategy, it is still being actively researched. The heart of this weapon is a small advanced nuclear bomb, which is surrounded by metal rods. If the system is deployed in space and surveillance satellites detect an enemy attack with nuclear missiles, the bomb will be detonated and all of the metal rods, acting like gun barrels, will instantaneously shoot powerful X-rays at the attacking missiles just before the rods are destroyed by the explosion. The powerful X-ray radiation, traveling at almost the speed of light, produces a destructive shock wave inside the missile and blows it up near the end of the boost phase. The X-ray laser is the most destructive of the proposed laser weapons because it has the shortest wavelength, and the shorter a laser's wavelength, the more powerful destructive energy it can deliver to a target. Current research includes attempts to find ways of powering X-ray lasers without nuclear devices.

Teller, to get around the 1967 agreement that bans nuclear weapons in space, proposed a pop-up version of the X-ray laser that could be launched from submarines after an attack was confirmed by intelligence satellites. The X-ray laser battle station would quickly climb to an altitude of 650 miles (1,046 kilometers), aim its laser rods, and explode, sending the deadly radiation beams toward the attacking missiles.

While the emphasis of SDI research is now on

nonnuclear weapons, 10 percent of the multi-billion-dollar research phase is still used for nuclear research. It is difficult to say if the deemphasis on nuclear space-based weapons is more public relations than a real criterion of the program. Even though the X-ray laser has fallen from prominence, primarily because it would appear irrational to achieve SDI's major goal of making nuclear weapons "impotent and obsolete" by using new space-based nuclear weapons, its size and weight are an important advantage. It could be easily carried to orbit in the space shuttle's cargo bay. For other, larger systems, including chemical lasers, entire new launch systems would have to be designed and built. No one should be at all surprised to see detailed engineering drawings of the X-ray laser published some time in the twenty-first century, after one of the major declassifications takes place.

Excimer and Free-Electron Lasers. Both these lasers have short wavelengths, making them more powerful and destructive than lasers with longer wavelengths. Unlike some of the chemical lasers, however, both are still in early stages of development and require extensive research.

The excimer laser (an abbreviation for "excited dimer," which means two linked atoms) would be ground-based because its generating apparatus would be extremely bulky. Large, lightweight mirrors in space, like those studied in the Lamp program by the Defense Advanced Research Projects Agency, could reflect and direct laser beams from the ground to enemy targets.

This laser is powered when gas molecules are excited by an electrical discharge, forming unstable compounds. These compounds then break down and give off light in the process.

Because such lasers would operate at short wavelengths and therefore be extremely powerful, they could destroy a missile in a fraction of a second and therefore operate in a pulsed mode. For an effective defense capability, there would have to be enough of these lasers on the ground to overcome the problem of cloud cover. Another problem to be solved by researchers is the ef-

fect of atmospheric perturbations on the beam. The atmosphere is so turbulent at certain altitudes that the beam of an excimer laser would lose its power over long distances. Research is underway to correct this by using what is called deformable optics to compensate for the deformation of the wave front that is caused by the atmosphere.

Free-electron lasers, if development proves them to be as extremely efficient at short wavelengths as theory predicts, will be good candidates for deployment in space. The power of this type of laser depends on fast-moving electrons that are violently agitated by so-called wiggler magnets. In this state of agitation, the electrons give off light. Progress on this laser has been encouraging.

It will be an equal challenge to find an economical way to feed these energy gluttons of the laser world. The tremendous power needs of the excimer and free-electron lasers may block their way into the real weapons world.

Particle Beam Weapons

Physicists at the Lawrence Livermore National Laboratory, Los Alamos Laboratories, the Naval Research Laboratory, and Sandia National Laboratories are experimenting with particle beam accelerators as space weapons of the future. Some experts consider beams of electrically neutral particles to be the ultimate space weapon because they are potentially more damaging than lasers, and, traveling at the speed of light, they would destroy their targets almost instantaneously. Some scientists also contend that they are useful only for defensive purposes. The other important advantage with particle beam weapons is that they would be extremely difficult to shield against because of the tremendous penetration power of their streams of accelerated subatomic particles. Even low-beam currents, experts believe, could probably disrupt or damage the sophisticated electronics of a missile or warhead, and high-beam currents could melt the warhead and detonate its high explosive. Scientists are also considering the use of such particle beams to discriminate real warheads from the decoys.

Inside a space-based particle beam accelerator, electrons would first be added to hydrogen atoms, creating charged particles (ions). They would then be put through electric fields, which would clump them together and speed them up before they received another magnetic boost in the linear accelerator. With more magnets and lenses, the beam would be focused and bent. Last, the beam of charged particles would be sent through a rarefied gas that would strip away the extra electrons making the beam neutral. If this were not done, the path of the beam would be strongly affected by the earth's magnetic field.

Giant machines such as the Advanced Technology Accelerator at Lawrence Livermore, the particle-beam fusion accelerator (PBFA-1), renamed Saturn for its SDI research function, at Sandia, and the newer RADLAC-2 linear accelerator, also at Sandia, shoot powerful pulses of electrons (sometimes delivering as much as 50 million electron volts at 1 second intervals) at test targets. For just a fraction of a second, this equipment can deliver more energy than the total electrical generating capacity of the United States.

The future of particle beam weapons faces some serious problems. More compact accelerators must be created, as it would be wholly impractical to boost the present-day mammoth

hardware into orbit. The huge power needs of such weapons is also a real concern, and possible enemy countermeasures require serious study. However, one important breakthrough in beam weapons research was accomplished in 1985 when the problem of beam guidance was solved at Sandia National Laboratories. To keep the beam from arcing and snaking as it makes its way through the earth's atmosphere or an equivalent test-gas medium, scientists employed a low-power laser to guide the electron

Another SDI weapons system under research is the space-based electromagnetic rail gun, which shoots non-explosive projectiles (smart rocks) to destroy enemy targets. This is the most developed of any of the so-called Stars Wars weapons systems. Courtesy LTV Corporation.

beam to its intended target. This success may speed up the development of particle beam and other directed-energy weapons that are undergoing intensive research in the first phase of the Strategic Defense Initiative. By the early 1990s, we should know if space-based particle beam weapons can or will be built.

Electromagnetic Rail Guns and Smart Rocks

This hypervelocity space gun has the great advantage of a basic and simple design. Along with the nonexplosive projectiles that it would rapidly shoot one after the other, this space weapons concept is the most developed of any of the Star Wars weapons systems.

By creating an electromagnetic force between two parallel rails, where an electric current interacts with a magnetic field, this rail gun could shoot off projectiles (called "smart rocks" because of their built-in sensors and mini-computers) at speeds of 23,356 miles (37,580 kilometers) an hour. Firing rates could be as high as sixty shots a second, and experts in this technology believe that projectile speeds of 223,000 miles (359,000 kilometers) an hour are possible and can eventually be obtained. The range of these space guns could be as far as 3,000 miles (4,827 kilometers), and they therefore could destroy enemy missiles and warheads in the boost or mid-course phases of flight.

Gerold Yonas, the chief scientist of the Strategic Defense Initiative Organization, keeps an example of this technology on his desk: a 1-inch (2.54-centimeter) thick aluminum slab that was pierced through by a small, 10-gram plastic pellet traveling at 4,471 miles (7,194 kilometers) an hour. Whatever the shape and size of the smart rock projectiles (and Robert Jastrow has suggested that they would resemble a 2-pound, 0.9-kilogram, coffee can), they would deliver more nonexplosive impact force than their equivalent weight in TNT.

If the smart rock—with all its sensors and microelectronics—weighed almost 18 pounds (8 kilograms) and was catapulted by the electro-

Another nonnuclear, kinetic-energy weapons system is shown here. It would use chemically propelled interceptors to destroy missiles in the boost phase. Courtesy LTV Corporation.

magnetic space gun at a velocity of 22,356 miles (35,971 kilometers) an hour, then the force of its nonexplosive impact would be equal to at least 220 pounds (100 kilograms) of dynamite. The space rail gun could fire a steady stream of computer-guided smart rocks to their target missiles at these extremely high velocities.

Such in-orbit weapons would be large and vulnerable, however. Guardian (so-called porcupine) satellites would have to protect these weapons from enemy attack, but this space weapon is politically sweet because such rail guns use no real explosives—just the impact force of the smart rocks. Even the protector porcupine satellites would use high-speed rockets against any enemy attack on the space guns. These rockets, also considered kinetic-energy weapons, could defend the space battle satellites against the enemy by their impact force alone, without explosives. Called kinetic kill vehicles (KKVs), they could also carry explosives to use against certain targets, but they are a very different class of space weapons from the laser and other directed-energy forms currently being researched.

While the electromagnetic rail gun and its smart rock projectiles are further along in development than most of the directed-energy systems, there still are some serious engineering problems to solve. The extremely small guidance and control system must be developed to withstand tremendous acceleration forces, and

the space-based power system must be able to deliver hundreds of megawatts of electricity to the electromagnetic rail guns, enabling them to shoot their smart rocks at the "threat cloud" of enemy missiles.

Experiments with rail gun weapons are being conducted by the Air Force and the Defense Advanced Research Projects Agency. Officials estimate that a ground-based demonstration system can be built by 1990.

Space Battle Management

Operational space weapons for strategic defense against an attacking ICBM force, no matter what the final mix of space-based and ground-based systems, also require complex systems to support them. Three essential supporting systems, some portions of which are currently beyond the state of the art, must be developed and perfected if a strategic defense system is ever to be deployed.

Launch Vehicles. Because of the large scale of such a space-based weapons system, a launch system to supplement the space shuttle fleet, with a larger payload capacity, would have to be developed. Many such unmanned expendable rockets are now under consideration or development as a result of the *Challenger* disaster. An expendable launch vehicle, more powerful than the late Saturn 5 and able to orbit a payload of about 300,000 pounds (136,000 kilograms), would be an essential part of the deployment phase. The new launch vehicle would have to cut payload costs by a factor of ten. If the space shuttle delivers 1 pound (0.45 kilogram) of payload to orbit at a cost of $3,000 (only one low-end estimate among several higher ones), then the new vehicle would have to deliver it to orbit for only $300. A presidential directive has ordered the study of such a rocket. If SDI research results in cutting the payload-to-earth orbit costs by a factor of ten, it could have a profound effect on space commerce and tourism, especially if the weapons are never deployed.

Computers. An integrated system of space weaponry would require supercomputers whose abilities far exceed those of the fastest available today—the Crays and the Cybers. Some five billion operations per second for the spaceborne imaging systems will be necessary, which is roughly twenty times faster than many of today's supercomputers. The fastest supercomputer in the world, the Cray-2, which is designing the next generation of airplanes and spaceships, has a computational rate of two billion calculations every second—three billion a second short. Other space weapons systems may require speeds that are equal to two thousand older-generation Cray supercomputers. It has also been estimated that there would have to be ten million lines of error-free code for the software to coordinate the complex second-by-second operations of the entire system. The last nine minutes of the computer-controlled space shuttle countdown, in comparison, has some 88,000 lines of code. Just ten years ago, such software would have been impossible, but the computers of today will help us create the software of tomorrow. By 1999 there will be more than one billion computers on or above planet Earth, and a lot more software to run them. But when it comes to space battle management, the software will have to be perfect: "We don't want a few lines of bad code mistakenly setting off a nuclear weapon," one SDI scientist said seriously, while some members in the audience laughed. Some computer scientists believe that the software challenges can never be met because they can never be tested except in a real war.

Space-Based Radar. Another essential support structure for the weapons system in space would be large radar platforms, perhaps the size of several football fields. Aerospace corporations have been studying such space-based radar systems for over a decade for the Defense Advanced Research Projects Agency, and it is hoped that small-scale versions can be tested on the space shuttle by the early 1990s.

Such large radar structures, which would be capable of imaging thousands of distant targets simultaneously, are essential for any space-based defense system. The size is necessary to provide the quality resolution needed to track what could be thousands of warheads, real ones and decoys.

Power in Orbit. There are big power requirements for energy-directed weapons such as lasers and particle beams, and for kinetic weapons such as the electromagnetic rail gun. If the battle management system does not have an abundance of reliable space-based energy, the most advanced space weapons become impotent.

Both nuclear and nonnuclear power sources are needed to support SDI technology. Such systems must be capable of supplying 5 to 10 megawatts of continuous energy and as much as 100 megawatts in brief bursts for certain beam and kinetic weapons.

The SP-100 nuclear space power research program, the system most advanced in research and development, is jointly sponsored by the Department of Energy, the Department of Defense, and NASA. In 1985 the decision was made to use thermoelectric conversion technology, which proved successful in earlier nuclear space power systems, but the 100 Kilowatts is probably beyond the capacity of proven thermoelectric material, and the engineering challenges are formidable. A functioning SP-100 system for ground testing should be completed by 1991. The megawatt power systems will follow about five years later, and it is projected that qualified hardware would be available by the mid-1990s.

Whatever the future of SDI, all these various research efforts will stimulate technology in general and eventually be transferred to the private sector, where there will be beneficial spin-offs, as there were during the Apollo years.

Strategic Defense Beyond 1990?

As the initial research years of SDI continue until 1990, the debate over the future of strategic defense weaponry will remain heated and in full view of people around the world, including the Soviets, who have their own active weapons research program. Opponents will cite all the possible countermeasures that the Russians could take to make a deployed space weapons system dangerously ineffective: increase their ICBM booster production; build more cruise missiles; develop highly reflective paint for their warheads that would reflect laser beams; spin their missiles so that laser weapons are not able to quickly burn through the missile skin; or shorten the boost phase of their ICBMs, which could make destruction impossible during this critical phase, before the warheads and decoys are released. They will also talk about the mind-boggling costs of an operational system, on the order of $1 trillion, which even the United States cannot afford, and the arms race in space that weapons development would ensure. Carl Sagan, one of the most well-known and outspoken critics of the Strategic Defense Initiative, has succinctly summed up the opponents' position: "In short, strategic defense cannot protect the U.S. in a nuclear war, can be overwhelmed, underflown, and outfoxed, flagrantly violates treaties America has solemnly ratified, is ruinously expensive, jeopardizes space exploration, and increases the chances of nuclear war. Other than that, it's a great idea."

The proponents of a space-based defense will emphasize that it is only a research program at present and will cite successful tests of the space-based technology, such as the space shuttle experiments with ground-based lasers that proved their tracking ability of spacecraft in orbit or the mirror relay of laser beams to test targets in space. They will talk about the ultimate goal of President Reagan's initiative: to rid the planet of nuclear weapons so that humankind no longer has to live under the threat of nuclear madness and the extinction of the human species and all other life on earth. As President Reagan said in his speech at the National Space Club in 1984: "Let us move on to a happier chapter in the history of man." No one, of course, will argue with that goal; too many nightmares have been a part of living under today's military strategy of mutual assured destruction (MAD).

Edward Teller, the man behind Reagan's initiative, believes that a space-based defense system, even moderately efficient, will have a strong deterrent effect when combined with some retaliatory ability. If the West had effective defense weapons in space, Teller argues, this would force the Soviets to reduce the burn phase on their thousands of ICBMs, which in essence would make their entire rocket arsenal obsolete, forcing them into extremely costly expenditures. Many experts disagree and are convinced that this would lead to a dangerous arms race in space.

Proponents believe SDI will lead to a safer planet; opponents believe it will lead to a more dangerous one. Both sides make, for the most part, convincing and intelligent arguments, but there are still many people who are not listening at all. Why bother, after all, if what we read in the media is just a watered-down, unclassified, public-relations version of reality? This view is reinforced by statements such as that by physicist Lowell Wood of Lawrence Livermore Laboratory, the intellectual heir of Edward Teller, who has said, "The things most discussed in public are the ones the government is least interested in." Some people ask: Why state an opinion on a partial truth?

There is, however, a common ground for both the opponents and proponents of the Strategic Defense Initiative: negotiations and political solutions.

Politics and the Silver Wings of Science

While the historical record up to the Iceland meeting in October 1986 keeps SDI proponents extremely skeptical about any such political breakthroughs, they certainly would welcome them. President Reagan's first science advisor, George A. Keyworth, is convinced that U.S. strategic defense options are necessary for any real progress in arms control, and the head of the Strategic Defense Initiative Organization, Lieutenant General James Abrahamson, has stated that the United States must support the antiballistic missile treaty with the Soviets during the research phase of SDI and keep open political channels such as the superpower summit meetings in Geneva and Reykjavik. The opponents of SDI, including Edward Teller's colleague in the Manhattan Project, Hans Bethe, are convinced that only a political solution will allow for nuclear arms reductions and a safer world.

Donald Kerr, Director of Los Alamos Laboratories, which is always at the forefront of weapons research, expresses the concern and opinion of many on both sides of the SDI fence when he says: " . . . the issues which are really at the core of this confrontation aren't amenable to technical fixes. Science provides new opportunities, but it doesn't get to the heart of the political and social issues." Today's issues, he believes, are the same as they were twenty years ago, in the late 1960s: missile defense and the modernization of strategic forces. But the leadership's understanding of these complex issues is no greater than it was two decades ago. "We don't," Kerr says, "deal with the fundamental issues that cause us to deploy these arms."

Global politics and the outcome of negotiations between Russia and the United States, Mikhail Gorbachev and Ronald Reagan, in the last half of the 1980s—before a U.S. decision is made to begin development of space weapons—will decide the fate of weapons in space in the twentieth century. We all must hope that the growing prospect of costly, advanced weapons technologies deployed in space, systems that no nation or group of nations can afford, will encourage a political breakthrough at the bargaining table, one that addresses the fears and needs of both superpowers and their allies. The result could be the dismantling of nuclear weapons and other weapons of mass destruction. If this hopeful scenario comes true, then the human spirit may breathe a deep sigh of relief at last. We can get on to more productive human ventures, using some of the newly forged technologies from SDI research to help alleviate human suffering on planet Earth. And if it is challenges the scientists and engineers want, then let them begin planning and launching some international expeditions into the solar system, first to Mars, then outward toward Jupiter and Saturn.

Is this another naive dream? No more so than going to the moon was in the early 1960s. Wars, be they fought on earth or above earth, by soldiers or robots, are fast becoming unacceptable to the majority of people on earth; more importantly, they are becoming unacceptable to those who hold the political power. If wars cannot be won today, why fight them?

Perhaps the 4.8 billion "soft target" people on earth can someday escape and feel safe from the constant "threat cloud" of thousands of nuclear missiles storming down on their countries and cities and bodies. Perhaps the space above their heads will also be clear of weapons capable of launching smart rocks or blasting laser and particle beams across their skies. Can humankind keep its space weapons and wars limited to the special effects of film and TV fantasies?

"The next stone age may come on the silver wings of science," Winston Churchill once remarked. The leaders and people of planet Earth must prevent this from coming true.

11

Settlement
& Colonization
Of Space

The Spaceward Expansion

T hose thousands of human footprints on the moon's surface, which will remain for some ten million years, are only the first steps in humankind's exploration of the solar system.

"We will return to the moon," said Neil Armstrong on the fifteenth anniversary of the historic moon landing in July 1984. He is not alone in his forecast. James M. Beggs, NASA's head administrator during the first half of the 1980s, also believes that we will return to the moon within the next twenty-five years. And Buzz Aldrin, the second human to set foot on the moon, has been working hard for years to promote a permanent manned moon base early in the next century.

A return to the moon after the space station is built is only the beginning of a well-planned and integrated expansion into the solar system during the next one thousand years. Mars still beckons, as it has for one hundred years, and ships and crews will set off for the Red Planet sometime in the twenty-first century. It is very likely that humans will walk on the red Martian soil by the year 2035. And after Mars, peopled rockets will thrust toward the outer planets of Jupiter and Saturn and explore their many moons until one or more are found that will sustain a base that can be self-sufficient—another foothold in the solar system.

Besides the manned missions to select planets and moons, there may be built the large space habitats, often called space colonies, that will be, depending on their size, the productive villages, towns, and cities on the frontier of space in the third millennium. Then there is the concept of planetary engineering, which may be realized in the more distant future. Even today studies indicate that humankind may someday control the technology and power to change the face of uninhabitable planets or moons into more earthlike places, new worlds where people can live and work and bear children who will continue the migration to even more distant worlds. The spaceward expansion is limitless. Today, with hundreds of exciting and imaginative ideas, there is enough work to keep our species busy for several thousand years.

The Moon's Bottom Line

Why return to the moon, to what Buzz Aldrin called "magnificent desolation" when he first stepped out of the *Eagle* mooncraft and became the second human to walk on the surface?

There are many valid and important scientific reasons to build a permanent manned base on the moon, but after the Apollo program, expensive science, no matter how worthy and important, does not have the persuasive power that good, bottom-line economic reasons do. Such a bottom-line argument is found in one of the most important resources of the moon, the oxygen trapped in the lunar soil and rock. It alone offers a strong incentive to construct a productive moon base that can eventually be largely self-sufficient. Oxygen is needed to fuel the rockets, create the atmosphere for space stations and factories, and fire up the industrial furnaces that will be important to the expanding above-earth economy.

The Apollo missions told us what the moon was made of: anorthosite, ilmenite, plagioclase, and silicon, to name a few minerals. Such knowledge was one of the most important legacies of the Apollo voyages. We now know that lunar rocks and soil can be processed to produce aluminum, titanium, iron, magnesium, silica, and oxygen. Most lunar rocks are composed half of oxygen, and oxygen is the precious commodity that life itself depends on—life on earth and life in space.

To produce the oxygen, moon base operations will focus on mining the mineral ilmenite, which will be separated out from other strip-mined lunar materials. It will then be chemically reduced by heating it in the presence of hydrogen, which must be imported from earth. By this method, the automated refining plant on the moon will produce iron, titanium oxide, and

ABOVE: *"We will return to the moon," said Neil Armstrong on the fifteenth anniversary of the historic moon landing in July 1984. He is not alone in his forecast.* Painting by Pat Rawlings/EAGLE ENGINEERING. Courtesy NASA.
PRECEEDING PAGES: *Whatever the design, giant wheels or cylinders, colonies could rotate to produce artificial gravity on their interior surfaces. Zero-gravity areas would be at the hub of the wheel or the center of the cylinder. For recreation, there might be low-gravity swimming pools, where one could swim in a complete circle and see others swimming above one's head or, if diving, below one's feet.* Courtesy NASA.

water. Some of the water will be pumped into an electrolysis facility, which separates the oxygen and hydrogen. The oxygen will then be liquefied and stored in tanks, ready to be exported to the space stations in earth or moon orbits.

Why go all the way to the moon for oxygen when there is plenty available right here on earth? Because it will be cheaper to produce and transport from a moon base. The weak gravity of the lunar surface and the development of inexpensive transport rockets that will take the oxygen to where it is needed above the earth will make it less expensive than if it had to be rocketed off the earth in huge heavy-lift vehicles. The same economies will hold true for other construction materials processed at the moon

base. Those that are used to build the moon facilities and make them more self-sufficient will also be exported to support the growing above-earth, space-centered economy. Metals that are abundant in lunar soil and rock—aluminum, magnesium, chromium—can be processed in a moon-base smelting plant and transported to the centers of commercial space industry for use as construction materials. Our return to the moon in the first decades of the twenty-first century will not be primarily to seek answers to the origin of the solar system; instead we will be looking for economic resources and products to supply the expanding high frontier. And when the moon base becomes a reality, it will have been the lack of gravity that made it all possible.

Earthtown, Moon

Studies by NASA and independent space experts show that the moon base is the next logical step after the permanent space station is orbiting the earth in the 1990s. Much of the same technology developed for the space station can be used to develop the lunar base. The pressurized modules designed for the space station can, with certain modifications, house the workers and scientists on the moon. The same is true for the space tugs, now called orbital transfer vehicles by NASA. The same rocket engines designed to take satellites to high geosynchronous orbit from low earth orbit will be able to ferry supplies and people to and from the moon. These economies of scale in space will eventually make the earth-moon space basin, with its well traveled routes of commerce, a busy and profitable place.

Just as the space station will grow in size and function year after year, so too will the moon base evolve from an outpost base of operations to a village and then a town on the moon. The first moonsteaders will lead a spartan life. After the initial exploratory mission, during which the astronauts and their equipment will land and explore the surface for thirty days to choose the best site for the moon base, cylindrical living modules and large soil movers will be brought to the lunar surface by unmanned descent rockets. The soil movers will first be used to bury the living and laboratory modules under more than six feet (about two meters) of lunar soil. This will protect the inhabitants from solar radiation, including the dangerous solar flares that can erupt at any time. Except for their activities on the surface, the moon workers and scientists will live underground. In the harsh, brilliant sunlight of moon's daytime, their buried habitats, thickly covered with heaped moon soil,

A lunar-supply space tug in orbit above the moon. It has transported start-up consumables for the moon base. Painting by Pat Rawlings/EAGLE ENGINEERING. Courtesy NASA.

will look like mysterious Indian mounds of a new world.

One of the first important tasks will be to set up the power system for the moon base. Most experts agree that the SP-100 space nuclear reactor program currently under development by the Strategic Defense Initiative will provide the technological base for the power station. Once power is plentiful, the mining operations will begin. Soil movers will scoop up the regolith, leaving shallow pits, and pilot facilities for oxygen extraction will be built. Additional pressurized modules will be brought to the lunar surface and filled with moon soil, and corn and wheat will be grown. Eventually the moon base will become more self-sufficient, but nitrogen and water will still have to be imported from earth. Yet some scientists suspect that even water may be discovered on the moon, trapped as ice in the regions near the lunar pole that are permanently in shadow. While it is hard to believe, right now there is more complete photo documentation of the planet Mars than of our own moon, and sophisticated satellites and unmanned probes during the first years of the moon base may discover useful resources in the unexplored regions of the moon that will cut down on earth imports.

As Earthtown grows and its population increases to several dozen, new, more spacious quarters could be provided by melting mounds of lunar soil with focused sunlight and creating a pliable building material that could be sculptured into larger habitats—moon adobes blending into the gently rolling hills of Earthtown.

Moon Science Beyond 2000

Ongoing scientific activities will be integrated into the daily life of the moon base once it is built, just as they will be during the site selection missions and the construction phase. The sun, for example, will be constantly monitored. If it throws off dangerous radiation, base activities will be restricted and people will remain inside their soil-covered moon mounds.

The more ambitious science projects and ex-peditions will probably begin after the moon base reaches an economic break-even point from its exports of oxygen and other lunar resources. One of these will be a wide-ranging geological exploration of the lunar surface that will answer many of the questions remaining after the Apollo expeditions. Scientists at the U.S. Geological Survey believe that a 2,500-mile (4,023-kilometer) expedition across the great Imbrium basin by a team of geologists would provide valuable scientific results. They would travel in a sophisticated moon rover containing equipment that could perform on-site chemical analyses. This mobile field laboratory would also be outfitted with high-resolution remote sensors and a deep-drilling rig with coring capabilities that could drill down as far as 1,000 feet (305 meters). The team would also deploy seismometers, heat-flow sensors, and other instruments along the way that would monitor this large area of the moon over several decades. Such a trek could solve the question of the origin of the moon, but more important for the growth of the base, it could also discover some important new resources that could affect the future of the base.

An exciting prospect for astronomy is to build lunar observatories on the moon that will house both optical and radio telescopes. While the Hubble Space Telescope in orbit above the earth will revolutionize astronomy, opening up the universe, state-of-the-art telescopes on the moon's far side may replace it during the first few decades of the twenty-first century. New limits for radio astronomy would be reached with such an observatory location because the site would be completely free from radio interference from the earth. Ultrasensitive instruments can for the first time be fully utilized in the search for extraterrestrial intelligence. A very low frequency radio telescope could receive radio transmissions that cannot even penetrate the earth's atmosphere. One astronomer has proposed that a huge optical telescope, with an aperture some 82 feet (25 meters) in diameter, be assembled on the moon. Such an instrument would have a light gathering power one hundred times that of the Hubble Space Tele-

An advanced moon base would have mining operations to extract the aluminum, titanium, silica, oxygen, and other resources from the rock and soil. Mass drivers (electromagnetic catapults) could be built to economically launch millions of tons of material that could be processed in orbiting smelting factories. The earth-moon basin, with its active trade routes, could become a busy and profitable place. Painting by Pat Rawlings/EAGLE ENGINEERING. Courtesy Lunar and Planetary Institute, © 1985.

The moon base may eventually grow into a town. Perhaps it will be named Earthtown, Moon. Here is an advanced moontown with its own bank—a branch of Lamar Savings. Building materials can be provided by melting mounds of lunar soil with focused sunlight, then pouring it to create any desired shape. Painting by Pat Rawlings/EAGLE ENGINEERING. Courtesy Lamar Savings, Houston, Texas, © 1985.

scope. It would have the capacity to see details in the cores of faraway galaxies for the first time and to directly observe planets orbiting distant stars. Cosmic ray and neutrino astronomy would also gain a decided advantage if conducted on the moon's surface. If such sophisticated instruments are on the lunar surface, it may be that evidence for other life in the universe will be discovered by scientists on the moon rather than on the earth.

Lunar Sports

As self-sufficiency increases and the moon base becomes larger, leisure-time amenities will be added. One of the first of these, no doubt, will be the lunar sports arena, where all activities will take place in the one-sixth gravity of the earth. New sports could include muscle-power flight, in which large artificial wings would be fitted to the arms. Tennis, if it survives at all, will have new rules, since most hit balls will travel in a straight line at least 200 feet (61 meters) before touching the ground, and the players will not be able to change direction quickly. Swimming would remain familiar, except that in doing certain strokes, for example, the breaststroke, the swimmer will be able to thrust the upper body clear of the water and probably improve earth time records.

The only sports activities on the moon during the twentieth century took place when Alan Shepard of *Apollo 14* threw a makeshift javelin and sent golf balls sailing through the lunar vacuum.

The Stepping-Stone Moon

If the spacefaring nations decide to build large space habitats, space colonies, or power-producing satellites in the earth-moon basin by the middle of the next century, demand for lunar resources will increase tremendously. The moon base, which will have become a moon town by then, will grow dramatically into an industrial lunar city. Mass drivers, electromagnetic catapults already designed on paper and tested in the laboratory, will be constructed to economically launch millions of tons of lunar material that could be processed into orbiting smelting factories.

The moon will become an industrial center if such grandiose space ventures are undertaken. But even if such large-scale space structures must await the twenty-second century for their share of the human activity pie, a lunar base will be essential for other space expeditions and can provide the operational experience needed before some of the planets and moons of our solar system are colonized. The early moon base would be an ideal training environment for the next feat of space travel—a manned mission to Mars. The moon will supply much of the fuel for the Mars ship—liquid oxygen extracted from the gray lunar soil. What will the rust-colored surface of Mars have to offer the space folk from earth on their decades-long, stepping-stone way to the moons of Jupiter and Saturn?

The Backup Planet: Mars

Red Mars has always fascinated earthbound eyes, and its celestial beckonings continue to this day. Even after two robot Viking spacecraft landed their electronic eyes, arms, and microlaboratories on its surface in 1976 and surveyed the landing sites, Mars still invites us to come and see for ourselves. The invitation has been accepted. The plans are being laid. Only the date and time remain to be confirmed.

Why will we go to Mars? Because the Red Planet is the most earthlike of the other planets in our solar system, the "most user-friendly" as one planetary scientist put it. Unlike the moon, Mars is rich in essential elements such as oxygen and nitrogen, and its land area is almost equal to the earth's even though it is only half the size, because there are no oceans or other large bodies of water. We can survive there and build a permanent colony that will eventually become

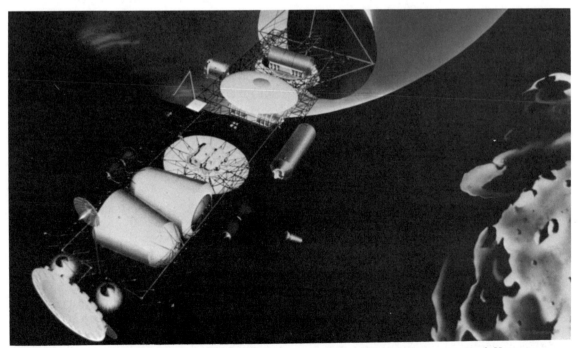

One Mars expedition scenario calls for launching two spaceships, one manned and one unmanned. Here an unmanned Mars freighter is arriving at Mars. This nuclear-electric-propelled freighter can be launched ahead of the manned spaceship and can be waiting for the colonists when they arrive. Painting by Mark Dowman/EAGLE ENGINEERING. Courtesy NASA.

self-sufficient. And because survival is the name of the game, a manned mission to Mars is ultimately a survival mission for humankind—a cosmic insurance policy that could be cashed in if our home planet is ever mortally wounded by human error or natural cataclysm. Within just a few decades, Mars can become earth's backup planet.

It will not be easy to journey to Mars and then set up a permanent base on its surface. It is a "friendly" planet only in comparison to the hostile and hellish environments of Mercury and Venus. But the fact that it will be difficult will make its accomplishment all the more valuable for humankind. Footprints in the rusty Martian soil will symbolize the success of a tremendous human endeavor, but it will be what we learned along the way to Mars, the new technology we mastered, that will yield the treasures for the future.

Our descendants will become the first Martians. Some of today's toddlers will be walking across the rust-colored soil and sand of Mars before 2050. And some of their children will be born there. In the future, hopefully, these Martians will grow up looking forward to visiting an earth that is still beautiful and habitable.

The Journey to Mars

How will we send our first Mars-bound astronauts across the more than 40 million miles (64 million kilometers) that, on the average, separate the two planets? It is no short hop to the moon this time—the distance is more than 160 times greater. The quantities of consumables that must be carried on the Mars mission are staggering. Such a Mars mission ship could easily weigh 1,000 tons. And the outward journey, the *shortest* leg of the mission, will take at least six months.

Despite the challenging problems that must be solved, many and various solutions have already been worked out in detail on paper by space organizations and individuals who believe that the next great space venture will be a manned spaceflight to Mars. Even back in the 1960s, when great Apollo was reaching for the moon, NASA awarded sixty contracts to firms for studies on how an American space effort could head for Mars after the moon. Such studies were halted, however, after Apollo's hasty retreat, caused by the blunt-edged program axes that were swung in Washington, D.C., in the early 1970s. NASA wanted to survive, and so it emphasized missions close to earth. Once the permanently manned space station is operational, in the 1990s, the manned Mars mission will be launched from earth orbit toward the fourth planet from the sun. It could even carry an international crew, including Americans and Russians.

What will the Mars spaceship look like and how many crew members will it carry? Several different mission concepts have been worked out, but one of the more interesting came out of a University of Colorado conference held in July 1984 by a group called the Mars Underground. In this scenario, three self-contained spacecraft, each with its own shuttle vehicle attached, would be assembled in earth orbit. They would each be as long as a twenty-story building is high (some 197 feet, 60 meters) and would carry five crew members each. All three would be launched into their trajectories toward Mars by expendable rocket boosters. Once their escape velocities and navigational headings were established, the spacecraft would jettison the expendable rockets, and all three would rendezvous and join together, forming a Y-shape configuration called a trefoil. Thrusters on the Mars spaceship would then be fired to spin the entire vessel. This motion would create artificial gravity in the living and working quarters equal to one third that of the earth's surface gravity—very close to the gravity conditions on the surface of Mars. This artificial gravity, besides preparing the crew for the surface conditions on Mars, would also have the physical benefits of retarding such negative physiological effects of weightlessness as bone loss, which could be extremely harmful during a long spaceflight.

The spinning, triaxial Mars ship would always have one side facing the sun. This sunward side would contain the fuel tanks, the engines, the power and heating equipment—most of the spaceship's mass. The habitat modules would be on the opposite side, in shadow. Such a design would protect the fifteen crew members from the sun's dangerous radiation, and if a solar flare ever erupted, the astronauts could seek temporary protection in the three shielded shuttle craft attached beneath the three living quarters on each of the ship's spokes. Lethal radiation bursts from the sun remain the biggest safety concern for a trip to Mars, and special radiation shelters may have to be designed and built for the Mars spaceships.

After a journey of six months, the spaceship would arrive, and the three shuttle craft and crews would descend to the surface of Mars, to the Martian village that was built by earlier expeditions using expendable spacecraft to transport the nuclear plant, the components of the refinery plant, the surface habitat modules, and other one-time surface hardware that would eventually make the Martian settlement self-sufficient. Once the refinery was operational and manufacturing rocket fuel, the shuttle craft from the last mission would be refueled, would rendezvous with the in-orbit ship, and would head back to earth with those crew members whose tours on Mars were completed.

Another Mars mission plan has been studied by the consulting firm Space Applications, Inc. It calls for a 500-ton spacecraft with a crew of four to be launched from earth orbit. Separate launches would carry the Martian lander craft and the return-to-earth craft. The return ship would arrive later than the crew and would remain in orbit above Mars. When the crew had completed its investigation, they would blast off from Mars, rendezvous with their spaceship, climb aboard the *Earthward Bound*, and head toward the home planet. The interior of the return ship may be painted in several shades of green, a color these first explorers on Mars would not see much of during their initial exploration of the Red Planet. That would come later, after the Martian dome farms were established. In the meantime, the returning Martian explorers could look forward to seeing the many and varied greens of planet Earth.

A Space Station Named Phobos

Some space experts believe that Phobos, the larger of the two Martian moons, would make an ideal space station from which to launch the landing expeditions to the surface of Mars. While Phobos is small, less than 15 miles (24 kilometers) in length, studies indicate it may be rich in water, carbon, nitrogen, and oxygen. If future space probes prove that the moon con-

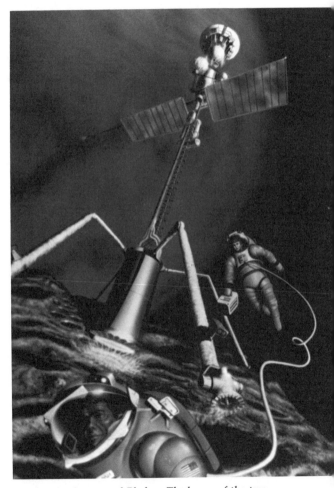

A space station named Phobos. The larger of the two Martian moons, Phobos may be worth its weight in gold, because studies indicate it is rich in water, carbon, nitrogen, and oxygen. It could become a resource-rich space station from which to stage expeditions and base operations on Mars. Painting by Pat Rawlings/ EAGLE ENGINEERING. Courtesy NASA.

tains these resources in abundance, Phobos may turn out to be an almost priceless piece of celestial real estate. If water is bound to the dark surface soil and could be baked out through the use of solar furnaces, this unattractive Martian companion could almost literally be worth its weight in gold because the estimated cost of transporting a pound of water from Earth to Mars is about $10,000.

The potential wealth in resources, along with the fact that, unlike Mars and Earth, Phobos has no gravity well for rockets to climb out of, has spurred scientists to think that this dark-surfaced moon could become a refueling station for Mars missions and could cut mission costs by billions of dollars. A Russian space probe is scheduled to go into orbit around Mars and Phobos in 1989 and to fire a laser beam into the moon, which will vaporize some of its surface and allow instruments to analyze it. If this Russian mission is successful, it should return enough information to earth to help scientists decide if Phobos will become the first interplanetary refueling station in the solar system. If so, Mars will become the gateway to the asteroid belt. And cosmic environmentalists should not be upset because Phobos's orbits are numbered. This little Martian moon has already lived out 98 percent of its life. It is expected to crash into the Martian surface in about one hundred million years.

Choosing the Mars Settlement Site

The settlement site on Mars will not be selected for its scientific value, but rather for access to resources such as water, nitrogen, and oxygen that will make the first human population on Mars self-sufficient and permanent. Because of the two or more years needed for the round-trip to Mars, most studies conclude that a permanent base on Mars is essential—no hit-and-run approach like the Apollo moon landings. This means that there must be a dependable and stable fuel and energy supply, which makes the settlement's location and its geological and chemical composition all important.

Where should the first Mars explorers begin to build their settlement? Some experts think that regions such as the Candor Chasm in the central section of the great canyon Valles Marineris, the largest known in the solar system, offer the layered deposits and geological features likely to hold valuable resources. Such a ridge-and-gully topography with layered sediments is also found in other Martian regions such as the Mangala region, which may have been formed by ancient surging rivers. Another site mentioned is the Olympus Mons region, the largest known volcano in the solar system, which rises 79,000 feet (24,000 meters) above the Martian desert. Different geological layers would be accessible, and magma flows are iron rich. The Mars orbiter probe to be launched by the United States in the early 1990s will certainly return a wealth of detailed information that will narrow down the final settlement site, but chances are it will be near a dry river bed or channel. If the settlement cannot be established near a flowing river or ocean coast as most towns and cities have been on earth, the next best place is near a spot where water once flowed and spread out over the ancient surface of Mars for the benefit of the early settlers.

The First Martian Village

There will be no golden shovels at the first ground-breaking for Mars housing; such ceremonial tools are better left on earth and the equivalent payload weight used for equipment that will help make the settlement self-sufficient. The first expedition to Mars will carry at least two habitat modules that will contain advanced recycling technology to make them less dependent on consumables. At least one pressurized heavy construction vehicle will also be in the cargo hold of the Mars ship. It will be used to dig trenches in the surface in which the habitats will be buried under at least 10 feet (3 meters) of red soil. This will protect the explorers from solar radiation when they are not working on the surface, and it will keep the life-support system away from the temperature extremes of

A Mars base may be built in the Candor Chasm, a central region of the great canyon Valles Marineris. Habitat modules will be buried under the Martian soil to protect the Martians from solar radiation when they are not working on the surface. Notice the pressurized greenhouses below. painting by Pat Rawlings/EAGLE ENGINEERING. Courtesy NASA.

the planet's surface. For example, at the *Viking 1* lander site in the northern hemisphere, temperatures ranged from -126 degrees F. (-88 degrees C.) before dawn to -10 degrees F. (-23 degrees C.) in the mid-afternoon—and these were summer temperatures. The first earth folk on Mars will, therefore, spend much of their time living under the surface. Each new expedition, and perhaps even some unmanned cargo spaceships, will carry additional habitat modules for the expanding Mars underground housing.

The Power Supply. After the habitats are installed and functioning, the next major project will be to make operational the nuclear power station upon which other facilities and settlement growth will depend. Because the sunlight falling on Mars is only 43 percent as strong as it is on earth, solar power collectors will not be able to supply the main power source for the resource refining and recycling plants. Solar energy will be used to grow food in pressurized greenhouses, however, and it has been estimated that a cultivated area equal to six football fields would be necessary to support fifteen people on Mars. For food production and other specific requirements, solar energy would be utilized. Cooling towers for the nuclear plants could be visible from the first village on Mars, but with luck a geological formation will be able to conceal them and retain the natural salmon-pink skyline of Mars.

Farming the Resources. Besides the seeds from earth that will be planted in the greenhouses for food—and there are solid practical reasons to believe that the first people on Mars will be vegetarians—the most important consumables on Mars will be water, breathable atmosphere, and rocket fuel.

Consensus among scientists who have studied the data from the Mariner and Viking space probes is that Mars is wet, perhaps not as wet as it once was, but wet nevertheless. The permafrost may provide water for the first Mars settlement, and some scientists even suspect that Mars has underground rivers. And although the atmosphere of Mars is extremely rare, it is made up of carbon dioxide, nitrogen, argon, and water vapor. Where the water is, and how much there is of it, must be determined before the first expedition sets off from earth orbit. The first supply of water will probably be produced by advanced fuel cells during the six-month journey to Mars. With large-scale recycling technology on the surface, this water will last until Mars water can be extracted and produced from the natural planetary environment. If water is found in sufficient quantities on Mars, then given adequate power, it can be hydrolyzed into oxygen and hydrogen for rocket fuel.

The extremely thin Martian atmosphere, similar to what we would find at an altitude of about 20 miles (32 kilometers) above the earth, is composed of 95 percent carbon dioxide. NASA and university scientists have devised plans for an automated refinery that utilizes an electrolytic pump to process the carbon dioxide into carbon monoxide and oxygen. After the carbon dioxide is pumped out of the thin atmosphere, it would be heated to high temperature until it breaks down into two gases. Bring hydrogen from earth (in the form of methane for easy transport), combine it, and presto . . . rocket fuel.

Nitrogen makes up only 2.7 percent of the thin atmosphere of Mars, so while there will eventually be oxygen aplenty for the Mars settlers to breathe, the nitrogen may have to be imported from the home planet if an earthlike atmosphere is considered essential. Advances in recycling technology will allow the original atmosphere brought from earth to fill human lungs again and again until methods are devised to produce large volumes of breatheable atmosphere from native Martian resources. Such atmospheric wealth could, over time, fill large domelike structures that could become the first parks and recreation arenas in the evolving Mars village. Will the village ever grow into a town and then a city named New Earth, Mars? Will New Earth ever become a favorite port for the sailors of interplanetary space? This all depends on what the first explorers, the robots and the humans, find in the valleys and on the volcanos of the Red Planet during the first half of the twenty-first century.

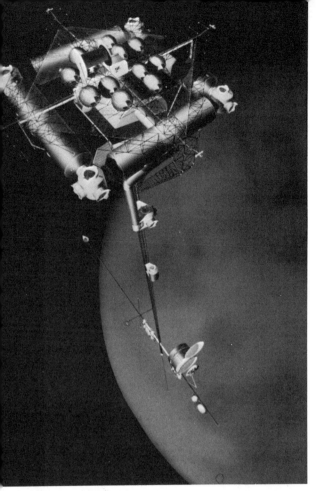

After the Mars base evolves, a space station may be built above Mars—a way station for new arrivals and those returning to earth. This design provides two varieties of artificial gravity by rotating: normal earth gravity and normal Mars gravity, which is one-third the earth's. Painting by Mark Dowman/EAGLE ENGINEERING. Courtesy NASA.

Explorers on Mars

Once the essentials of life are being dependably produced on Mars, scientific expeditions can set off to those corners of the planet that require more intensive study than the orbiting satellites can provide. A four-legged rover that can roll like a tank over smooth surfaces or step over rocks in rough areas is under development by the Jet Propulsion Laboratory. Advanced designs of this vehicle can be both manned or remote-control, or more fully automated for expeditions up to several hundred miles (kilometers) from the settlement. Concept studies have also begun on airplanes and dirigibles that can fly to more faraway locations on Mars. Because of the thin Martian atmosphere, the planes would require wide wingspans, but because the gravity is only one-third of the earth's, their power requirements would be modest. Hydrazine engines could be used, and solar cells mounted on the top of these Mars planes could supply additional power. The same solar power could be used for the Mars dirigible.

On manned expeditions the explorers would require protection while working on the surface. Mars may be the best backup planet we have in the solar system, but its surface conditions are still hostile to human life. If an explorer left his pressurized surface rover without protection, he would suffocate in only half a minute from lack of oxygen. If he carried an air supply but wore only light clothing, he would freeze to death in just a few minutes. And if he did not wear a pressure suit, the air pressure on Mars, which is equivalent to 1 percent of that on Earth, would cause his blood to boil at once.

The pressure suits designed for the surface of Mars will not be as cumbersome and awkward as the space suits used in earth orbit. They will not need a sophisticated design to hold the pressure; the body's skin and the elastic suit itself will become the pressure wall, so the suits can be designed for ease of movement and flexibility in the one-third-earth-gravity environment. It is therefore likely that the Mars surface suit will be a skin-tight garment that resembles a diver's wet suit. The freedom of movement that such a Mars suit provides will make work on the surface of Mars much less physically demanding than the spacewalks in earth orbit or on the surface of the moon. This means that more can be accomplished in less time during these exploratory expeditions that set off from the Martian village.

What Treasures Found?

What will the more extensive manned explorations of the Martian surface find? It is doubtful they will find life or evidence of extinct life forms, but it is still not impossible. It will be the complete geological findings gathered over time that may provide the most far-reaching rewards of such planet-wide expeditions. The discovery of significant ore deposits could begin an economic boom on Mars that would lead to many other human settlements. It could also prove to be the primary mover in speeding up the timetable for colonizing the moons of Jupiter and Saturn. What the first Martian settlers discover beneath the surface of their new planet will directly influence how the human expansion into the solar systems treats the asteroids on the way to the planetary giants, Jupiter and Saturn.

Mining the Asteroids

More than three thousand large asteroids have been discovered, and most of them travel orbits between those of Mars and Jupiter. Hundreds of other move in orbits that come close to the earth and moon, and still more are wanderers between the outer planets. The resource riches of these celestial rocks may prove to be the foundation for the industrial revolution of space within the next century. It has been estimated that the resource reserves in the asteroid belt alone could construct a land area that is some three thousand times that of the earth's.

By astronomical prospecting techniques that obtain the spectra of these bodies and interpret them, their composition can be determined. As a result, scientists know that certain types of asteroids have resources that are rare on the moon and Mars: water (up to 20 percent by volume); carbon; small amounts of nitrogen; and many precious metals. What this tells us is that what we cannot find on the moon and Mars, we can find on the asteroids. Almost all the resources of earth can be found off earth, and they don't have to be hauled out of the deep gravity well that earth, like other more massive objects, has. This fact may more than compensate for the distance the majority of the asteroids are from earth.

There are more of the so-called earth-crossing asteroids than estimated a few decades ago.

There are between one thousand and two thousand of them, and one such asteroid, designated 1976 AA, travels closer to the earth than any other large object except the moon. Its diameter is about 2 miles (3.2 kilometers), just a cosmic speck compared to the planets, but this rocky speck has a mass of over ten billion tons! With a mining operation on its surface, it could yield tremendous amounts of rocket fuel, precious metals, and other resources. The solar system has, therefore, supplied us with mines in the sky—some close, some far—to keep the new frontier in space expanding.

And this is just one example of a small asteroid, of which there are thousands. Ceres was the first asteroid to be discovered—in 1801. It is also the largest, with a diameter of about 621 miles (1,000 kilometers). Although Ceres is located beyond Mars in the asteroid belt, its surface composition indicates that it holds riches for the spacefarers who get there first, set up their solar furnace and other mining equipment, and use low-cost space transports powered by the mass driver propulsion system to deliver the ore and other resources to either Mars or the earth-moon system. Its carbon, water, and precious metals would make Ceres (or perhaps one of its richer neighbors) a resource island on the way to the moons of Jupiter and Saturn, moons that will contain their own unique resource wealth. While the asteroids appear to be unimpressive chunks of cosmic rock, they may become the building blocks of the new solar system—the one humankind begins to reshape over the next few centuries.

A Choice of Moons

After Mars and the asteroids, the giant planets of Jupiter and Saturn and their families of moons await the spacefarers and their machines. State-of-the-art robot probes will perform the scouting missions and discover in more detail what resources and energy sources are there for use by the interplanetary industrial revolution that may begin in the late twenty-first century. It is even possible that a new generation of self-replicating robots will journey to the diverse worlds that orbit Jupiter and Saturn during the twenty-first century, and the human space folk will remain busy in the inner solar system with large-scale projects in the earth-moon basin, on Mars, and on the asteroids. What is wonderful about the two largest gaseous planets in our solar system is that in many respects they are mini-solar systems in their own right, each with its own family of unique moons cruising along its orbits.

Jupiter claims amazing Io, the most volcanically active spot in our solar system. Io's tug of war with Jupiter and the other large moons creates its molten core, which constantly erupts, creating vast lava flows. This tremendous heat energy might be harnessed in the future. And then there are the icy worlds of Europa and Ganymede. The moon Ganymede, the largest known in the solar system, has a thick outer layer of ice believed to equal half of its entire mass. Heat up Ganymede, and half this moon would evaporate. But future spacefarers would prefer to utilize the ice that several of the moons of Jupiter and Saturn contain. One space scientist has worked out methods to use these vast quantities of ice as a source of oxygen and hydrogen for life-support consumables and for rocket propellant.

Besides Saturn's snowball moons, such as Iapetus, Rhea, Tethys, and Dione, which could provide water and its component gases, there is the magnificent red-veiled Titan. This big moon, second only to Ganymede in size, is the only moon in our solar system known to have a substantial atmosphere. As future spacefarers travel into the Saturn system, Titan will be there with its thick atmosphere, composed mostly of nitrogen—a very helpful gas resource for replenishing or creating earthlike atmospheres in large space colonies and mining operations on the moons of the outer planets. Add these outer moon resources to those of the inner planets and asteroids, and humankind has enough to reshape the solar system for the benefit of our species.

Colonies in Space

Since the Space Age began, the most exciting vision of humankind's future in space has been that of large man-made structures populated by hundreds, thousands, even millions of people who work, play, and live out their years in these mini-worlds.

"We can," wrote Isaac Asimov in the *National Geographic* of July 1976, "build space colonies in near space [which] would fulfill the functions that are now fulfilled by cities on the surface of the Earth." And the late Wernher von Braun, one of the men most responsible for getting to the moon on time, believed that building large space habitats would be difficult, but that such strong challenges to the best nations and the best minds would help forge an expansive future for humankind.

The difference between future space stations and space colonies is merely one of scale. How big is the structure? How many people live in it and for how long? How self-sufficient and regenerative is its ecology? Answers to these questions help determine if a space structure is a station or a settlement or a colony or a city in space. A few dozen space people could inhabit the second-generation space stations, but when the population goes over one hundred, it certainly should be called something other than a station. Just as the earth has villages, towns, and cities, so too will space have distinct names for the small, medium, and large habitats and workplaces above the earth.

A Dream Revisited

When visions of space colonies abounded in the latter half of the 1970s, especially under the guiding influence of Gerard K. O'Neill and many space advocacy groups, the concept of having large groups of people living and working in space goes back to the writings of the Russian space pioneer Konstantin Tsiolkovsky in the early decades of this century. Many of the

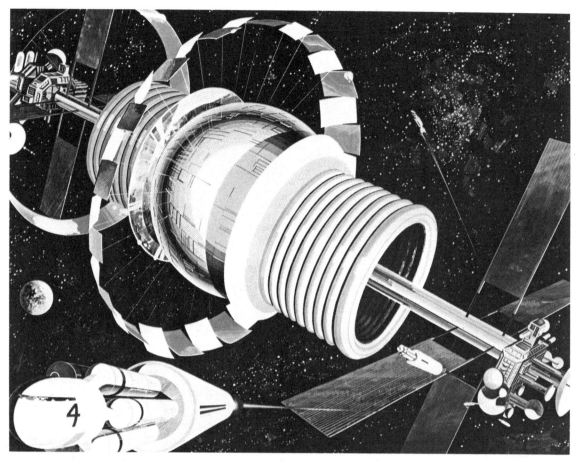

ABOVE AND FACING PAGE: *Space colonies, large-scale structures in space, captured the public's imagination in the late 1970s, when the U.S. manned space program was on hold, waiting for the space shuttle to fly. The large colonies were envisioned to hold tens of thousands or even millions of people.* Courtesy NASA.

other important concepts of space colonization have been covered by such science writers as Arthur C. Clarke, Krafft Ehricke, and Dandridge Cole.

The contemporary concepts of space colonies, often very detailed in their presentation and with solid mathematics to back them up, began their gestation in a freshman physics course at Princeton University in 1969. The course was taught by Gerard K. O'Neill. To challenge some of the brightest students in the class, he began a special seminar that put several problems before these bright students. One problem they considered was how large could a pressurized vessel be that would rotate to provide gravity and hold an atmosphere in space. The answer they came

up with was fascinating: several miles in diameter. It was these calculations that began the modern space colony movement which was extensively covered by the media in the 1970s and captured the imaginations of tens of thousands of people around the world (including the Russians), who believe that the human future lies beyond the planet Earth.

Designs Gigantic

The incredibly large scale of the space colony designs is what initially boggled the minds of all who became acquainted with O'Neill's ideas. Famous space station designs such as the torus wheel of *2001* were dwarfed to insignificance by comparison. The ideas were shocking at first, even to people who were space activists and made it their main avocation. At a time when the United States was between the Skylab missions and the test flights of the space shuttle, when no American astronauts were in space, the smallest space colony designs called for huge rotating aluminum cylinders that were about 1,600 feet (488 meters) in diameter, with circumferences of about 1 mile (1.6 kilometers), containing populations ranging from ten thousand to fifty thousand people. And these so-called Island One designs were the smallest space colonies; the calculation told of medium- and large-sized space habitats with dimensions that seemed like utter fantasy.

In his book *The High Frontier: Human Colonies in Space*, O'Neill writes of moderate-size colonies with diameters of 4 miles (6.4 kilometers) and lengths of 20 miles (32 kilometers), with to-

tal interior surface land areas of 500 square miles (1,295 square kilometers) supporting populations of several million people. Again, these sizes are based on math and technology that is available today. The largest colonies possible, using today's construction materials and building technology, could be some 15 miles (24 kilometers) in diameter, 75 miles (121 kilometers) in length, with total land areas of some 7,000 square miles (18,135 square kilometers), and supporting tens of millions of people. And most of the materials to build them could be refined from moon rock and soil that would be catapulted from the moon's surface, with its very low gravity, by a mass driver at relatively low cost, especially when compared with what they would cost if launched from the earth's deep gravity well.

Talented space artists created images of these large colonies that often looked like cities in giant bottles as the houses and wooded lands and

streams climbed the steep cylinder walls and continued full circle. Miniature San Franciscos and other familiar places were rendered by these artists in huge space structures. Their images depicting the interiors of these future colonies in space were not strange, but earthlike and nonthreatening. And in many cases, the paintings were rendered with photographic realism; they made the concepts and numbers more credible. The same can be said for the many books and articles on the subject of space colonies that appeared in the last half of the 1970s. In T. A. Heppenheimer's *Colonies in Space*, for example, the details of everyday life were described: what the houses would look like; what the colonists would eat; what kind of new sports and leisure-time activities the colonists would engage in. Such details certainly did lend credence to such grandiose space ventures, but many people still looked upon colonies in space as far-out fantasy.

Even now, a decade later, as the United States plans and designs in cooperation with several other nations a space station that will become operational in the 1990s, a large colony in space still seems incredible to many people. Why? Because the space station will accommodate only about twelve people, not the thousands predicted for the space colonies, and by the time the space station is operational, the Space Age will be more than three decades old. But will the huge colonies ever be built?

Time's Way

It does well to remember that the first human orbited the earth in 1961. About thirty years later there will be a permanent space station orbiting the earth with twelve people aboard, and more than one thousand people will have experienced life in space, twelve of whom walked and worked on the moon's surface. The size dif-

RIGHT AND PAGE 342: *Talented space artists created images of these large space colonies that looked like gigantic shopping malls or miniature San Franciscos.* Courtesy NASA.

ference between the interior of a Mercury capsule and the interior of the space station will be tremendous; and the same will hold true when a larger space habitat of the future is compared with the first permanent space station. No one should underestimate the power of human energy and purpose over time.

Also, if we emphasize a few basic but not-so-obvious facts, large space colonies suddenly move from what seems like the realm of fantasy to a realistic scenario for future space activities. First of all, there is no gravity, practically speaking, above the earth. This means that the same large-scale structures that are unfeasible to build in earth's gravity become possible in space. And there is a constant source of powerful energy from the sun, and no breakthrough technology is necessary to capture and utilize it. Indirect sunlight from large mirrors could fill the colony with light through large windows, thus protecting the population from the direct radiation of cosmic rays. People would live and work on the interior surface of the colony—an inside-out world—and the entire habitat would rotate at a rate that would create an artificial gravity through centrifugal force. This would be true of any colony design—cylinder or wheel. Special areas of the colonies would be set aside for growing crops. By completely controlling the agricultural environment—the humidity, the sunlight, the nutrients—food production would be maximized. Even now NASA is conducting space food research, including a program to evaluate the sweet potato for long duration space missions. Energy, food, shelter, challenging jobs, unique recreational opportunities—it sounds like the good life in space.

Why Build Colonies in Space?

Why build these celestial pyramids? Why would even the wealthiest nations or groups of nations on earth want to spend tens of billions of dollars to establish self-sustaining space communities in stable orbits in the earth-moon basin? What would be the economic incentives for building these floating cities in space? Without such incentives, colonies would never be built. And what other motivations would push humankind into space on such a large scale.

There have been many answers. For one, planet Earth has a finite surface area and limited resources. As the world population continues to increase, international stresses will also increase. The risk of a worldwide holocaust becomes greater. But space is almost infinite, and the resources in our solar system alone are vast and can offer a future of continued expansion and diversification for humankind. Perhaps the surface of our planet, or any other for that matter, is not the best place for an expanding technological civilization. Isn't space the only place left for the next industrial revolution? Why not clean up planet Earth by moving all the dirty work and its pollution into space? And by building colonies in space, we cast off the constraints of planet Earth and open up new paths of evolution and diversity for our species. It would also be humanity's evolutionary insurance policy. "Wouldn't it be terrible," wrote Ray Bradbury, "to wake up one morning and discover, without remedy, that we were a failed experiment in our meadow-section of the universe ... Man imprisons himself in his own bastille, not noticing he has the keys in his hands. Any time he wishes he can unlock, step forth, fly, be free."

All well and good for the long view, but what about the immediate economic incentives that will be needed for such a tremendous undertaking? The magic word becomes "energy," the constant solar energy from the sun. Above the earth, beyond the fluctuations of the planet's weather and day-night cycles, sunlight streams forth twenty-four hours a day. There it is, ready to be taken, immense quantities of clean, undiluted energy from our star, enough to satisfy the energy needs of thousands of civilizations more advanced than any the earth has yet produced. If such tremendous amounts of energy could be economically delivered to the surface of the earth, there would be an economic rationale for

building space colonies. The colonists could build these huge power satellites, larger than the island of Manhattan, that would deliver 10 million kilowatts to earth receiving stations through microwave transmission. Many thorough and serious studies have been completed. The engineers' numbers work. This new generation of spacefarers would utilize lunar resources to build the powersats, just as they would do to build their colonies in the first place. The calculations present their objective tale: Power satellites are a promising new energy source and colonies in space can build them. Time will have its way. When energy becomes too expensive on earth, the colonies could provide it for less from space. And an old dream realized—journeys to the moon—will turn the new dream of space colonies into a reality. We now know what the moon is made of, and we know what we can build with it.

Life on High

A Choice of Gravities. Whenever colonies are built (probably in the last half of the twenty-first century) and whatever design is finally chosen, they will rotate to create artificial gravity through centrifugal force. The need to imitate an earthlike gravity in the space colonies is essential to the health of the colonial pioneers. In zero gravity, bones weaken, circulatory changes occur, and the entire body undergoes a profound adjustment to the weightless environment. All these changes can negatively effect performance and mental outlook, and the experts agree that an artificial earthlike gravity must be present for any long-term living in space.

Artificial gravity may mimic the real thing, but there are some differences. Gravity, for example, will not be the same everywhere in the colony. As a person goes toward its center of the colony, toward its axis of rotation, gravity will become less. Just by walking up the stairs of a home or office, the colonist will lose some weight. For all practical purposes, there will be no gravity along the axis of rotation, and this space can provide exciting zero-gravity recreational activities. A colony could even be designed to have its older population live in an area of weaker gravity, which could lengthen their lives.

Although the size and design of the colony structure help to determine the best rotation rate, one revolution per minute is considered to be the ideal by many experts. If the rotation rate is too fast, a side effect called Coriolis acceleration could cause problems. This effect, found in all rotating environments, tends to change the direction of moving objects and can also confuse the mechanism of the inner ear. If the rotation rate is too fast, it can cause dizziness, motion sickness, and rolling eyes. But in very large space colonies, and in small ones with a rate of about 1 rpm, people should not experience these problems. Still, if the rotation rate in a smaller colony sped up without being corrected, there would be some curious effects, such as curving shower sprays or baseballs and golf balls departing from their straight-hit trajectories. And pizza makers would have to be extra careful when they tossed their pies into the air!

Future Hearth and Home. The colonists' homes will literally be built from the moon up because that is where the majority of building materials will come from. Housing will be built from plenty of aluminum, titanium, steel, and glass—all extracted and produced from the same lunar soil that will also cover most of the colony to protect the population from harmful radiation. Furnishings will be made of ceramics and aluminum derived from the moon. Even the draperies and carpets can be made from moon material, a type of Beta cloth fabric woven of glass fiber. The Apollo moon program used glass-spun beta cloth because it was fireproof. Because silicates are so common in the moon soil, stained glass wall and window art will probably be common, its colors dancing as the huge mirrors let in the indirect sunlight.

Brick, glass, and metal will make up the exteriors, and the homes will be built on terraces. The largest homes will be limited to about 2,000 square feet (185 square meters), and might contain only half that area—small but efficient. The living rooms might have a window that looks out onto the immensity of space. If the earth passes by as the person looks out, a pleasant warm emotion might flow; if, however, there is only the immense blackness of space at that time, the colonists might be overwhelmed by the darkness, as some astronauts have been.

As the colonists tend to the greenery of their small yards, just a motion of the head can remind them where they are. By looking up, especially in the smaller space colonies, they can see their neighbors' homes and streets and kids hanging down about 1,600 feet (488 meters) above their heads.

The Work of People and Robots. Hard work can be a blessing both on and off the earth, but some occasional relief is always welcome, and the space colonists can count on getting more than a little help from their robot friends, who will help them with their jobs and in their homes. And when the robots get tired and have problems, their human friends, specialists all, will be there to give them a new part or a new program to make things right.

Teleoperators for the remote-controlled robots will be one group kept very busy building new colonies or the huge power satellites. They and their robots will build large space structures together. Human hands will be on the sophisticated controls that will guide the robots in their building of new colonies, processing plants, or power satellites. NASA has already designed and tested some of the automated machines that will become the robot construction workers in space. One, called the Space Spider, can produce aluminum structures in space by unrolling and fabricating them along a preprogrammed path and continue by following and adding new material along the path formed by its previous pass. The Space Spider robot can advance about 20 feet (6 meters) a minute, which means that under ideal conditions it could fabricate a structure 1,600 feet (488 meters) in diameter in 144 hours, or about 6 days.

Gerard O'Neill believes that there will be few scientists in the first few colonies. Other than construction specialists, many of the first colo-

SETTLEMENT AND COLONIZATION CALENDAR

Event	Date
First lunar settlement	2010
Mars base	2035
Small colonies in space (ten to thirty thousand people)	2050
Large colonies in space (more than one hundred thousand people)	2150
Regional terraforming of Mars, the moon, and Venus	2100
Global terraforming of the planets and moons	2300+
Man-made biosphere around the sun (The Dyson Sphere)	4500 to 5000

nists will be sophisticated repair personnel to maintain and upgrade all the robots, the equipment of the industrial plant that processes moon material, and so forth. Other than a large number of these specialists there will be many worker specialists who are familiar in today's society—food service and medical personnel, testing and quality control specialists, traffic controllers, chaplains, and so on. Like earthbound workers, they will be happy to have a day off or weekend of rest and relaxation after a hard day or week on the job.

Leisure-Time Fun and Games. The lack of gravity at the center of the colonies, whether their designs are great spoked wheels with a central hub or large cylinders, will afford some wonderful free-time activities. The low-gravity swimming pool will be like nothing experienced on earth because it will be in the shape of a large cylinder and water will be in a complete circle. At only about one-fiftieth normal earth gravity, the waves will be high and in slow motion. A swimmer will be able to swim the entire circle of the pool and return to the starting point, at which time he or she can look straight up and see some friends swimming or splashing directly above. When a diver enters the water, a hole

will be left for a second or slightly longer before filling in. And with a powerful enough push off from a shallow area, the swimmer could swim through the waterless low-gravity area in the center of the pool; small hand-held paddles or feet fins moved against the atmosphere would help the swimmer move freely and avoid getting stuck in zero gravity at the exact center without anything to push against. The ultimate challenge might be the vertical dive, where a powerful jump off a springboard would send the diver through the zero-gravity center and then "up" to the opposite water of the cylindrical pool. With all these extra possibilities, some new water sports spectaculars will no doubt be performed.

Once only a dream, human-powered flight could become common in the low- to zero-gravity areas of the colony. People could strap wings on their arms and legs and fly, or they could use a small hang-glider type aircraft that would become airborne with human leg power turning bicycle pedals to drive the propeller. Even without any strap-on wings or other contraptions, low- and zero-gravity areas will offer both challenges and entertainment for people trying out new body motions and learning how to control them. And the colonists may vote for some special getaway suites in the zero-gravity areas, so they can experience the new marvels, freedoms, and some challenges of weightless sex.

Redesigning the Planets

Scoffing at way-out ideas and seemingly unattainable dreams of men and women has been a regular pursuit of some well-intentioned but ignorant people for thousands of years. That an idea or dream vision should be scrutinized and thought about by as many people as possible is healthy for human affairs in the long run, and such is often the case of many of the contemporary democracies. When this world is at its best, ideas have their fair hearings by audiences large or small, without any burnings at the stake or jailings or banishments. No one has to be reminded of the many major technological advances in the modern world that were thought to be so much hogwash while they were still in their conceptual stages. Large numbers of people dismissed them as the idle speculations of eccentrics. But again and again, human ingenuity, individual or group, has brought a dream into reality, often for the benefit of many people on planet Earth. Perhaps terraforming the planets is one such idea.

Cosmic Tools for Terraforming

Seething Venus did not get furnace-hot overnight, nor did the surging waterways of Mars evaporate or freeze into permafrost in a few hours. On the cosmic clock, a million years is just a tick, as powerful forces sculpted the planets of our solar system.

Colossal amounts of energy would be needed to change the present environments of Venus or Mars or the moon in time spans that approach a few human generations. The enormous energy of our sun is there to keep us alive and help us, once we learn more about how to harness it on a large scale. Big mirrors in orbit around the earth and other planets could have many uses, such as increasing or decreasing or concentrating sunlight on the planetary surfaces. For the hot inner planets of Mercury and Venus, the amount of sunlight falling on their surfaces could be reduced by giant parasols, artificial rings, or shadow-casting dust clouds. The changing temperatures that would result could then open the way for other long-lasting changes in a planet's ecology. This could include the introduction of genetically engineered organisms that are capable of creating a more earthlike atmosphere over time.

A Cool Venus? If the rotation rate of Venus could be increased from its slow spin of 243 earth days, a magnetic field would be created that would ward off some of the sun's radiation, which is responsible for the hellish, inhuman greenhouse effect. One proposed method of changing a planet's rotation rate is to use nuclear explosions to change the course of comets or asteroids, which would then plunge into off-center regions of Venus. Once our "sister planet" was spinning more rapidly, genetically engineered algae or other microorganisms, able to thrive in the hellish conditions, could transform the rich carbon dioxide atmosphere into a more earthlike one and remove the sulfuric acid that contributes to heating up the planet. Huge dust clouds could be put into orbit around the planet, which would shade it and allow oceans to form on its surface. Another method of decreasing the amount of sunlight reaching the surfaces of the inner planets of Mercury and Venus would be to build huge reflecting mirrors and put them into orbit above the planets.

Mars Remade

Because Mars is too cold and needs more sunlight to bring its surface temperatures up, it has been suggested that its two dark moonlets, Phobos and Deimos, could be mined for their dark material, which could then be pulverized and spread like soot over the surface of Mars. This would absorb more sunlight and melt the polar caps and what scientists believe are permafrosts beneath the surface. Additional heating, if needed, could be provided by giant mirrors concentrating sunlight onto surface regions. Water could run again in the riverbeds that had been dry for millions of years. With the help of genetic engineering, perhaps on the blue-green algae of earth, the rarefied Martian atmosphere could become earthlike, and the greening of Mars would begin.

Such brief scenarios do not begin to suggest the tremendous complexities of such grandiose schemes of planetary manipulations. For Mars, even after the Viking missions, more detailed information is needed about its water, its surface soil chemistry, and the concentration of salts in the soil before any serious model can be created and studied. Many experts admit that much of the knowledge and technology needed to accomplish planetary engineering is beyond the capabilities of humankind and may remain so for some time.

Selective Terraforming

Even the moon has been seriously considered as an object of terraforming, and preliminary studies suggest that even its small mass could hold an engineered atmosphere for thousands of years. But the problem of importing such a mass of atmosphere, even if some of the gases within the lunar rock and soil could be released through nuclear excavation, appears insurmountable today. At least one scientist has concluded that it is not feasible, with current technology, to create a dense atmosphere on the moon that compares with that on earth.

Perhaps the global approach to terraforming of the planets and moons of the solar system should be reconsidered, with more emphasis placed on terraforming regional pockets of Mars and the moon. True, work in these areas is being done, but a more concerted effort could produce results that would actually be used on missions, manned and unmanned, in the twenty-first century. Studies to develop technologies and methods for terraforming city-size regions on the surfaces of other worlds would have to face many of the same problems that global studies would, but they would be more in keeping with the large engineering projects of the world today, and they are more likely to be applied in the next century. Establishing earthlike ecological niches on the moon and Mars will be no small chore. After the space station is operational, this is where our extraterrestrial energies should be directed, with some breathing time, of course, for far future speculations—those that have a history of uncovering useful ideas that can be applied in the next few decades of the Space Age.

The Dyson Sphere

Freeman J. Dyson, the famous American physicist at the Institute for Advanced Study at Princeton, speculates as a hobby, and his creative imagination has produced ideas that are authentically mind-boggling.

In one such thought experiment, Dyson asked himself what an advanced civilization would need to push its colonization to the limits, and his answer was total resources, first of the solar system, then of a galaxy, then perhaps of several galaxies. All the exploitable resources are either the matter of the planets, comets, immense dust clouds, and so forth, or energy in the form of starlight. One type of advanced civilization would in theory be able to control the total light energy of its star. How would this be done? Dyson decided, with some high math to back up his speculations, that such a civilization could build an immense biosphere shell around the star, which would capture and utilize its entire photon energy output. This amount of energy is almost inconceivable, but it helps to know that the earth receives only two parts per billion of the sun's total output.

Given sufficient time, Dyson believes, any long-lived technological species, including humankind if we survive into the far future, can construct such an artificial biosphere. In our solar system, the resource could be the giant planet Jupiter. To dismantle our giant planet, we would need energy equivalent to the sun's total output for eight hundred years. With this amount of vaporized material, the artificial biosphere—either solid or composed of millions of separately orbiting habitable balloons—could be constructed around the entire sun at a distance equal to the radius of the earth's orbit.

The inside surface area of such a biosphere around our sun would be one billion times greater than that of earth's surface area, and the solar energy captured would be more than one hundred thousand billion times that of today's world production of energy, enough perhaps to take us to the next level of advanced civilization—controlling the resources of the entire Milky Way galaxy, so immense that light takes some one hundred thousand years to cross its spiral disk, filled with hundreds of billions of stars.

When would humankind be capable of building such a gigantic sphere around the sun—if we survive? Some 2,500 to 3,000 years from now, Dyson believes, which would put this outside date at about A.D. 5000.

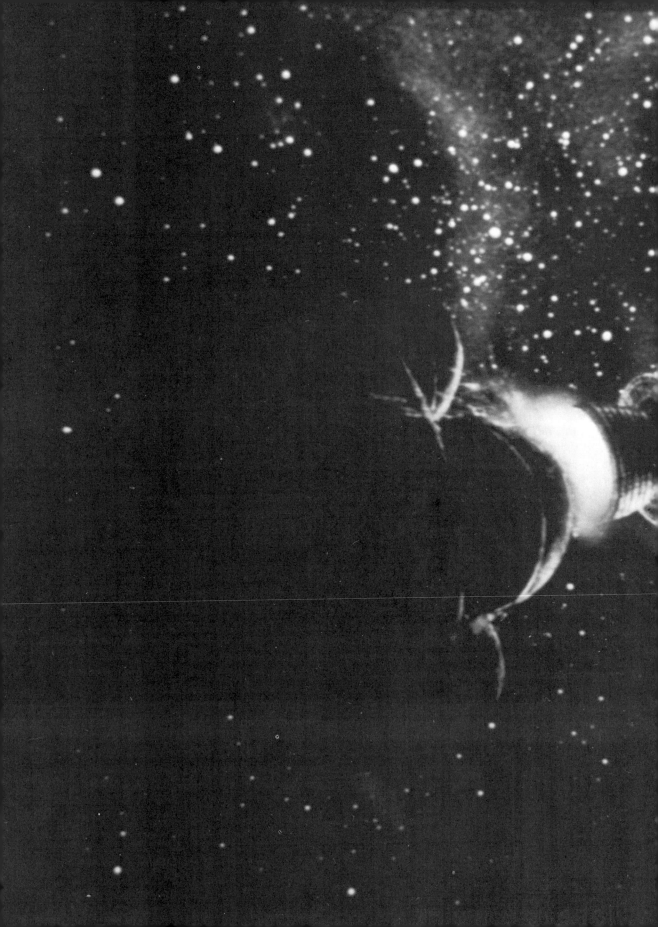

12

Extraterrestrials & Star Trips

Lost in the Cosmic Haystack

Are we alone in the universe? If not, where are they? Do intelligent extraterrestrials (ETs) really exist among the hundreds of billions of stars in our Milky Way galaxy or the tens of billions of galaxies beyond, or is ET life found only in wonderful film fantasies or books produced on planet Earth? These are some of the cosmic questions that humankind asks itself. But as we close out the twentieth century, no one knows the answers. There are many impressive opinions, presented in hundreds of scientific papers, some complete with equations, but it still comes down to a cosmic guessing game.

Whatever position a scientist may take or strongly prefer, there is an act of faith behind it. Reasonable arguments are offered by well-known and respected scientists to persuade us that there are at least one million advanced technological civilizations in our Milky Way galaxy alone. Other equally respected astronomers, biologists, chemists, or physicians convincingly argue that we are alone in the galaxy, the only intelligent life that has evolved to the point of attempting to find our extraterrestrial neighbors in the cosmos.

These are grand-scale differences of opinion, but one does well to remember that the search for extraterrestrial intelligence (known as SETI) is only three decades old—an almost imperceptible microtick on the great cosmic clock. In its first few decades, the search has been modest in scale and funding. Only a few straws in the great cosmic haystack have been examined. The search is still in its infancy; we are just learning how to listen.

"The probability of success is difficult to estimate," wrote Giuseppe Cocconi and Philip Morrison in their historic paper "Searching for Interstellar Communications," published in the magazine *Nature* in 1959, "but if we never search, the chance of success is zero." Little did they know that a young astronomer, unaware of their work, had reached the same conclusion. He would soon point his instruments toward distant stars and listen for signs of extraterrestrial intelligence. This was twenty-two years before Hollywood's *E.T.* rigged up a transmitting antenna from an umbrella, a saw blade, a record player, batteries, and a Speak and Spell® control panel, and called his "people."

Green Bank and Project Ozma

Frank D. Drake was the first person on planet Earth to peek into the cosmic haystack and search for electromagnetic signals from intelligent life elsewhere in the universe. It was April 8, 1960, a cold and foggy morning, when Frank Drake got up at 3:00 A.M., and went to the 85-foot (26-meter) radio telescope of the National Radio Astronomy Observatory at Green Bank, West Virginia. The radio telescope is located in a valley flanked by the Blue Ridge Mountains and is well protected from radio interference. After an hour of fine-tuning, the big dish antenna was pointed at the first prime target, a sunlike star about 11.9 light-years away—Tau Ceti. Project Ozma (named after Princess Ozma in the Oz books) was born when this big metal ear at Green Bank, with its specially designed equipment, listened for intelligent signals from space for the first time in human history.

A Mysterious Signal

A second star target was also observed during the first observing day of Project Ozma. The star was Epsilon Eridani, some 10.8 light-years from earth, and when Tau Ceti set in the west at about noon, the big dish pointed to Eridani and listened for signs from cosmic neighbors.

Suddenly the chart recorder went off the scale and rapid bursts of noise, about eight a second, came from the loud speaker. Could it be? Was it that simple? Was this an intelligent signal from the vicinity of Epsilon Eridani? No one could find an explanation.

About ten days later the signal appeared again, and with it the answer. After the mysterious signal was picked up the first time, a simple horn antenna was pointed out the window and attached to a second receiver. When the big dish picked up the signal again, the horn antenna also picked it up. This proved that the signals had to be man-made radio signals. It was later learned that these signals were from a secret military experiment; no intelligent signal was coming from Epsilon Eridani.

Project Ozma continued for three months, through the month of June 1960, and accumulated a total of 150 hours of observing time on these two nearby stars. No other signals were heard, not even false alarms, but the search for extraterrestrial intelligence had begun in earnest. And it had begun at the beginning of the Space Age, a year before the first men from planet Earth rocketed into space.

Estimating ETs

In the early 1960s, a few years after the completion of Project Ozma, Frank Drake had formulated an equation that helped to estimate the number (N) of technological civilizations in our Milky Way galaxy. This equation, with N equaling the product of multiplying a series of probabilities, is known as the Drake equation or the Green Bank equation. In the following years, Drake, Carl Sagan, and other extraterrestrial enthusiasts arrived at a value for N as one technological civilization per million stars in the galaxy. Other scientists homed in on certain elements of the equation, broke them down into several subunits, and analyzed them more thoroughly. Without much real data, their work sometimes appeared to be no more than an interesting and challenging intellectual exercise.

Today this value for N (a single civilization for every million stars) stands out as an optimistic prediction of other intelligent life in the Milky Way—estimated at between one hundred thousand and one million other galactic civilizations, depending on the more accurate dimensions of the galaxy and its number of stars.

ABOVE: *Where are the ETs? Enrico Fermi, Nobel Prize-winning physicist, surprised his luncheon companions at Los Alamos in 1950 by asking them, "Where are they?" Fermi's calculations suggested that evidence of an advanced galactic civilization should be everywhere around us, given the age of the galaxy, and that ETs should have visited us long ago. But since there is no such scientific evidence, perhaps they do not exist.* Courtesy American Institute of Physics, Niels Bohr Library. PRECEEDING PAGES: *Humankind's first starship, a fully automated, unmanned interstellar spacecraft, has already been designed in amazing detail by the British Interplanetary Society as Project Daedalus, in 1978. Here, the second stage of* Daedalus *ignites in deep space about two years after launch. The second stage will burn for 1.76 years and attain 12.8 percent the speed of light, or about 86 million miles (138 million kilometers) per hour.* Courtesy Don Dixon, © 1983.

What is important to emphasize is that the right side of the equation is composed of probabilities, and when these probabilities are multiplied together, they equal the number (N) of technological civilizations in the galaxy. What are these elements on the right-hand side of the equation? They are:

• The average rate of star birth in the Milky Way galaxy each year
• The number of sunlike stars that are single

and not members of binary or multiple systems
- The number of stars with planetary systems
- The number of earthlike, habitable planets per planetary system
- The number of planets on which life starts
- The number of life forms that evolve to intelligence
- The number of intelligent civilizations that have developed an advanced technology
- The average lifetime in years for the civilizations with advanced technologies.

Today it is impossible to fill in all these blanks with any degree of certainty or completeness. Perhaps the best-known element is the rate of star formation in the galaxy. At the other extreme is the completely unknown value of the average lifetime of an advanced galactic civilization. If the average life of an advanced technological civilization were only a few hundred years because of self-destruction brought about by the technology, then the odds against any two galactic civilizations ever communicating would be long indeed. The gaps in knowledge about how life arose and the random nature of biochemical evolution itself make the prediction of the evolution of life impossible—another unknown in the equation.

Wishful Equations?

Princeton physicist Freeman Dyson, whose heart is certainly open to the possibility of intelligent extraterrestrial life, casts a cold eye on all such formulas that lack data. In his book *Disturbing the Universe*, he writes: "I reject as worthless all attempts to calculate from theoretical principles the frequency of occurrence of intelligent life forms in the universe." His attitude is wait and see. And Jesco von Puttkamer, a program manager at NASA, notes that the value of N is obtained by "multiplying many conditional probabilities . . . . But there is a catch: in multiplying odds we are also multiplying possible goofs in our estimates, and so the likelihood of deriving a good judgment decreases as we proceed."

The optimism of the 1960s and 1970s, when leading scientists from the United States and Russia such as Carl Sagan and I. S. Shklovskii generally agreed on a "best guess" of one million advanced civilizations in the galaxy (averaging out to one every three hundred light years) that would be at or beyond our present level of technology, has been countered more recently by several rigorous arguments that present more pessimistic interpretations based on the extremely limited information.

For example, in his 1975 article "An Explanation for the Absence of Extraterrestrials on Earth," in the *Quarterly Journal of the Royal Astronomical Society*, Michael H. Hart argues that because there are no intelligent beings from outer space on earth now, it is quite implausible, as the optimists would have us believe, that intelligent life in our galaxy is very common, with thousands of advanced civilizations scattered throughout it. Hart concludes, given his thesis, that it is probably a waste of time and money to do an extensive search for radio messages from other civilizations about the galaxy. That there are no extraterrestrials on earth now, Hart argues, is strong evidence that we are the first intelligent civilization in our galaxy, although we do not know why. If this is so, then the descendants of humankind may occupy most of the habitable planets of the Milky Way galaxy in the future. We may become those we seek: the extraterrestrials. Wishful thinking, equations or no, helps make dreams come true.

Where Are Our Galactic Neighbors?

The optimistic searchers for ET intelligence remind us that our Milky Way galaxy is an immense spiral carousel of stars, so large that it takes light (traveling at 186,000 miles, 299,274 kilometers per second) 100,000 years to cross its diameter. Considering the size of the galaxy and considering the fact that the search for extraterrestrial intelligence has just begun with a few quick peeks into the cosmic haystack of radio frequencies, no one should be surprised that intelligent signals from ET life have not yet been recorded. This viewpoint does not seem at all unreasonable when we know that searching for a radio signal of one hertz (a cycle a second) out of a total radio band of one hundred billion hertz is no small chore. In fact, if astronomers spent just one second to tune their radios to each of these one hundred billion frequencies, that alone would take three thousand years—and this would be just one direction in space, just one star or planet. If one million stars were covered in this way, multiply three thousand years by one million and see how much time would go into the project! With such numbers in mind, one can see that any quick (ten years or less) results from SETI activities appear next to impossible. It also proves the need for powerful high-technology signal processors that can scan the heavens faster, more thoroughly, and automatically. The future promises to have such sophisticated search machines monitoring the cosmic cacophony on selected radio frequencies.

In contrast to the ET optimists, there are many respected scientists who seriously question the existence of ET life. Enrico Fermi, the famous Nobel Prize–winning physicist, was the first to question the work of the ET optimists when he surprised his luncheon companions in 1950 by asking the now famous question "Where are they? Where are the extraterrestials?" Even today, the answer to this mystery is still being sought. No one on planet Earth today knows where the other extraterrestrials are.

ET Optimists and ET Pessimists

Why is the question "Where are they?" valid when put up against the fact of the immense distances of our home galaxy? Aren't the cosmic distances just too much for even an advanced spacefaring civilization to navigate across? Many experts would answer no. Fermi's question is valid because of time—the ages of the universe and the Milky Way galaxy. In other words, given enough time, any interstellar space, no matter how vast, can be crossed.

The galaxy is believed to be ten to twelve billion years old. Assume that the galaxy evolved without life forms for half this time; this still leaves five or six billion years for life to evolve

and for advanced galactic civilizations to master interstellar flight and colonize the entire galaxy, which could easily be done, some experts tell us, in two to three million years. Eric Jones, a scientist at the Los Alamos Scientific Laboratory, wrote several papers in the late 1970s that strongly make the case that a stellar civilization would sweep across the galaxy, leaving unmistakable signs everywhere, in less than one hundred million years. Jones and other ET pessimists therefore join Fermi in seriously asking, "Where are they?" No scientific evidence of past visits from extraterrestrials has been found on planet Earth, and none of the tens of thousands of astronomical observations, more sophisticated and sensitive with each passing year, has yielded evidence of extraterrestrial intelligence. One does well to remember that more than 99 percent of all life forms that ever appeared on planet Earth are now extinct. Perhaps the same discouraging numbers hold true for ET life elsewhere in the universe. Perhaps technological civilizations do not last long. If this is so, humankind will learn this fact within the next one hundred years.

Both the ET optimists and the ET pessimists have done good work considering the fact that they have very little data to work with. The truth is that they are all optimists by virtue of the fact that they even attempt to solve the extraterrestrial mystery from their earthbound perspective. Time will tell. Ten thousand years should easily do it, if humankind survives that long.

The SETI Majority: On With the Search

Most SETI experts agree, no matter what their positions are on the various aspects of extraterrestrial intelligence, that the search must go on. Thinking of life among the stars and neighbors in the skies, of course, has been going on for hundreds and thousands of years. The Epicurean philosopher Metrodorus of Chios expressed a very contemporary viewpoint in his book *On Nature*, written in the fourth century B.C. "To consider the earth as the only populated world in infinite space is as absurd as to assert that in an entire field sown with millet only one grain will grow." And the same concepts were expressed in China more than 1,500 years later when Teng Mu wrote *The Lute of Po Ya* during the thirteen century A.D.: "How unreasonable it would be to suppose that, besides the heaven and earth, which we can see, there are no other heavens and no other earths?" Perhaps these similar expressions, separated as they are by centuries and continents, time and space, present an authentic intuition of humankind that has as much merit in approaching the truth about ET intelligence as do the Greenbank equation and all the fiddling with its elements. But then we must ask: Is human intuition usually optimistic? These historic thoughts about ETs are certainly on the side of the optimists.

The search for ET signals gained a legitimate standing in the scientific community only in the 1980s. The International Astronomical Union established a new commission in 1982: Commission 51, Search for Extraterrestrial Life. In the same year, The International SETI Petition was signed by seventy-three prominent scientists, including seven Nobel Prize winners, from fourteen countries. Carl Sagan prepared the petition, and it had the support of The Planetary Society. Although the opinions of these scientists on ET life and our possible communication with it varied widely, they all strongly believed that the search should continue and that much would be learned no matter what the eventual outcome.

The introduction to The International SETI Petition said in part: " . . . we are unanimous in our conviction that the only significant test of the existence of extraterrestrial intelligence is an experimental one. No *a priori* arguments on this subject can be compelling or should be used as a substitute for an observational program. We urge the organization of a coordinated worldwide and systematic search for extraterrestrial intelligence." Everyone agrees that a long-term program of listening is important. It will collect real data over time, which is worth more than all the probability equations combined. The future program, under development by NASA, will

cover some ten million times the cosmic space that was covered in the dozens of search programs through the mid-1980s.

The U.S. Congress was persuaded to approve a modest $1.5 million in the NASA budget for 1983 to begin the development of the long-term SETI project, which would last five years and cost a total of about $7.5 million. It took Carl Sagan and other prominent scientists years to coax SETI critic Senator William Proxmire, bestower of the Golden Fleece awards, from a position of hostile criticism into one of skeptical neutrality. Perhaps they had to remind him that Columbus, after all, discovered America after starting out on an impossible goal under several wrong assumptions.

SETI After Ozma

Some four dozen separate radio searches have been carried out since 1960; the antennae have been aimed at more than a thousand stars, some globular clusters, galaxies, and quasars. More than 3,500 hours of observing time have accumulated during these searches carried out by seven countries: Australia, Canada, France, West Germany, the Netherlands, the Soviet Union, and the United States.

One such search, Ozma II, took place at the National Radio Astronomy Observatory at Green Bank, West Virginia, from 1972 to 1976. The project, headed by astronomers Ben Zuckerman and Patrick Palmer, tuned in on 674 stars within 80 light-years of the sun for a total of 500 observing hours. In the wavelengths covered, no signals were received from extraterrestrial life.

Astronomers at Ohio State University Radio Observatory have been conducting a radio search for ET life since 1973. It is an all-sky search, largely a voluntary effort on a shoestring budget, but it is the longest-lasting effort to date, thanks to the dedicated work of Robert S. Dixon and John Kraus.

In 1978, American and Australian astronomers turned their dishes to twenty-five globular clusters of stars that orbit the galaxy's center and

The world's largest radio-radar telescope at Arecibo, Puerto Rico, is responsible for dozens of discoveries about our solar system and the distant universe. Arecibo's radar transmitter sends out the strongest signal yet to leave planet Earth. The advanced Suitcase SETI was tested on 250 nearby stars at Arecibo in 1982 and became Project Sentinel, a major step forward in the search for extraterrestrial life. Courtesy Ames Research Center, NASA.

listened for signals from intelligent life, but nothing was heard. Toward one of these globular clusters, M-13, however, a signal was transmitted by the huge radio telescope at Arecibo, Puerto Rico, during a dedication ceremony in 1974. Just because no intelligent radio signals were heard from M-13 in the 1970s does not mean that they will not be heard in A.D. 51,974—the approximate future date that we could hear an answer to our message to M-13 if it is received and answered by ET life. Because M-13 is some 25,000 light-years away, the round-trip message time would be about 50,000 years, so if such life did not have the technology

now to be sending messages, it very easily could when the message arrives in 25,000 years.

Paul Horowitz of Harvard University conducted the Project Sentinel search that monitored the sky from 1983 to late 1985 using the 84-foot (25.6 meter) dish of the radio telescope at Harvard's Oak Ridge Observatory. The project used an apparatus called the "Suitcase SETI" because it is portable and can be used on any radio telescope around the world to search for signals at any special frequency, such as the one emitted by neutral hydrogen, the most common element in the universe.

The Suitcase SETI is still hooked up to the refurbished dish of the Oak Ridge Observatory and continues to search thousands of stars in more than two-thirds of the northern sky. The equipment was upgraded in late 1985 with a new antenna, which is now capable of scanning up to 8.4 million radio channels. Now known as Project *Meta* (for megachannel extraterrestrial assay), it was largely funded with a grant from the famous filmmaker Steven Spielberg.

New frequencies, the so-called magic frequencies such as 1,420 gigahertz (1 gigahertz, abbreviated GHz, is equal to 1 billion cycles per minute) of neutral hydrogen, are monitored each year after a search is completed. While this particular radio telescope cannot match the ultimate sensitivity of the Arecibo antenna (the world's largest and therefore in demand for many research programs), it nevertheless is covering a larger part of the sky than ever before in the search. The older digital signal processor was capable of monitoring 128,000 channels; the upgraded system can handle about sixty-six times this number—over 8 million channels. If there are a small number of very powerful signals from ET life, Horowitz and his equipment at Oak Ridge Observatory should record them.

Horowitz is one of the optimists: "With four hundred billion stars in the galaxy and ten billion galaxies, there certainly is life," he says. He will be carrying the SETI torch for a few more years, until the new generation of high-tech equipment begins scanning the skies for the new NASA program in the late 1980s.

Future Search: NASA's SETI Program

SETI searches in the next few decades will be far more ambitious than anything that has been done before, and they will be capable of listening to a great number of frequencies on very narrow channels. A coherent signal on a narrow wavelength would almost certainly come from an intelligent ET source because no natural source known at this time can create them.

An instrument called a multichannel spectrum analyzer, which is able to filter a wide-band signal into thousands of narrower bands, has been developed by NASA and Stanford University. The prototype had 74,000 channels, and each one was only 1 hertz wide. But electronic designs advanced in the next few years, and the state-of-the-art analyzers continued to grow to accommodate more than 8 million channels. When the megachannel extraterrestrial assay was initiated at the Massachusetts site of Paul Horowitz's SETI project in the fall of 1985, it represented a major advance in SETI research. Yet this 8.4-million channel receiver, as exciting as it was, still covered only about one-nineteenth of the microwave window that the future NASA megachannel analyzer will.

NASA's high-technology radio receiver will be capable of simultaneously searching these millions of discrete channels in the microwave portion of the spectrum. Once these signal analyzers are connected to the large radio telescopes around the world that make up the Deep Space Tracking Network, the configuration will become the most powerful dedicated computer ever made, generating one billion bits of information each second. This mountain of data will be processed automatically, and even if an unusual signal is heard, it will be sorted out from the background signals. Such equipment connected to the great radio telescopes of the world will be able to examine in minutes more interstellar data from a broad portion of the radio spectrum in which the background noise is low than did all the SETI programs combined over almost three decades with several thousand hours of observations.

NASA's future SETI program includes two distinct searches—a general all-sky search and a

This radio telescope at Harvard, Massachusetts, houses the world's most sophisticated ET search equipment. Called META (for megachannel extraterrestrial assay), it can receive 8.4 million channels simultaneously. The system has some 500,000 solder joints that were all done by hand and 25,000 microchips. Paul Horowitz, the leader of Project META, is convinced that it can detect signals from an advanced civilization located anywhere in our galaxy. The film producer Steven Spielberg donated funds for the project to The Planetary Society and turned it on September 29, 1985. Courtesy Paul Horowitz, Harvard University.

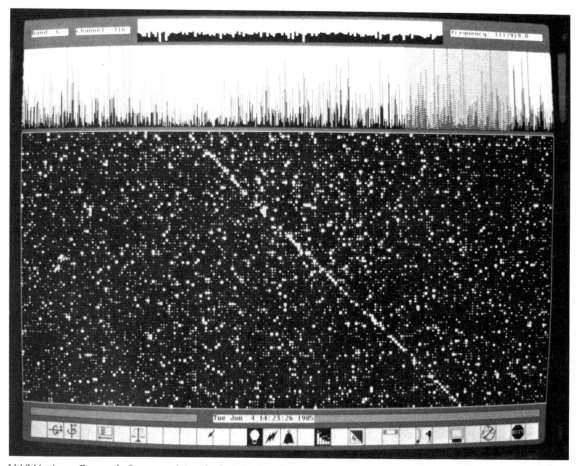

NASA's Ames Research Center and Stanford University helped develop the state-of-the-art multichannel spectrum analyzer that accommodates more than eight million frequency channels. Eventually such equipment will be connected to the great radio telescopes of the world and will generate one billion bits of information each second. The prototype equipment captured Pioneer 10's 1-watt signal in 1985 from a distance of 3.3 billion miles (5.3 billion kilometers), when the spacecraft was beyond our solar system. Courtesy Ames Research Center, NASA.

search that targets more specific celestial objects. The all-sky survey will cover the frequency range from 1 to 10 gigahertz, almost the entire microwave window that falls between the infrared and radio broadcast portions of the spectrum. This sweep of the entire sky will be conducted over a three-year period with the advanced signal analyzer attached to the antennae at the Jet Propulsion Laboratory. Even though this search observes each direction in space for a few seconds, giving the search low sensitivity compared to the more focused search, it does increase the chances that all possible sites for intelligent life will be observed. This search strate-

gy also includes some monitoring of spot bands between 10 gigahertz and 25 gigahertz.

The other search objective is to examine a narrower frequency range, from 1.2 gigahertz to 3 gigahertz, in much greater detail. It will include 773 stars within 80 light-years of the sun and solar system that have been selected on the basis of their spectral type, which determines such traits as temperature, life span, mass, and so forth. Alpha Centauri A, a star in the star system nearest to the sun and earth, will be one of those stars thoroughly studied. The center of the Milky Way galaxy and a few external galaxies will also be listened to intently. A few spot bands between 3 gigahertz and 25 gigahertz will also be covered. With this focused SETI search working in tandem with the powerful Hubble Space Telescope, there could be an ET surprise by the year 2000. Even the discovery of a habitable, earthlike planet by the turn of the century would bode well for the search for extraterrestrial life in the twenty-first century.

The Cosmic Waterhole

The searchers for extraterrestrial life know that the challenge is in the search itself, among the billions of narrow frequency bands, not in the actual detection of the signal. Astronomers have likened the problem to trying to find a friend in a large city without knowing where to look, without any prior meeting place. There would be no problem in recognizing an intelligent signal from ET life. But where to look for the friend or the ET signal is a serious challenge.

The more focused SETI search being conducted by NASA includes the frequency range known as the "waterhole," from 1.4 gigahertz to 1.7 gigahertz. Since SETI activities began in 1960, this frequency range has always been considered a good prospect for detecting intelligent signals. The reason for this is that it is located in the center of a radio-quiet region, which contains the natural frequencies of two natural radio emitters: the interstellar hydrogen Atom H; and the interstellar hydroxyl (hydrogen-oxygen) molecule OH (which is oxygen chemically bound to hydrogen)—the elements of water upon which our known life is based. The 1.4 to 1.7 gigahertz frequency range has thus been considered one of the so-called magic frequency ranges, one of several that SETI experts have considered most likely to broadcast messages from other advanced civilizations in the galaxy.

The waterhole range of a frequency is still only a very small fraction of the electromagnetic spectrum, and some astronomers suggest that this magic frequency has been chosen more for philosophically attractive reasons than for scientific ones. The more scientific approach brings forth a wider frequency band that is the most efficient for electromagnetic communications, the microwave window between 1 and 10 gigahertz, which the all-sky SETI search will cover from earth-based observatories. Above the earth's atmosphere, this microwave window increases tenfold, to an upper range of 100 gigahertz, and future satellites are expected to cover this range before the end of the century.

The ET Signal: Will We Hear It?

On the surface of the earth, sophisticated search machines will eventually be hooked up to the large radio telescopes around the world. So predicts John Billingham, chief of NASA's Extraterrestrial Research Division at Ames Research Center in Mountain View, California. They will collect and analyze the hundreds of billions of bits of data automatically. After they are in place, he believes, there is a chance that a detection will be made during their first decade of operation. Beyond the year 2000, because advances in SETI search technology are so unpredictable, Billingham considers it impossible to make a forecast. The sensitivity of the listening instruments will no doubt increase over time— if they are funded and built. Radio telescopes in earth orbit or in orbit on the far side of the moon could extend coverage to perhaps several thousand light-years.

Michael Papagiannis, an astronomer at Boston University, thinks that, one way or the other, we will have an answer in the next ten to twenty years: " . . . we will either have detected some sort of signal from SETI, or we will begin to accept the fact that we are alone in our galaxy." Such negative results from an extended SETI search would still be important for humankind, Papagiannis says, because we would realize "that we are the torch bearers of the flame of cosmic consciousness in our entire galaxy." Even if we conclude we are alone in *this* galaxy, there is no cause for despair because there are some one hundred billion galaxies in the universe. The odds are good that we have intelligent living neighbors somewhere in the cosmic realm and some life forms in our own galaxy, even if they may not be intelligent enough to communicate with us during a particular decade or two.

If we come to be convinced that humankind is presently the only intelligent life in the Milky Way galaxy, we may find it more imperative to adopt an attitude of cosmic responsibility and do everything within our power to hedge against extinction, to preserve ourselves and other species on planet Earth for future evolution, migration among the stars, and perhaps contact with some future intelligent life in our galaxy.

The Planets of Distant Suns

In the quest to know if there is other intelligent life in the Milky Way galaxy, hundreds of other questions must be considered, such as: How many stars have planets and how many planets are suitable to sustain life as we know it? But like so much else in the search for extraterrestrial intelligence, there are very few facts and a great deal of theory based on probability. The last decade of the twentieth century will change this, however; in fact, discoveries in the late 1980s began to replace probability theory with astronomical evidence. The Hubble Space Telescope, the Space Infrared Telescope facility (four thousand times more sensitive than the Infrared Astronomical Satellite, *IRAS*), and other in-orbit instruments capable of detecting any earth-size planet within forty light-years of our solar system will begin the actual count of planets around the other stars and start to determine their surface conditions. Another unknown in the ET equation will soon be replaced with real numbers.

The first breakthrough in planetary detection took place in 1983, when the Infrared Astronomical Satellite detected a disk of gas and dust around the bright star Vega—the kind of disk that astronomers predicted would exist around a young star in the process of forming planets. A similar disk was found in 1984 around the star Beta Pictoris, a star visible only from the southern hemisphere and some fifty light-years away

from earth. This time the disk was actually photographed by astronomers, who believed, based on planet-formation models, that there were planets around the star. This could not be proved conclusively, however. Soon after this discovery, still another celestial surprise: Astronomers presented evidence for a massive planet-sized body in orbit around a dim star known as Van Biesbroeck 8, which is about twenty-one light-years from earth. Designated VB 8B, the object is slightly smaller than our planet Jupiter, but as much as eighty times the mass and is about nine times hotter. While it may be hotter than the gaseous giant planets of our solar system, it is still too cool to be a star, so astronomers give it a place between a planet and a star, a position that has existed in theory for many years, and label it a brown dwarf.

NASA has its own Planetary Detection Project and has agreed to work with the University of Arizona to study the planetary systems around distant suns. This new branch of astronomy will be one of the most exciting scientific disciplines in the next few decades, as the new technology extends humankind's senses even farther into space and time. If the rate of planetary formation around other stars in the galaxy bears out the optimistic predictions of such people as Isaac Asimov and Carl Sagan, who believe that there could be as many as seven hundred million to one billion habitable planets in the universe, then the optimists will probably once again dominate the search for extraterrestrial intelligence.

Between one hundred and two hundred stars will be scrutinized for planets before the end of the century, and some real data about other solar systems will replace what has been complete ignorance during the first twenty-five years of the search for ET intelligence.

A possible solar system was found around Beta Pictoris, a star visible from the southern hemisphere, in 1984. As the technology for detecting planets advances dramatically in the next few decades, the question of how many habitable planets there are in the galaxy may be answered, and the search for life elsewhere in the universe will include intensive study of actual planets around distant stars. Courtesy NASA.

Of UFOs and Ancient Astronauts

There are many people today who would deny that the Fermi paradox even exists. The paradox juxtaposes the concept of abundant life throughout the Milky Way galaxy and the fact that extraterrestrials are not among us—Fermi's famous question, "Where are they?" After all, think of all the people who believe in UFOs and the works of Erich von Daniken. What do we tell these people who would tell us that there is plenty of evidence for ancient astronauts and the extraterrestrial origins of UFOs? We tell them that we want very much to see evidence that proves humankind has been visited by extraterrestrial intelligence sometime in our past or that UFOs are piloted by ETs from another star system. Then we rationally argue *against* their conviction, with facts at hand, that there is no scientific evidence for such ET visits. There may be intriguing hearsay evidence, such as the case of Robert Sarbacher, currently head of the Washington Institute of Technology, who was told by government officials in 1950 that the Pentagon had recovered a crashed UFO. Still no hard evidence. No matter how much we wish we could find such evidence, there is none.

Many big names in science—for example, Arthur C. Clarke, Robert Jastrow, and Carl Sagan—have argued against such convictions based on faith alone by refuting the majority of such ancient astronaut stories and UFO experiences point by point. No indisputable evidence of visiting ancient astronauts or their ships has been found in the earth's geologic strata of three-plus billion years. And no verifications of UFOs has ever been made by the high-technology space surveillance network that has been in place, in one form or another, for almost three decades, even though this network often tracks fast-moving meteors as they strike the earth's atmosphere.

Robert Jastrow represents a healthy scientific skepticism about UFOs. "Why would a being with the ability to cross the distance between stars bother to stop down in New Hampshire and pick up some woman and give her a physical examination?" Jastrow asks. Such typical reports are suspect, he believes, because the UFOs always land in remote places where only one or two people see them or come in contact with them. Like so many others, Jastrow awaits some scientific evidence and believes that, in the meantime, we must be prepared for the first ET contact.

Low Life in the Galaxy

Advanced galactic civilizations created by intelligent extraterrestrial life are at one possible extreme of evolution in the universe; another extreme would be primitive life deep within the immense gas and dust clouds ebbing and flowing throughout the galaxy. That galactic low life (or high life, for that matter) could be found anywhere other than a planetary surface or atmosphere at first seems ridiculous, but the fact that the chemical precursors of life, the polypeptides and amino acids, for example, have been detected in interstellar space, makes the possibility much more reasonable.

Radio telescopes have detected more than fifty naturally forming organic molecules in interstellar space over the past two decades. They include water, ammonia, carbon monoxide, hydrogen sulphide, formaldehyde, hydrogen cyanide, methyl alcohol, and silicon monoxide. These large molecules are composed of the most common atoms, the so-called building blocks of life: carbon, hydrogen, oxygen, and nitrogen. When radio astronomers discovered the molecules methylamine and formic acid in the galaxy, the scientific community became excited. Why? Because these molecules can combine to form glycine and amino acid, and amino acids combine to form proteins so important for life. Then extraterrestrial amino acids were discovered for the first time in the Murchison meteorite, which fell in Australia in 1974. Even though the biological jumps from such molecules to amino acids to DNA to living cells are tremendous, the presence of such interstellar chemistry gives the ET optimists more cause for hope. Indeed, such discoveries have become the basis for entire new theories of the origin of life on earth and in the universe.

Seeding the Earth

Francis Crick's concept of panspermia is one such theory—that life was seeded on earth by an extraterrestrial civilization many aeons ago, when primordial earth had the conditions favorable for the development of life. The idea of panspermia was first put forward by Lord Kelvin; he suggested that life came to planet Earth on the back of a meteorite. Crick's version, presented in his book *Life Itself*, was formed in collaboration with Leslie Orgel, a biochemist at the Salk Institute, and was in large part an intellectual exercise that grew out of an international meeting on the topic of communication with extraterrestrial intelligence held in the Armenian Republic in 1971.

Many people believed that Crick had cracked the scientific limb he had been out on when he seriously presented his panspermia ideas, but this winner of the Nobel Prize felt the concept should be fully explored as an alternative to life evolving from the primordial chemistry of ancient earth—an explanation that Crick seriously doubted and considered improbable because of the numerous conditions and complex sequence of events that had to be met before it could happen.

Cyril Ponnamperuma, director of the Laboratory of Chemical Evolution at the University of Maryland, has wondered why Crick presented the improbable explanation of an extraterrestrial civilization seeding planet Earth at just the right time in geological history as an alternative to what the scientist considered another improbable history—life on earth from scratch, from the primordial soup.

"There's no way of disproving Crick's idea," Ponnamperuma told *Omni* magazine, "but I feel uncomfortable with it . . . . Sometimes I wonder whether he really believes what he wrote."

Francis Crick is not alone in presenting some unusual and controversial ideas about extraterrestrial life and the origins of life on earth. The famous astronomer Fred Hoyle has also gone out on what many peers consider to be a scientific limb by arguing that life originally evolved (and continues to evolve) in the same vast molecular clouds from which stars are born, once the condensed gas and dust builds up pressure and temperature until the fusion process begins. At some point in the evolution of these great interstellar clouds, they become biologically active.

Hoyle has also presented a theory that diseases on earth have come via comets from space. While Ponnamperuma goes so far as to admit that organic molecules can exist on comets and that, under special conditions, some type of cometary life could evolve, he dismisses Hoyle's theory as "bizarre": "To get a virus, specific to a human, evolved completely away from the earth is very, very hard to accept," says Ponnamperuma. "You've got to throw away all of modern biology."

As skeptical as Ponnamperuma is about such Johnny Appleseed ideas of spreading life about the galaxy, he is a believer in extraterrestrial life and even gives the intelligent type a chance of existing. He refuses to accept the idea that we are alone in the universe. The SETI searches, he contends, have just begun to scratch the surface. "In order to detect a signal, you probably have to look for at least thirty years."

If humankind ever makes contact with ET life, there will be a whole shift in consciousness. We will, Ponnamperuma believes, "feel less freakish, part of a magnificent cosmic plan." Even if contact is not made in the next few thousand years, this expert on chemical evolution believes, we humans will be leaving the solar system and inhabiting other worlds orbiting other stars in our vast Milky Way galaxy. Humankind will evolve from terrestrial to extraterrestrial life. We will become the extraterrestrials, the members of the galactic club, perhaps its founding members. What, after all, is a few thousand years on the cosmic calendar? Approximately 1/5,000,000th the age of the universe.

To the Stars?

Given enough time, so the speculation goes, even the stars can be reached by earthlings and their interstellar ships. But is this so? Certainly it would be somewhat easier for humankind to

Pioneer 10, which flew by Jupiter in December 1973, became the first human artifact to leave our solar system. It is heading toward a point in the sky near the constellations Taurus and Orion and may come "close" to another star (a few light-years or tens of trillions of miles) once every one to two million years! An average distance between two stars equals about 228 million times around the earth's equator. Courtesy TRW, Inc.

send automated starships to the nearby stars than a peopled starship whose descendants would reach a star, perhaps at the expense of their descendants never seeing their home solar system again. Whether such far-future starships are piloted and tended by humans or robots is a small question resting on the shoulders of a giant one: Why travel to the stars at all, especially knowing the staggering quantities of energy and resources needed to cross such deserted interstellar voids? On the basis of what we know, or even imagine, today, why would our technological civilization want to weigh anchor from earth and set off toward the stars, a journey of decades and centuries across trillions of miles (kilometers) of interstellar space? There is plenty, after all, to keep us busy right here in our own solar system.

The distances between the stars, based on today's spacecraft speeds, appear impossible to cross. An Apollo spacecraft at its average earth-moon and return speed, would take some 850,000 years to reach the nearest star to our sun, Proxima Centauri, one of the three stars of the Alpha Centauri system, 4.3 light-years away. *Pioneer 10* was the first man-made object to leave the solar system, in June 1983, and it carried the first cosmic postcard in the form of the famous gold-plated aluminum plaque with its naked man and woman and other engraved message elements designed and drawn by Frank Drake and Carl and Linda Sagan. The spacecraft left the solar system at a speed of about 25,000 miles (40,000 kilometers) an hour, heading toward a point in the sky near the constellations Taurus and Orion. Astronomers estimated that this human artifact might come "close" (a few light-years or tens of trillions of miles or kilometers) to another star once every one or two million years. For it to enter the planetary system of a distant sun, it was estimated, ten billion (one thousand million) years would have to pass!

If we scale down the average-size star to a diameter of 1 inch (2.5 centimeters), about half the size of a golf ball, then the next star would be 100 miles (161 kilometers) away. Our sun's real diameter is 109 times that of the earth's, or some 865,000 miles (1,391,785 kilometers), equal to about 36 times around the planet Earth. An average distance between two stars, then, is equal to some 228 million times around the earth's equator. Expressed in a more personal way, this distance between the stars would be like having your next-door neighbor in the suburbs living about 40,000 miles (64,000 kilometers) away!

Talking about the nearest star to our sun is one thing, but when galactic distances are described, the human scale of space and time is swallowed up in one great cosmic gulp. Both the Voyager spacecraft will leave the solar system by the end of the century. But at their average speed relative to the sun, it would take them six hundred million years to travel to the center of the Milky Way galaxy. The nearest stars, perhaps, may be within our reach given enough time, say a century or two, but what about starship highways around the Milky Way, perhaps through black holes, at the hyperspace speeds of Hollywood film fantasies? Forget it. That's as close to "never happen" as anyone can reasonably predict for the next few thousand years.

Across the Light-Years

Why would humankind ever travel across the light-years to other planetary systems? Even assuming that technology can solve the tremendous problems of interstellar flight and that the international community approves the expenditure of huge resources, we still need a reason to go. Human curiosity or sense of adventure alone cannot justify such an immense human undertaking. Because they, the stars, are there is not the answer as it was for Mount Everest and the mountaineers who conquered its summit. The human spirit has proved itself over and over

again, and it does not have to challenge the light-years to prove that it continues to endure.

There are two reasons earth folk would build their starships in the next few hundred years and point them in the direction of the nearby stars. If an intelligent signal is received from extraterrestrial life within about ten light-years (the odds are against this occurrence), then there would be a powerful motivation to travel to them for an encounter, or at least to meet them halfway. The other reason for such a migration would be survival of our species, a threat to our solar system or our sun that would be known for a long time in advance. It is conceivable, for example, that new knowledge about the sun could predict a shortened solar life span and make interstellar human migration essential for survival of the species. Assuming that there are no such cosmic upheavals in our corner of the cosmos, it would then seem that only a communication with ET life within just a few light-years would provide the impetus for starships setting off across the interstellar gulfs. We would have to know that intelligent beings were within our reach, even it if was a long reach of several decades. Interesting planetary systems that might, after careful study with state-of-the-art astronomical instruments, give every indication of harboring life would not be enough to make a commitment to such a human endeavor. From what we can imagine today, the will to build and pilot starships would come from ET life calling planet Earth. Instead of having "ET phone home," one of ET's fellow creatures would phone earth over the light-years and create a cosmic bond that would draw us starward.

Interstellar Speed Limits

The year in which the extraterrestrial call is received—if it ever is—will in large part determine how long it will take to build and launch an interstellar expedition, and how far and how fast such a starship can travel. The technological level in any given century beyond 2000, as well as the degree to which solar system resources are utilized, will be the key to attempting a voyage to nearby stars. Whether the starship is guided by artificial intelligence alone or whether it is supplemented by star-bound people from planet Earth—this does not matter much. The biggest factor is: When? If the cosmic call came in tomorrow, there would be some real problems in putting together an expedition in a few decades. The decision might be to wait for an advanced technology and make up the time in flight.

Freeman Dyson tinkered with several starship designs during Project Orion and got one, on paper, to fly at 3 percent the speed of light—somewhat over 20 million miles (32 million kilometers) an hour. While this speed is 33 times slower than the speed of light, it is 5,760 times faster than Apollo's average round-trip speed to the moon. It would take such a "slow" starship about 130 years to make a one-way trip to the nearby Alpha Centauri system. At 10 percent the speed of light, it would take about 43 years; and at 40 percent, about 11 years to travel the 4.3 light-years—one way. Any consideration of speeds beyond 40 percent the speed of light is taking freewheeling speculation into the twilight zone, which was already visited in Chapter 7 with reference to the photon rocket.

For a spaceship to get 25 light-years out from earth at 10 percent the speed of light, the one-way travel time jumps to 250 years. It quickly becomes clear that at 10 percent the speed of light (more than 66 million miles, 106 million kilometers, per hour), almost all interstellar travel becomes a journey of several human generations—whether we actually fly the starships or send sophisticated robots and wait for the information to return to earth. If the extremely long cosmic odds were with us, of course, and the ET life with which we made contact was on a similar technological level of spacefaring skills, then perhaps we could exchange cultures at some midway point by sending cosmic arks, interstellar cultural exhibits, that could have living crews or sophisticated artificial intelligence to run the missions and conduct the diplomacy. Two galactic civilizations traveling toward a common island (star or navigational point) of

interstellar space at 10 percent the speed of light would double the speed and cut the journey time in half.

One fully robotized starship has already been worked out in amazing detail by the British Interplanetary Society, which published the seminal work *Project Daedalus* in 1978.

Humankind's First Starship

The project to design a *practical* starship, unmanned and controlled by advanced robotics, began in 1972. The captain of the first starship design team on planet Earth was Dr. Alan Bond, a propulsion engineer and former scientist at the rocket division of Rolls Royce. It was Bond who suggested in that year that members of the British Interplanetary Society form a working group and do a detailed feasibility study of a one-way, unmanned starship that would fly by Barnard's Star, which is some 5.9 light-years from our sun and solar system. The members enthusiastically supported the concept, and it was decided to proceed in January 1973—the same year that the United States launched Skylab, its first space station.

Propulsion. The design was based on state-of-the-art, available technology, and the group decided to use the nuclear pulse rocket as their propulsion concept. The engine would expel small spheres of frozen deuterium and helium-3, about half the size of a Ping-Pong ball, which would then be exploded by electron beams behind the ship. These spheres would explode at the amazing rate of 250 each second, each one releasing energy equal to about 90 tons of TNT. The grand total of energy released each second would therefore equal 22,500 tons of TNT. The expanding gases from these explosions would eventually propel the starship to a top velocity of about 13 percent the speed of light, some 86 million miles (138 million kilometers) per hour, which would fly it past Barnard's Star about fifty years after launch.

The Daedalus starship would use some 30,000 tons of helium-3 and 20,000 tons of deuterium to fuel its nuclear pulse engine. But because helium-3 is almost nonexistent on earth, it would have to be extracted from Jupiter's atmosphere, which is perhaps one of the most difficult logistical features of the Daedalus project. A base on one of Jupiter's moons such as Callisto or a large orbiting space station would have to be established for the helium-3 mining operations. Alan Bond did suggest an alternative way of obtaining the helium-3: artificially breed it on the surface of the moon. This too has its problems; the waste heat generated by the breeding process would be equal to the world's energy consumption, at today's rate, for some seven hundred years.

Size and Weight. After several design revisions, the unmanned starship had an on-paper launch mass of 104.8 million pounds (47.5 million kilograms). Its combined two stages and payload were to have a total length of some 650 feet (200 meters) and a total weight of 68,000 tons, including fuel, with a modest 400 tons (about equal to the total weight of five Skylab space stations) devoted to the payload star probes that would be activated during the flyby of Barnard's Star. The primary engine reaction chamber was designed to be 330 feet (100 meters) in diameter, but very thin, like a huge foil dish.

Thrust and Rocket Stages. Each small fuel sphere would be injected by a magnetic piston at about 27,000 miles (43,000 kilometers) per hour into the combustion chamber, where it would be bombarded by high-energy electron beams and exploded, creating a magnetic spring reaction in the magnetic field to give the starship constant acceleration.

Some eight and a third months after launch from an orbit around either Jupiter or the moon, two of the first-stage fuel tanks would separate, and the other four would separate during the first two years of the mission, at which time the entire first stage would fall away and the second stage would ignite. The second stage would burn for 1.76 years, at which time it would cease firing and the probe would coast the rest of the

way to its destination star at 12.8 percent the speed of light—almost 86 million miles (138 million kilometers) an hour!

Star Ahoy! Some fifty years after launch from our solar system, this first starship conceived by earthlings would release an armada of seventeen targeted scientific probes of all sizes and containing different sensors as the sophisticated artificial intelligence center came alive in the mother ship and used all of its computing power, which had remained dormant during the journey. Each probe would transmit its information back to the starship for relay to earth. The data stream for a single image frame would take some three and a half hours to transmit, and the data for as many as one thousand images could be stored on the starship.

Perhaps for the first time in human history, close-ups of another star's planets, some of them actually showing surface features, would be seen after the data-bit signals traveled for another 5.9 years at the speed of light back to planet Earth.

Interstellar Ports of Call

While Barnard's Star is not as close to our solar system as is the Alpha Centauri system (5.9 light-years versus 4.3 light-years), it was chosen by the British Interplanetary Society study group as the target star for the unmanned Daedalus starship. It is easy to forget, when tossing around light-years as if they were miles or kilometers, how much farther Barnard's Star is than Alpha Centauri: the 1.6 extra light-years are equal to forty million times the earth-moon distance of about 240,000 miles (386,000 kilometers). The Daedalus study team believed that if they designed their unmanned starship to travel the almost six light-years to Barnard's Star, their design work would be more adaptable to other interstellar journeys and could probably make journeys out to about nine light-years from our solar system. This would include the closer Alpha Centauri duo as well as stars such as Luyten 726-8 A/B, Wolf 359, and Lalande 21185.

Barnard's Star, named for astronomer Edward Emerson Barnard, who discovered it in 1916, is a faint red dwarf, which has the fastest motion across the sky of any star. It is, in fact, moving toward the sun and solar system at about 24,000 miles (39,000 kilometers) an hour. Over the next ten thousand years, it will have moved some two light-years closer to earth and will replace the Alpha Centauri system as the closest star to earth. If the first starship is launched in the next two centuries, however, this star's motion toward the sun will not in any way lessen the tremendous challenge of designing, building, and launching it.

Another reason that Barnard's Star holds such interest as a target star for an interstellar voyage is that several studies, in particular Peter Van de Kamp's work at Sproul Observatory, have strongly indicated that this red dwarf has planetary companions orbiting it. The Hubble Space Telescope will study this and other neighboring stars and accumulate definite observational evidence of planetary systems around them. Any one of several stars could eventually replace Barnard's Star as the destination sun for humankind's first interstellar voyage.

Project Daedalus Star Ranking

Star	Distance (l.-y.)	Rank
Proxima	4.25	6
Alpha Centauri A/B	4.3	1
Barnard's Star	5.9	3
Wolf 359	7.6	8
Lalande 21185	8.1	11
Sirius A/B	8.6	10
Luyten 726-8 A/B	8.9	7
Ross 154	9.4	12
Ross 248	10.3	13
Epsilon Eridani	10.7	4
61 Cygni A/B	11.2	2
Epsilon Indi	11.2	9
Tau Ceti	11.9	5

Star Search

As other sophisticated astronomical instruments follow the Hubble Space Telescope into orbit above the earth in the next few decades, the ranking of neighboring stars as targets for interstellar voyages will change. It is, nevertheless, all but certain that the destination star for the first starship voyage will be among the list that was published by the British Interplanetary Society in 1978 as part of their *Project Daedalus*. Even though Barnard's Star was chosen for the study, it had a ranking of third on the target star list. The group considered it important to design their robot starship for the extra distance. The star list, by distance from the sun, is shown below. One of the three stars, Proxima Centauri, in the Alpha Centauri system, is listed separately because it is separated from the Alpha Centauri A/B by some 930 billion miles (1.5 trillion kilometers)—almost fifty-eight light-days.

The rankings of these nearby stars were determined by considering many factors, including the type of star, its temperature and life span, and the probability of planetary systems and the evolution of life forms on the planets. All known stars up to a distance of 10.7 light-years (Epsilon Eridani) were included, and the three stars beyond that distance were similar in type to our Sun or had a high ranking. One of these thirteen stars will probably be scrutinized in the next few centuries by the far-flung technological neophytes *Homo sapiens*.

Peopled or Roboted Starships?

One of the more attractive destination stars defined by the Daedalus study is Epsilon Eridani, a sunlike star some 10.7 light-years away. Assuming that the next few decades of advanced in-space telescopes tell us that this star has the most earthlike planets around it, will we send a robotized starship toward it or a space ark with a human crew? In this hypothetical instance, when there is actually no extraterrestrial communication taking place, either a new generation of in-space sensing technology would be designed and built or a robotized starship would weigh anchor and head toward interstellar space. A peopled space ark would depart the solar system only if ET communications had been established, and even then a starship with advanced artificial intelligence might be chosen to navigate the interstellar gulf. Why send people to Epsilon Eridani, after all, when it would take about 110 years to make a one-way trip at one-tenth the speed of light, a technological possibility in the next century? This is the big question: Why send people to do an advanced robot's work? Eye-to-eye contact may be a positive force here on earth, but if the ETs we communicate with have no eyes as we define them, what is the point of a creature-to-creature meeting when computer enhanced data will create as many detailed images as we care to see?

There are alternatives to physically sending a starship, peopled or not, to Epsilon Eridani or any other star that proves attractive in the twenty-first century. If it is only habitable planets around another nearby star that create the motivation, advanced above-earth or above-moon instruments may satisfy our curiosity and sufficiently increase our knowledge. A so-called ultimate telescope, proposed by Princeton University physicist Eric Hannah, could in theory use the entire sun as a gravitational lens to focus a distant star's image on a large, flat array of photodetectors in space. If ever built, such an ultimate telescope could amplify the light from a faraway star two hundred billionfold. It could show detail one hundred million times finer than the Hubble Space Telescope can, and it could even show features the size of houses on the planets that orbit Barnard's Star almost six light-years (36 million million miles, 58 million million kilometers) away.

The more ambitious plan would be to send an unmanned Daedalus-like (Daedalus in Greek, by the way, means "cunningly wrought") starship to explore such a planetary system, which could harbor evolving life forms that had not become spacefaring, interstellar-communicating species. An advanced on-board artificial intelligence could probably do as much or more

than a trained human crew. Still, at only 10 percent the speed of light, the travel time to and information return from Epsilon Eridani would be some two hundred years—about the time it took two industrial nations to acquire the rudimentary means of leaving their cradle, planet Earth.

Without an ET message in hand, it is doubtful that a peopled starship will ever fly and deliver its crew or their descendants to a new world orbiting a distant star. Our answer may be: Let us let the robots take those interstellar risks.

An ET Message to Earth

If our sophisticated, computer signal analyzers do make contact with extraterrestrial intelligence, the response could eventually be a space ark filled with earth people and their fellow creatures. Such a response, however, would take several decades if such ET contact were made in the year 2019, and it would be the most improbable response.

After confirmation of the initial contact (an exacting, scientific procedure that could take weeks), communication with the ET intelligence could be initiated through a transmission program and then dramatically increased over the next few years. Or earth could decide not to respond for some reason. If thorough analysis leads experts to the conclusion that there is no danger in responding to the message, transmission will begin. The data stream from earth would include whatever information is both scientifically valid and internationally approved and screened for transmission by a multinational forum such as the United Nations.

Our ability to understand and interpret the first extraterrestrial message will in large part determine earth's response. And assuming that such an analysis is fruitful, then the location and distance from earth of the ET intelligence is also tremendously important. If the signal comes from a planet near a star one hundred light-years away, humankind's answer in the next century will be limited to radio or other electromagnetic radiation traveling at the speed of light—the speed limit of the universe. If, however, an intelligent signal comes from a star such as Epsilon Eridani (10.7 light-years from the sun), then an entire range of responses from earth may be possible in the next few hundred years: radio contact only; radio silence or censorship (if the contact message implies malevolence); launching a roboted starship; or launching a peopled space ark.

Radio contact. Radio contact with ET intelligence is astronomically more probable than physical contact. If the ET civilization is any more than twenty light-years from earth (the distance many experts believe to be the limit of possible exploration), it is doubtful that a starship will be launched to such a faraway celestial island because a one-way journey would take about two hundred years at 10 percent the speed of light. But with various wavelengths of the electromagnetic spectrum, humankind can communicate a tremendous amount of information *at the speed of light* and can do so over a distance of a few hundred light-years in fewer than a dozen human generations, which represents a short period of time when compared to the sixty-six thousand human generations that took us from the trees of the jungle to the surface of the moon.

That we or other galactic civilizations can send our messages via electromagnetic waves at the speed of light is luck of a cosmic order; but just as important as this is the fact that almost any kind of information can be sent in the message—including the genetic codes of the human species and its fellow earth creatures, which could be recreated by intelligent ETs somewhere in their solar systems far, far away. With DNA technology going through a revolution, the time is near when complete instructions on how to build a human being could be sent at the speed of light to intelligent creatures among the distance stars. In this way, the cybernetic seed of humankind could be broadcast throughout the galaxy. If advanced life forms on distant worlds catch our molecular DNA secrets, humankind could become a new species on those distant planets. Of course, such ETs may not want to

have anything to do with us; they may consider us too low on the evolutionary spiral to bother with. If such galactic life receives a DNA message from us several decades after it was transmitted from earth, they may accurately predict on the basis of analysis that we no longer exist—just another of the millions of galactic species that die out each year.

Broadcasting humanity's seeds indiscriminately throughout the Milky Way galaxy does have its own risks, however. Some would argue that we would be forsaking our long evolutionary heritage by almost nonchalantly giving away the secrets of our species. How do we know that some evil ET alien would not recreate us and commit atrocities against our fellow creatures? We must practice caution in deciding what we broadcast into the universe. Perhaps our application for membership in the galactic club has already been turned down on the basis of ET life having heard some of our earlier radio broadcasts, such as "I Love Lucy" or "The Shadow," that were not intended for ET ears.

Radio Silence and Censorship. An immediate transmission after receiving a radio message from ET life could turn out to be a grave error. Painstaking analysis should be completed, and a profile of the ET life should be projected based on its message. Planet Earth's and our species' survival could be at stake. Scientists as well as dozens of science fiction writers have speculated on various outcomes of our species coming into contact with malevolent aliens. Common themes include earthlings being made slaves, harvested as food, used as pets, or put in zoos. Even the theory that, unknown to us, the earth and humankind are part of an ET zoo has been proposed. Our zookeepers want to preserve our natural habitat, and, therefore, do not reveal themselves.

A Robotized Starship. The Daedalus project, as already summarized, was the first detailed feasibility study of the design of a starship and an interstellar voyage. Because the life spans of advanced robots will be much longer than the natural lifespans of humans, and because future robots will have the ability to reproduce themselves, our first interstellar voyage probably will be fully controlled by a high order of artificial intelligence. Even Gerard O'Neill's concepts of large colonies in space could be adapted to interstellar travel and populated with a robot crew. Such an extraterrestrial starship was the focus of Arthur C. Clarke's 1973 novel, *Rendezvous with Rama*. In it, the solar system is visited by a huge spaceship, some 31 miles (50 kilometers) long, populated by biological robots. It turns out that this spaceship from a mysterious galactic civilization was only passing through to fill up with energy from our sun. They were not interested in earth or its creatures.

A Peopled Space Ark. J. Desmond Bernal, the British physicist, wrote his prophetic book *The World, the Flesh and the Devil* in 1929. Often quoted by such visionaries as Arthur C. Clarke, Freeman Dyson, Gerard O'Neill, and Olaf Stapledon, the book describes the expansion of life into space and includes a space ark in which generations of people live and die, never knowing the earth from which they came, nor the destination planet on which their descendants will land and live. In 1929, the concept of a space ark seemed the only way of transporting people across the vast distances of interstellar space and time.

The fact that several human generations would be required to journey at a fraction of the speed of light from star to star was not encouraging. This is why space arks are usually envisioned as large vessels able to accommodate attractive earthlike environments to keep the interstellar void at bay and out of the spacefarers' psyches. But the size of the space ark was directly interrelated to its propulsion and speed. If an O'Neill-type space colony were adapted to an interstellar ark and left the solar system traveling at 1 percent the speed of light (about 6.7 million miles, 10.8 million kilometers, per hour), it would take more than 13 human generations—about 430 years—to reach the nearby Alpha Centauri star system. Would such a long voyage attract enough volunteers? Who would want to go and why?

There is always that extraterrestrial rub: Why travel to the stars? What would be humankind's motivation? Beyond the survival of the species or a physical encounter with extraterrestrial intelligence, such motivations are secrets held by the future. Assuming for the moment that there is an important reason for earthlings to physically meet ET intelligence at a distance of between ten and twenty light-years and at a velocity of no more than 20 percent the speed of light, what are the alternatives to a voyage requiring several human generations? A deep and almost dreamless sleep.

As the Silent Stars Go By

An imaginative science fiction genre has often supplied a method of reaching the stars within a human lifespan: Suspended animation has been a way of slowing down human biological time to better match the cosmic time scale of interstellar journeys. Some type of human hibernation, in which the human body and brain are frozen in time, may become a real alternative in another one hundred or two hundred years for spacefarers journeying between the stars. A California research team has recently discovered, for example, that many organisms are able to survive dehydration and exist in a dormant state by producing a sugar called trehalose. This and other such discoveries may eventually lead to safe techniques that will suspend human biological processes over long periods of time.

In the film *Alien*, the crew members were held in suspended animation while their spaceship crossed the light-years. Although artificial hibernation or hypothermia for the human body is beyond present capabilities, human embryos have been frozen, thawed, implanted in the female uterus, and brought to a successful birth. It is conceivable that a nursery of frozen human life forms, shielded from radiation at the center of a starship (let us name it *Earthark*) could survive an interstellar journey of hundreds or thousands of years, and then complete their gestation and formative years under the guidance of programmed robot nannies before reaching their destination star and planetary system, in which they will meet the ET life or colonize a habitable planet.

If we can suspend the human biological clock in some way, and at the same time cut down tremendously on on-board consumables, physical travel to the stars becomes more realistic. If this cannot be done, be it with the embryonic human seeds ready for gestation or fully grown earthlings ready for the challenges of new worlds, then a one-hundred member crew on board the starship *Earthark* would need the following consumables for a twenty-year journey:

• Water: 7.9 million pounds (3.6 million kilograms)

• Food: 3.3 million pounds (1.5 million kilograms)

• Oxygen: 1.1 million pounds (0.5 million kilograms)

These three life-sustaining consumables add up to some 12.3 million pounds (5.6 million kilograms), a total weight that would greatly influence the starship's size, propulsion design, and flight time to the stars.

If our descendants board the starships and set sail on the interstellar oceans, they will probably choose, at some early point in the journey to chemically suspend their lives in time with future life-suspension techniques. Perhaps these deep-sleeping adults will be accompanied by an embryonic nursery of frozen fertilized eggs from various earth species on this first great voyage—an ark from planet Earth. As they fly to the stars, suspended in a deep and dreamless sleep, their robot friends will stand watch and tend to them during the cosmic journey spanning decades. For these star-bound humans, this will be a dreamless road to the stars, but a road built by the persistent dreams of humankind over thousands of years. Images of new and distant worlds may lie dormant during this cosmic slumber, but in essence they are what power this starship into the future. Such images will form again when the cosmic sleepers awaken to complete their lives and explore new worlds.

Is any of this possible? Will this happen to our descendants born on planet Earth? Will they become true extraterrestrials journeying from star to star? Human dreams tell us yes, this is the distant future of the Space Age.

APPENDIXES

The space groups listed below offer a variety of services and information for the serious space buff. Publications, products, activities, and trips are just some of the offerings. Call or write for more information about a specific group.

ACTIVE SPACE ORGANIZATIONS

Headquarters	Contact	Headquarters	Contact
American Astronautical Society 6212B Old Keene Mill Court Springfield, Virginia 22152 (703) 866-0020	Carolyn Brown	Planetary Society 65 North Catalina Avenue Pasadena, California 91106 (818) 793-5100	Louis Friedman *Executive Director*
American Institute of Aeronautics and Astronautics 1633 Broadway New York, New York 10019 (212) 581-4300	James J. Harford *Executive Director*	Space Foundation P.O. Box 58501 Houston, Texas 77258 (713) 474-2258	Nancy Wood *Executive Director*
		Space Studies Institute 258 Rosedale Road Princeton, New Jersey 08540 (609) 921-0377	Bettie Grebre *Development Coordinator*
L-5 Society 1060 East Elm Street Tuscon, Arizona 85719 (602) 622-6351	Greg Barr	Students for the Exploration & Development of Space 800 21st Street NW Washington, D.C. 20052 (202) 676-7102	Nick Dobelbower *Executive Secretary*
National Space Club 655 15th Street NW Washington, D.C. 20005 (202) 639-4210	Rory Heydon	Women in Aerospace 6212B Old Keene Mill Court Springfield, Virginia 22152 (703) 866-0020	Diana Hoyt *President*
National Space Society 600 Maryland Avenue SW West Wing Suite 203 Washington, D.C. 20024 (202) 484-1111	Dr. Glen Wilson *Executive Director*	Young Astronauts Council 1015 15th street NW Suite 905 Washington, D.C. 20005 (202) 282-1984	Jack Anderson *Chairman*

ON-EARTH SPACE EXHIBITS

The air and space museums listed below are a selected list by region and do not represent a complete guide. Many cities and states have science museums that devote some of their exhibit areas to spaceflight and other Space Age activities. For other museums, exhibits, and theaters of interest, ask for information from your travel agent or the chamber of commerce in your area of interest.

East

Spaceport USA
Kennedy Space Center, Florida
32899
(305) 452-2121

Over 2 million people visit the Kennedy Space Center each year. Two bus tours are available: an historical tour of Cape Canaveral Air Force Station, and a tour of the Kennedy Space Center, which includes the immense Vehicle Assembly Building (with a Saturn 5 in front), the giant Crawler Transporter, and several launch sites, including 39A and 39B from which the space shuttle is launched.

Many historic rockets are displayed outside the Visitor's Center. The Gallery of Space Flight displays several spacecraft, spacesuits and other equipment, and many models of unmanned spacecraft, including the Viking Mars Lander. Two theaters provide visitors with the ultimate in exciting visual entertainment.

National Air and Space Museum
Smithsonian Institution
Washington, D.C. 20560
(202) 357-1552

The Smithsonian Institution's National Air and Space Museum is the most popular museum in the world, receiving more than 12 million people each year. Twenty-three exhibit areas display the history and future of air and space flight. The serious buff could spend days observing and enjoying the thousands of artifacts. A few examples are John Glenn's Mercury *Friendship 7*, Apollo 11's command module *Columbia*, a Skylab Orbital Workshop, and an Apollo-Soyuz spacecraft.

Two auditoriums, the Albert Einstein Planetarium and the Samuel P. Langley Theater provide exciting visual and learning experiences using state-of-the-art projection systems. Two recent films shown here and elsewhere around the country are *Hail Columbia* and *The Dream is Alive*.

Florence Air and Missile Museum
Highway 301, North-Airport Entrance
P.O. Box 1326
Florence, South Carolina 29503
(803) 665-5118

The Florence Air and Missile Museum is located on a World War II training base for fighter pilots. Some 38 aircraft, missiles, and rockets are displayed, including jet fighters, bombers, and a Titan Intercontinental Ballistic Missile. The

space displays include Alan Shepard's Apollo 14 spacesuit, a Gemini capsule, space parachutes, and portions of the Saturn 5 moon rocket and the Apollo launch computers.

Goddard Space Flight Center
Greenbelt, Maryland 20771
(301) 344-8101

The Goddard Space Flight Center, like most NASA centers in the United States, has a visitor center and offers tours of the facilities. Several replicas of Earth satellites can be seen, and there is a special room displaying memorabilia of the American rocket pioneer, Robert Goddard. Visitors can hear their own voices bounced off an orbiting satellite and view a full-scale mockup of a shuttle cargo bay.

Central

The Alabama Space & Rocket Center
Tranquility Base
Huntsville, Alabama 35807
(1-800) 633-7280; in Alabama, 1-800-572-7234

The Alabama Space and Rocket Center has been called the Earth's largest space museum, and it also sponsors the famous U.S. Space Camp. In the Rocket Park at the rear of the museum, visitors can see rockets and missiles of the early Space Age, including the giant Saturn 5 moon rocket. Inside are displayed several spacecraft which actually flew such as the Apollo 16 command module, *Casper*, and the Mercury capsule *Sigma 7*. Replicas of Gemini and lunar excursion module spacecraft are also on exhibit.

The museum features more than 60 hands-on exhibits, which allow visitors to fire a rocket engine, guide a spacecraft by computer, and feel the sensation of weightlessness. Inside the Lunar Odyssey, three simulated spaceflights are offered, and the Spacedome Theater presents state-of-the-art cinematic adventures. Tours of the nearby NASA-Marshall Space Flight Center are also available.

The center runs the United States Space Camp which offers a variety of programs for children and adults, ages 11 and up.

The Johnson Space Center
Public Services Branch, AP4
Houston, Texas 77058
(713) 483-4321

Home of the Mission Control Center for actual manned space missions, the Johnson Space Center offers a variety of exciting tours and exhibits for the space buff. Tours of the Control Center are available when no mission is in progress (although visitors might witness a simulated one), and lunar rock and soil samples can be seen on a tour of the Lunar Sample Building.

The Teague Visitor Center contains an extensive collection of space hardware and exhibits, including Apollo and Mercury spacecraft. A large auditorium shows many NASA films throughout the day. A spare Saturn 5 rocket, positioned horizontally, can be seen outside.

In another building, visitors can walk through a Skylab space station trainer and see other exhibits on how astronauts live and work in space. Another facility houses a full-scale mockup of the space shuttle orbiter where the new astronauts constantly train and practice for their missions.

The Kansas Cosmosphere Discovery Center
1100 N. Plum
Hutchinson, Kansas 67501
(316) 662-2305

The Hall of Space includes spacesuits, spacecraft, and other space artifacts, and there are many interactive, hands-on exhibits. An OMNI-MAX ® projection system in the Cosmosphere Theater gives the audience the next best thing to actually blasting off the pad and experiencing weightless life in orbit.

McDonnell Douglas Prologue Room
McDonnell Douglas World Headquarters
P.O. Box 516
St. Louis, Missouri 63166
(314) 232-0232

Past and future aviation and space projects are presented with models, dioramas, and pictures. The Prologue Room includes exhibits on missiles and space, Mercury and Gemini manned spaceflight, Skylab, and the space shuttle—all programs in which this major aerospace company made significant contributions.

Experimental Aircraft Association, Inc.
EAA Aviation Center
Wittman Airfield
Oshkosh, Wisconsin 54903-2591
(414) 426-4800

The EAA sponsors the "world's largest fly-in" each year, and the EAA Aviation Center is one of the world's great aviation museums. While spaceflight and spacecraft are not the center's primary focus, there are many exhibits of interest to the space buff. The Parasev 1-A, for example, was an experimental vehicle built and flown by NASA in the early 1960s to evaluate a possible on-land touchdown for manned spacecraft. Famous astronauts such as Gus Grissom and Neil Armstrong piloted some of the more than 300 test flights of this sailwing vehicle.

The Henry Crown Space Center and Omnimax ®
Theater
Chicago's Museum of Science and Industry
57th Street and Lake Shore Drive
Chicago, Illinois 60637-2093
(313) 684-1414

This $12 million addition to the Midwest's most-visited cultural institution opened in July 1987. Besides many NASA artifacts, space science and technology exhibits (including a space station replica), and the theater, it offers simulated rides aboard a space shuttle mockup.

Mountain

Space Center
P.O. Box 533
Alamogorado, New Mexico, 88311-0533
(1-800) 545-4021; in New Mexico, 437-2840

The Space Center includes the International Space Hall of Fame and the Clyde W. Tombaugh Space Instruction Center. Outside the complex visitors can see the Little Joe II rocket used to test launch escape systems for manned Apollo flight and the famous rocket sled, Sonic

Wind II, which tested human response to the extreme acceleration and deceleration forces in 1954 before the Space Age began.

The indoor exhibits display such unusual hardware as the V-2 rocket guidance system and the capsule that carried Ham, the first chimpanzee in space.

Pima Air Museum
6000 East Valencia Road
Tuscon, Arizona 85706
(602) 574-9658

A full-scale X-15 rocketplane can be seen at the Pima Air Museum, as well as the third B-52 bomber that launched it. Besides the extensive aircraft museum that contains airplanes from all periods of the twentieth century, including a replica of the Wright brother's aircraft, a separate Titan Missile Museum, with a Titan ICBM in its silo, has recently been opened to the public. Tours of this Titan Missile Museum take visitors down to the command level of the silo and the missile can be seen from several perspectives, including a view from the top into the silo through a thick transparent cover.

West

California Museum of Science and Industry
Aerospace Museum
700 State Drive, Exposition Park
Los Angeles, CA 90037
(213) 744-7400

Exhibits in the Aerospace Museum include the Gemini 11 spacecraft, a lunar module descent engine, replicas of communications and military satellites, and a Viking biology experiment lander, the device used to search for life on Mars. Visitors can also tap into a national weather satellite and call up images of various American cities. A simulated space shuttle is also exhibited.

A multimedia program, "Windows on the Universe," is staged with 21 projectors on 9 separate screens. The script was written by Ray Bradbury and narrated by James Whitmore. An IMAX® projection theater is also part of the facility.

Museum of Flight
9404 East Marginal Way South
Seattle, Washington 98108
(206) 767-7373

The Boeing Company's original manufacturing plant, the historic Red Barn, was moved to Boeing Field and restored to become the new museum's first facility open to the public in 1983. Then the Museum of Flight Great Gallery was completed in 1987. It displays about 35 aircraft and spacecraft, including an Apollo lunar rover. The recently completed facility contains a 300-seat auditorium for large-screen movies and special events.

Reuben H. Fleet Space Theater and Science Center
Balboa Park, P.O. Box 33303
San Diego, California 92103
(619) 238-1233

The Reuben H. Fleet Space Theater, named for a famous pioneer aviator and aerospace industrialist, has been a model for other space theaters throughout the world. It features the OMNI-MAX® film format, which projects the largest motion picture frame ever developed. This technology allows the audience to experience the sensations of spaceflight on the large screen when viewing films such as *Hail Columbia* and *The Dream is Alive*.

The Science Center presents over 50 interactive exhibits that present scientific principles to young and old in entertaining ways.

INDEX